COMPUTATIONAL METHODS FOR LARGE SPARSE POWER SYSTEMS ANALYSIS

An Object Oriented Approach

THE KLUWER INTERNATIONAL SERIES IN ENGINEERING AND COMPUTER SCIENCE

Power Electronics and Power Systems

Series Editor

M. A. Pai

Other books in the series:

OPERATION OF RESTRUCTURED POWER SYSTEMS
Kankar Bhattacharya, Math H.J. Bollen and Jaap E. Daalder, ISBN 0-7923-7397-9
TRANSIENT STABILITY OF POWER SYSTEMS: *A Unified Approach to Assessment and Control*
Mania Pavella, Damien Ernst and Daniel Ruiz-Vega, ISBN 0-7923-7963-2
MAINTENANCE SCHEDULING IN RESTRUCTURED POWER SYSTEMS
M. Shahidehpour and M. Marwali, ISBN: 0-7923-7872-5
POWER SYSTEM OSCILLATIONS
Graham Rogers, ISBN: 0-7923-7712-5
STATE ESTIMATION IN ELECTRIC POWER SYSTEMS: *A Generalized Approach*
A. Monticelli, ISBN: 0-7923-8519-5
COMPUTATIONAL AUCTION MECHANISMS FOR RESTRUCTURED POWER INDUSTRY OPERATIONS
Gerald B. Sheblé, ISBN: 0-7923-8475-X
ANALYSIS OF SUBSYNCHRONOUS RESONANCE IN POWER SYSTEMS
K.R. Padiyar, ISBN: 0-7923-8319-2
POWER SYSTEMS RESTRUCTURING: *Engineering and Economics*
Marija Ilic, Francisco Galiana, and Lester Fink, ISBN: 0-7923-8163-7
CRYOGENIC OPERATION OF SILICON POWER DEVICES
Ranbir Singh and B. Jayant Baliga, ISBN: 0-7923-8157-2
VOLTAGE STABILITY OF ELECTRIC POWER SYSTEMS
Thierry Van Cutsem and Costas Vournas, ISBN: 0-7923-8139-4
AUTOMATIC LEARNING TECHNIQUES IN POWER SYSTEMS,
Louis A. Wehenkel, ISBN: 0-7923-8068-1
ENERGY FUNCTION ANALYSIS FOR POWER SYSTEM STABILITY
M. A. Pai, ISBN: 0-7923-9035-0
ELECTROMAGNETIC MODELLING OF POWER ELECTRONIC CONVERTERS
J. A. Ferreira, ISBN: 0-7923-9034-2
MODERN POWER SYSTEMS CONTROL AND OPERATION
A. S. Debs, ISBN: 0-89838-265-3
RELIABILITY ASSESSMENT OF LARGE ELECTRIC POWER SYSTEMS
R. Billington, R. N. Allan, ISBN: 0-89838-266-1
SPOT PRICING OF ELECTRICITY
F. C. Schweppe, M. C. Caramanis, R. D.Tabors, R. E. Bohn, ISBN: 0-89838-260-2
INDUSTRIAL ENERGY MANAGEMENT: *Principles and Applications,*
Giovanni Petrecca, ISBN: 0-7923-9305-8
THE FIELD ORIENTATION PRINCIPLE IN CONTROL OF INDUCTION MOTORS
Andrzej M. Trzynadlowski, ISBN: 0-7923-9420-8
FINITE ELEMENT ANALYSIS OF ELECTRICAL MACHINES
S. J. Salon, ISBN: 0-7923-9594-8

COMPUTATIONAL METHODS FOR LARGE SPARSE POWER SYSTEMS ANALYSIS

An Object Oriented Approach

by

S. A. Soman
S. A. Khaparde
Indian Institute of Technology Bombay, India

Shubha Pandit
Sardar Patel College of Engineering, India

SPRINGER SCIENCE+BUSINESS MEDIA, LLC

Electronic Services < http://www.wkap.nl>

Library of Congress Cataloging-in-Publication Data

Soman, S. A., 1967-
Computational methods for large sparse power systems analysis : an object oriented approach / by S.A. Soman, S.A. Khaparde, Shubha Pandit.
p. cm. -- (The Kluwer international series in engineering and computer science ; SECS 651. Power electronics and power systems)
Includes bibliographical references and index.

Additional material to this book can be downloaded from: //http: extras.springer.com

ISBN 978-1-4613-5256-3 ISBN 978-1-4615-0823-6 (eBook)

DOI 10.1007/978-1-4615-0823-6

1. Electric power distribution--Data Processing. 2. System analysis--Data processing. 3. Object-oriented methods (Computer science) 4. Electric power systems--Automatic Control--Mathematics. 5. Sparse matrices. I. Khaparde, S.A., 1949- II. Pandit, Shubha, 1961- III. Title. IV. Kluwer international series in engineering and computer science. Power electronics & power systems.

TK3091 .S64 2001
621.319'1--dc21

2001050363

Contents

Preface

I yield freely to the sacred frenzy; I dare frankly to confess that I have stolen the golden vessels of the Egyptians to build a tabernacle for my God far from the bounds of Egypt. If you pardon me, I shall rejoice; if you reproach me, I shall endure. The die is cast, and I am writing the book–to be read either now or by posterity, it matters not. It can wait a century for a reader, as God himself has waited six thousand years for a witness.

—J Kepler,V.M. Tikhomirov: Stories about Maxima and Minima

Why should one write a book? The answer to this question largely decides what appears in a book, if it is ever written! One of the authors (S. A. Soman) was introduced to large scale computing by Profs. K. Parthasarathy and D. Thukaram at Indian Institute of Science, Bangalore, India (IISc) in early 1990 when the computation power and memory capability of PCs was quite limited (at least by today's standards!). Consequently, a considerable amount of time had to be spent in learning sparsity techniques. While our initial attempts in sparsity exploitation were amateurish, the excitement caught on. Soon, it was realized that the initial impression that sparsity exploitation is quite a bit of an art in programming was not strictly correct. This brought us in touch with outstanding work in sparsity exploitation for linear system solvers. Initially, it was felt that parallel computing may be a gateway to fast computations for power system analysis. Over a period of time, our views have become more conservative. We now feel that finer innovations are the key to large scale computation programs.

A wealth of organized literature already exists in the domain of scientific computation, *e.g.*, SIAM publications, ACM transactions on Mathematical Software etc. Books and monographs have been written on direct sparse methods for sparse symmetric matrix, linear system solvers etc. This brings in two important questions (1) how much of it is relevant to us and (2) how much has the power system analysis community contributed to pioneering research in sparsity. The question of whether we are the leaders or the followers is in itself fascinating.

Our impressions, as they stand today, are that pioneering work in sparsity exploitation was done by power system researchers in 1960's and 1970's. For example, Tinney and Walker introduced Minimum Degree Algorithm in 1967. However, somewhere down the line, it appears that we have become followers and started specializing in application engineering of such techniques to power system applications. Simultaneously, the cause of matrix computation was furthered by work of leading numerical analyst. One can quote the application of least squares (LS) method [George and Heath, 1980] to power system state estimation, reflection of the work of Philip Gill of Systems Optimization Laboratory, Stanford on developing large scale OPF etc. Monographs on specialized topics like, LS [Lawson and Hanson, 1974], Matrix computation [Golub and Loan, 1989] eigen value computation by [Parlett, 1998] point to the vast amount of scientific knowledge that has been accumulated over a period of time.

With passage of time, it was felt that a book or monograph on sparsity exploitation techniques for power system analysis was worth an effort. To our knowledge, the work of [Wallach, 1986] was the first major contribution in bringing sparsity, numerical analysis, algorithms, data structures, parallel computation and power system computation in one place for a post graduate level student.

While sparsity is a major issue in power system computation, the EMS applications have another dimension of software engineering. Articles in IEEE Computer Applications in Power usually capture this flavor. Around 1996, we started exploring the utility of the OO paradigm to scientific computing. Specifically, we were looking was whether this much hyped paradigm was useful in the context of traditional computing. To begin with, we were a bit sceptical but soon the 'hype' cleared and we started seeing benefits in programming, debugging and the architecture of the programs (software). In our research we set some simple but achievable goals. We were not interested in redesigning or inventing new algorithms for power system analysis application which are based on message passing. We thought that if we can develop a library of classes that simplifies programming and write applications without any degradation in speed of computation, it should be sufficient. The following few years were quite interesting and eventful. Being FORTRAN programmers earlier, the benefits of OO programming and design were a revelation. Library classes were written for power system objects, sparse matrix etc.

In the year 2000, we had an interaction with Prof. M. A. Pai and the idea of book then took a concrete shape. While the initial idea was to write a monograph of OO applications to power system analysis, the idea was refined later so as to cover the overall framework of large scale computing for power system analysis.

Each book has its own style. In writing this book, we have tried to give a unified treatment of OO computing, sparse matrix techniques and its applications to power system analysis. The book is divided into 3 parts. The first part is on OO computing. It introduces the OOP paradigm and data structures like linked list which are used to implement a sparse matrix data structure. A few important classes like sparse matrix, set, graph, etc are introduced. A classification of objects and an architecture for power system analysis applications is developed. While the chapter 2 has introductory flavor, we expect that it will also be enjoyed by those who have done a lot of programming in power system analysis. The chapter 3 on data structures provides a systematic treatment of issues involved in storage, scalability and flexibility in accessing and modifying sparse matrix elements. New computation classes like graph, set are introduced to facilitate sparse matrix computing.

The second part of the book deals with large scale computing. In this part, we develop a large sparse symmetric positive definite linear system solver which is a workhorse in most power system analysis applications (chapter 4). In chapter 5, a sparse QR decomposition tool is developed which is popularly used for numerically stable implementation of state estimation. Chapter 6 and 7 introduce and develop sparse matrix methods in optimization required for applications like Optimal Power Flow. This concludes the part II of the book.

The third and the final part of the book deals with applications. Load Flow Analysis and Short Circuit Analysis modeling, algorithms and OO implementation is covered in chapters 8 and 9 respectively. A noteworthy aspect here is that by using templates, the same implementation of linear system solver is used for both load flow analysis and short circuit analysis. The chapter 10 introduces the state estimation. All the components of state estimation *viz.*, NTP, observability analysis, state estimation and bad data processing are discussed. The treatment of state estimation chapter is such that it emphasizes (1) optimality issues (2) effect of uncertainty (3) numerical methods and (4) OO design. Chapter 11 introduces the OPF application. The chapter provides a systematic review of various computational methods for OPF. The class of sparse optimization methods which directly compute load flow and optimal solution *viz.*, Newton method and SQP are discussed in detail. The chapter 12 introduces computational issues in power system dynamics.

This chapter is more in the nature of a brief overview. Usually, Power System Dynamics is covered as a separate graduate course and many of the intricacies of computation and modeling can be understood only after going through such a course.

The book can be used for teaching a one semester power system analysis course at the graduate level as well as an independent course on large sparse matrix methods. In writing this book, we have tried to keep treatment of the chapters fairly independent (the OO design principle of decoupling and cohesion). Thus, if power system analysis applications have to be emphasized, then the part 3 can form the bulk of a course. It can be taught without getting into the nitty-gritty of OO and sparsity issues in LSS, QR etc. (chapter 2-5). Alternatively, for those already acquainted with power system analysis, sparsity issues can be emphasized by covering chapter 3-5 in detail. In our experience, an optimization course is not usually taught at the undergraduate level. Hence, chapter 6 and 7 introduce optimization methods and the instructor can give a complete treatment of optimization through chapters 6, 7 and 11. The chapters on load flow analysis and short circuit analysis can be used in an undergraduate course and dealt in brief in a graduate course.

Most of the chapters are supported with review questions. To quote Herstein [Herstein, 1975], "The value of a problem is not so much in coming up with answer as in the ideas and attempted ideas it forces on the would-be solver." For number crunching, the programs provided on the diskette can be used. We recommend that the readers should dwell on class interface in the diskette. Not all the material presented in the book can be comprehended in first reading. Hence, we have marked * to those sections which can be skipped in the first reading. The applications discussed are rich in diversity, both from a computational perspective (numerically intensive, graph theoretic and dynamical systems) and domain of application (off-line applications e.g., in planning, and online applications e.g., in energy control centers). The thrust of the book is to provide tools that make large scale program development practical and enjoyable.

As this is the first edition of the book, it is possible that some aspects of the subject may not be adequately covered. It is also likely that there may be some errors, typographical or otherwise. We welcome feedback on such errors as well as suggestions for improvements.

To Mata, Pita and Guru

Acknowledgments

The book in its evolution went through many revisions. The authors benefited tremendously due to the comments, criticism and suggestion from friends, colleagues and research scholars. Dr. Rishikesh Joshi and Dr. B. Menezes, helped authors in developing the perspective on OO modeling. Ms. Madhuri Patwardhan and Mr. Devkar gave valuable suggestions on the chapter-2. The chapter-3 on data structures was reviewed by Dr. Milind Sohoni. His sense of humor made the discussions entertaining. Dr. Mahesh Patil, reviewed the chapter of Linear System Solver. His expertise in handling sparse matrix computations helped in revising the initial drafts. Ever smiling Dr. Hemchandra did a meticulous job in reviewing the optimization chapter, the end result being that one of the authors (S.A.Soman) has to teach a course of linear and non-linear optimization! The chapter on sparse QR decomposition was reviewed by Dr. A.M. Kulkarni. He was generous enough to approve the initial draft with minor changes. The chapters on Load Flow Analysis, Short Circuit Analysis, State Estimation and Optimal Power Flow went through a number of revisions. Significant improvement and changes were suggested by Prof D. Thukaram of IISc Bangalore, Dr. H. Mangalvedekar of VJTI Mumbai, and A. M. Kulkarni, K. N. Shubhanga Aithal of IIT Bombay. The chapter on OPF was reviewed by Prof. D. Thukaram. The authors are grateful to Prof. Thukaram to have come all the way to Mumbai to give suggestions and final inputs when time was running out!

The initial blueprint for the book did not have a chapter on dynamics. The inclusion of dynamics chapter with a computational viewpoint was suggested by Prof M. A. Pai, UIUC and Dr. A. M. Kulkarni drafted the chapter. Prof Pai was a source of inspiration all through the work. His prompt response to email queries, suggestions in writing style and keeping the overall tempo strong can not be forgotten. Authors are thankful to Prof. Milind Malshe, Humanities Department for correcting the chapter 1 and providing tips on good writing style. Mr. Nitin Joshi typed most of the manuscript and figures in TEX. His expertise, sincerity and hard work helped in maintaining the tempo of writing and made writing a pleasure. The book could not have been completed without the help and support of many friends and colleagues. In particular, we would like to acknowledge the support and encouragement received from the Profs. B. G. Fernandes, Kishore Chatterjee, Mukul Chandorkar, Madhav Desai, D. Manjunath, U. B. Desai, V. R. Sule, Dr. A. Pandian and Dr. Mukesh Taneja.

Mr. Alex Greene and Ms. Melissa Sullivan of Kluwer Publication were helpful throughout the project. The interaction with them on email helped in keeping the book at the top of our priorities.

Last but not the least, the book would never have been completed without the love and support of family members, who had to endure our excessive working hours because of the 'book'. S. A. Soman would like to place on record the love and moral support of wife Meghana, son Jay, parents, brother Dr. Harshavardhan Soman and Taranekar 'Vahini'. Shubha Pandit would like to acknowledge the love and support of her husband, children, in-laws and parents.

S. A. Soman
S. A. Khaparde
Shubha Pandit

30 August 2001
Mumbai

Chapter 1

INTRODUCTION

The story begins in the late 1960s when there was a concerted effort to develop new operating system techniques.

—Richard Petersen, Linux:The Complete Reference

1. History of Power System Computing

Use of digital computers for power system simulations gained importance during the early fifties. Load flow analysis [Ward and Hale, 1956] was one of the first computer applications in power system analysis. Before the digital computers era, analog computers and scaled physical models were used for power system computation and simulations. For example, analog computers were used to implement load flow like computations. They were called ac network analyzers and used to occupy a large floor area. The sixties and the first half of the seventies witnessed many important breakthroughs in computational methods for power systems like Optimal Power Flow (OPF) algorithm [Dommel and Tinney, 1968], the electromagnetic transients program [Dommel, 1969], power system state estimation [Schweppe and Wildes, 1970], transient stability [Dommel and Sato, 1972], and fast decoupled load flow [Stott and Alsac, 1974]. Since then, the field of power system computations has grown rapidly. In the last five decades, power system computational techniques have evolved dramatically and have achieved a high level of sophistication and maturity. In tandem with the digital computer revolution, many traditional barriers and constraints which existed earlier have been overcome. The advantages of digital simulations over analog simulations are:

- **flexibility:** Many complex scenarios can be modeled and simulated without extra hardware development. Further, it is easier to upgrade programs than the hardware.

- **reduced costs:** Once the computational facilities are established, the only overheads in digital simulation are computer time and programming effort which, far outweigh investment cost in analog systems.

- **large system simulations:** Today's programs can handle systems with thousands of nodes routinely by the exploitation of sparsity. The increasing size of power systems is due to the detailed system modeling and integration of smaller power systems into larger grids. The computation speed is suitable for on-line computations.

2. Present Scenario

Over the years the computational power (one measure of it is 'flops') of the computers has increased tremendously. At the same time, the computational requirements for analysis of power system problems has also increased. This growth in the computational requirements is mainly due to the following reasons:

- **complexity:** Power systems are inherently complex because of the dynamical nature of power apparatus (e.g., a generator), use of complex technologies (e.g., high voltage ac and dc transmission systems) for transmission of power and control (e.g., real time protection, Automatic Generation Control (AGC) and Energy Management Systems (EMS)).

- **automation:** The almost complete dependence of modern society on electricity calls for a secure and optimal operation of power systems. This can be achieved through automation. Historically, digital electronics was first introduced at the substation level as a Remote Terminal Unit (RTU). The aim was to monitor the transmission system through a Supervisory Control And Data Acquisition (SCADA) system which acts as backbone in the energy control center. Limited processing was envisaged at the energy control center. With the phenomenal growth in computation and communication, at an affordable price, *automation* at transmission, distribution and substation level has become a reality. We are witnessing a movement from uncontrolled to automated systems.

- **decision making by simulation:** Simulations play an important role in the decision making at both the planning (off line) and operation level (online). Considering the high level of reliability expected from electrical energy systems, it is imperative that simulations depict accurately the real life system. The quality of simulation depends upon (1) data (2) modeling and (3) solution methodology. To have a higher level of confidence in simulations, it is now common to model systems to a greater detail.

We may say that high level of belief in power system simulations increase the responsibility of the power system software professionals to provide more powerful, reliable and user friendly tools for large scale system simulations.

3. Purpose & Scope

This book is about *modern* computing as applied to power system analysis. In other words, the emphasis of the text is on computational methods which drive the application functions like load flow analysis, power system state estimation, Optimal Power Flow (OPF) etc. While writing a text book on power system analysis, usually emphasis is placed on model building (e.g., π model of a transmission line) and development of tailor made algorithms for specific applications (e.g., Fast Decoupled Load Flow). In this book, we endeavor to take a different route to learning; a path which has been far more instructive to us.

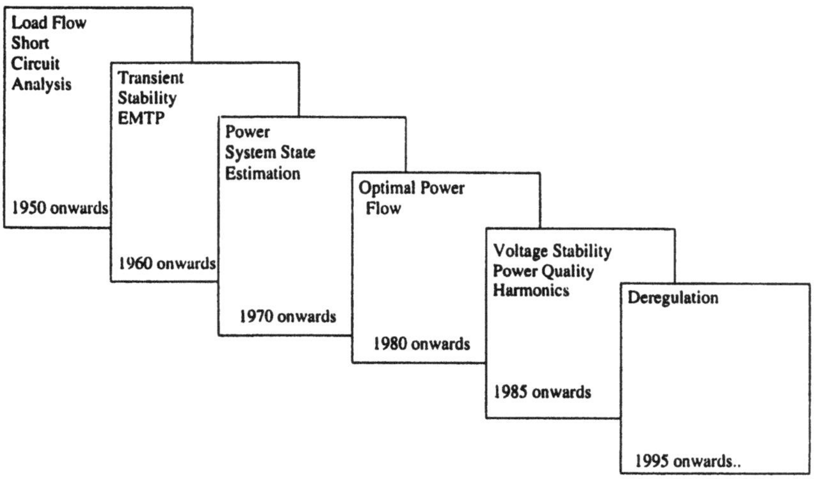

Figure 1.1. Evolution of Power System Analysis

The area of power system analysis and computing is rich with diverse application problems and tools to solve them (Figures 1.1)[1]. This richness is a result of assimilation of expertise over a long period of time, *i.e.*, *evolution*. New problems as well as new tools evolve over a period

[1]Years mentioned reflect when applications were widely accepted as technologically viable.

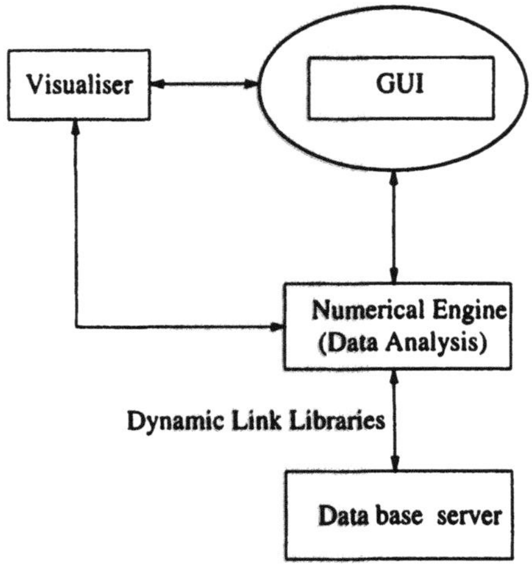

Figure 1.2. An Information Oriented Architecture

of time. All through the evolution, physical laws governing the application remain the same. This introduces an element of commonality in modeling and solution methods of the multiple power system analysis applications which can be exploited in design and implementation phase. To cope up with the complexity of the computations what is required is to *highlight* the important tools used to solve such problems. Considering the diversity of the tools from the perspective of numerical stability, sparsity exploitation, data structures, etc., we stress a unified approach to computations. This is in contrast to the traditional paradigm which primarily stresses applications. The approach followed in this book is to analyze multiple applications for *commonality* and then develop a broad computational framework that is amenable to large scale computing.

Large scale computing has two dimensions: (1) the complexity of the software, and (2) the complexity of computation. One measure of the complexity of the software is the number of lines of programming involved. Object oriented development is one way to handle this complexity. When diverse programming languages are used to develop a software, this complexity can be handled by *component* technology. The complexity of computation arises due to the need to handle large scale power systems. It is usually handled through exploitation of sparsity. In this text, the sparsity issues are not considered *in addendum*; rather, they are the heart of application functions. In this way, one can harness technology of sparse matrix computation [Sergio, 1984] that makes efficient large scale programming and application development possible (typically in thousands of nodes). By identifying a critical set of the computational tools, we control the complexity of computation and therefore the associated developmental effort and risks of programming.

The first generation power system computation tools were programs with limited capability. A program implemented an application functionality in isolation; while the earlier programs had an application oriented architecture, today's software is based on information oriented architecture.

An application has many facets, e.g., (1) that of the end user and (2) of a programmer. We refer to programmer as the one whose key responsibility is to develop application functions. The perspective of a programmer requires a discussion on the software design and programming languages. However, this perspective is quite different from that of an end user. Obviously, an end user expects the specified application functionality. He is also concerned with the interpretation of the information in a way conducive to human perception. He does not worry about the algorithms and computational methods. Usually the stress is on interpretation of results, data integrity, navigability between applications, visualization etc. This leads to a software architecture which has a database layer, an application layer and a visualization layer (fig-1.2).

Further Reading

[Elgerd, 1982] provides an excellent treatment of the electric energy systems theory. [Kundur, 1994] provides a comprehensive treatment right from the basics to the state of art of power systems technology. Both these books emphasize power system theory, *i.e.*, modeling, system interaction etc. [Stagg and El-Abiad., 1968] is one of the first books on computational methods in power system analysis. [Pai, 1979; Kusic, 1986; Wood and Wollenberg, 1996; Heydt, 1986] deal with the computational methods in power system analysis. Sparsity issues and programs/algorithms for large power system analysis are stressed in [Brown, 1975; Wallach, 1986; Alvarado et al., 1991].

Chapter 2

OBJECT ORIENTATION FOR MODELING COMPUTATIONS

By a process of abstraction we get behind the layers of body, mind and intellect and reach the Universal Self, "the true light which lighteth every man that cometh into the world".

—Dr. S. Radhakrishnan, Vice-President of the Republic of India

1. Introduction

A power system is composed of *objects* like generators and loads, which are interconnected by *objects* like lines and transformers. The physical laws provide a mechanism whereby objects collaborate to provide higher level behavior (system operation). This fundamental object based structure of power systems provides a natural mapping of real life objects to program objects. Traditionally, power system computation has been done using procedural programming languages like FORTRAN, PASCAL and C. The question as to whether a procedural programming or an Object Oriented Programming (OOP) paradigm[1] is better for the development of power system applications is debatable. The success story of the procedural methods makes a strong case for *status quo.* The aim of this chapter is to demonstrate the utility of object oriented analysis, design and programming techniques for power system analysis. We argue for a paradigm shift from procedural to object oriented computation. Basic concepts of object orientation and its application in analysis and design are reviewed. A design framework for power system analysis

[1] We use a single term OOP to include object oriented analysis, design and programming.

is developed. Let us begin by reviewing the procedural programming paradigm.

1.1. Procedural Programming

Many of the power system analysis programs and numerical methods have been written in FORTRAN (FORmula TRANslation). The choice of the language of implementation has a significant influence on how we go about tackling a computational problem. This is because the programming language provides nuts and bolts with which we build programs, the end products.

The benefits of procedural programming in power system analysis are as follows:

- Analysis problems are expressed through a set of equations (linear, nonlinear, differential etc). Languages like FORTRAN and C can cope with this level of abstraction with ease.

- The well established computational techniques are essentially algorithms which can be implemented with ease in procedural programming.

The focus in this traditional programming paradigm is on the desired computations. It can be said that the *procedural programming* is based on the following principle [Stroustrup, 1991]:

Decide which procedures you want; use the best possible algorithm you can find.

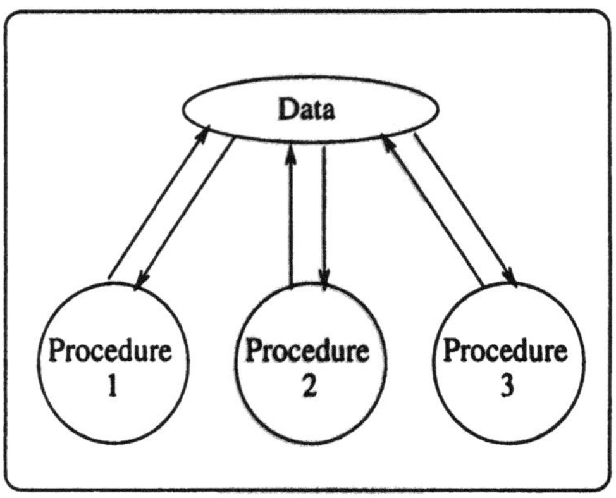

Figure 2.1. Procedural Programming

Figure 2.1 shows the generic structure of a program using procedural programming language, wherein:

- the data is more or less global;
- data is used, shared and modified by procedures that act on it. The temporary association is obtained by parameter passing;
- the limited modularity of the program arises from the functions or subroutine libraries.

The biggest drawback of the procedural programming paradigm is that it does not emphasize data abstraction. With every function or subroutine the data has to be passed through parameters. Limited amount of grouping can be obtained through use of COMMON blocks in FORTRAN, and structures in C and PASCAL. The data structure, models of real-life objects (like π-model of line) are all open. Also, there is no association of data and its methods. Almost everyone has a freedom to access and modify the data. No security checks such as whether the function has a privilege to modify the data are implemented. *Consequently any ripple of change during evolution carries over to many procedures, subroutine or functions.* This makes large scale programming fragile and unwieldy. Procedural languages are not geared to handle complexity of large scale systems. The next section gives a flavor of how power system computations can be elegantly handled in an Object Oriented (OO) framework.

2. Object Oriented Programming

This paradigm is meant for *programming on a large scale.* OOP languages provide a cohesive binding of data to the methods or functions or procedures that act on it. The data associated with an abstraction is well encapsulated in an object[2] and only prespecified methods in the interface can access it and/or modify it, *i.e.*, change the internal state of the object. The interface functions need not contain details of data structure and models. In other words, interface hides data structure and models. *Thus, the ripple of change due to changes in data structure and model is restricted to the corresponding object.* The principles of object oriented programming can be captured through few guidelines as given by [Stroustrup, 1991] as follows:

Decide which classes you want; provide a full set of operations for each class; make commonality explicit by using inheritance.

[2]Object: something that has crisply defined boundaries; an object has state, behavior and identity [Booch, 1994].

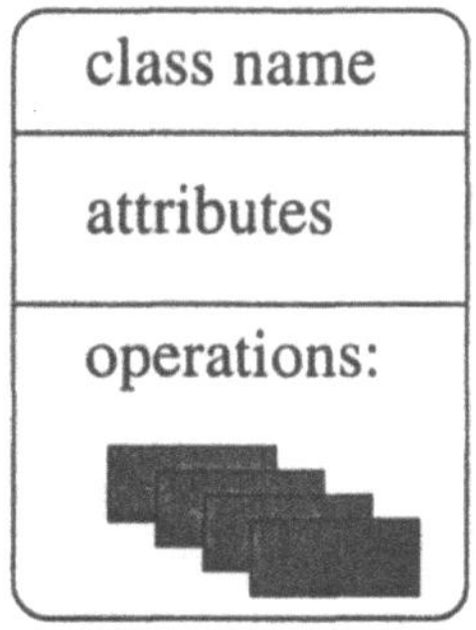

Figure 2.2. **Class Representation**

For example, a 'line' object encapsulates[3] the model (say the π-model) of a transmission line. A user of the object 'line' need not be made aware of these details. A user can simply query the object for say power flow across terminal node names. Such a functionality is provided by a *method*[4] defined on the object. Further, outside access to the object can be restricted, thereby, preventing unwarranted and unforeseen corruption of the data. In this book, we will develop a number of objects as we proceed. Structurally similar objects are specified through a *class* which is a user defined type. An object is then simply an instance of the class. For example with a class called *Line*, the declaration *Line* A[100] creates an array A of 100 line objects. Each object $A[i]$ is called an instance of the class *Line*.

2.1. Class

Abstraction is an important concept in OOP. It involves a process of focusing upon the essential characteristics of an object. In design stage, it is neither feasible nor desirable to capture all the details of any real life object. It is required to capture the desired perspective of an object's behavior. As an example, let us consider the interface of class *Line*.

In most of the applications that we come across a line is represented by a π model. Hence, we make *r, x, by2, from & to* as the attributes of the class *Line*. Similarly, each line has two terminal nodes which also form attributes of the class.

[3]Encapsulation provides compartmentalization of structure and the behavior. Therefore, encapsulation separates contractual interface from implementation.
[4]Similar to a function in C and a subroutine in FORTRAN.

```
Class Line {
private :
 float r, x, by2;
 char * from, to;
 ....
public :
  Line(); // states will be initialized to a default value
  Line(float r1, float x1, float by2, char * f1, char *f2);
  // attributes of a line required during instantiation
  ComputeFlow(const float & v1, const float & v2);
  ~Line(); // dtor
  .....
};
```

A method having the same name as the class is known as constructor (ctor). A constructor initializes the object *i.e.*, defines its default state. Usually, an object may be created by many ways and hence many constructors may be specified. Two ways of constructing a line can be found in the interface above. Declarations like *private* and *public* govern access from the outside world. An object needs a destructor (dtor) to destroy it cleanly when its use is over. Function *ComputeFlow* computes the flow across the line. A class interface as shown above is usually listed in a header file (.h file). *Line* class will be redefined and discussed in detail in chapters on Load Flow Analysis and Short Circuit Analysis. We now introduce the concept of parameterized class.

2.1.1 Templates

Many OO languages like *C++* support a kind of syntactic polymorphism, also known as *templates*. *Templates* permit implementation of an algorithm with variable data types. Quite often, one encounters situations wherein similar (or identical) operations have to be written for different data types. Consider the example of addition of two matrices of *real* numbers. The same operation needs to be performed on matrices having *integer* numbers. While the *actual* operations involved in both the cases are identical, the data type or objects on which they operate are different. This calls for a separate piece of code for each of them. This problem has been very conveniently handled in *C++* , by permitting *parameterized* classes. In such classes, methods are defined for an arbitrary data type or object. When the class is instantiated, a concrete data type needs to be specified. This provides *C++* programmers with economy of expression and generalization not available in traditional languages like C. A class *Matrix* implemented with templates appears as follows.

```
template<class T>
class Matrix<T>
 {
  private :
      int row, col;
       T  * value;
      .....
  public :
     Matrix () {row=col=1; value= new T [row*col]; };
                    // default ctor
    ~ Matrix () {delete [] value; }; // default ctor
     Matrix (int Row, int Col, T Val); //ctor
     Matrix(Matrix<T> & A); // copy ctor
     friend Matrix<T> operator +(const Matrix<T> &A ,
            const Matrix<T> &B); // addition operation
     Matrix<T> & operator =(const Matrix<T> &B);
                   // assignment operator overloaded
     ...
 }
```

Following matrices can now be created as instances of this class.

```
Matrix<float> A; // matrix A initialized with
         // elements having float data type
 Matrix<int> G;  // matrix G initialized with
         // elements having integer data type
```

In short, *template* is a specification for a family of related classes, which allows *family of functions* to be defined over a *family of data types.* Usage of templates involves no run time overhead. Class Matrix has been supported in the diskette along with this book.

2.1.2 Operator Overloading

The process of using the same symbol for more than one operation in a single context is called operator overloading. *C++* allows it to be extended even up to the user defined types or classes. Overloaded operators make the code look more natural as they conceal the mechanics of the operation but emphasize on the effect of the operation. Overloading functions and names serve primarily a notational convenience. Many standard arithmetic and logical operators like $+$, $-$, $\%, =$ and $<, >$

can be overloaded in C_{++} to provide similar interpretations over objects. For example, operator +, = can be found overloaded in the interface for class *Matrix* above. Use of these operators will let a program for addition of two matrices appear as follows:

```
    main()
    {
// this program creates 2 10X5 float matrices adds them
// to another matrix C. Two type of ctors are used.
      .....
      Matrix <float>A(10,5,1.0), Matrix <float>B(10,5,2.0);
      Matrix <float>C;
      C = A + B;
        .....
    }
```

A constructor for class *Matrix* can initialize a 10×5 matrix A, with each element having a value as 1.0. Matrix B, of the same dimensions, can be initialized to a value of 2.0. Another matrix C has been invoked by the default constructor of class *Matrix*. By overloading operator + to perform matrix addition, and operator = to assign the value, a very simple and elegant code, for addition of two matrices, can be written. In case of matrices, operator A < B can be used to order matrices A and B based upon matrix norm e.g., Frebenious norm. The feasibility of the expected operation *e.g.*, dimension checks can be performed as a part of the method relieving user the burden. Operator overloading is extensively used in this work and is discussed at appropriate places. Operator << (to read data) and operator >> (to print data) can similarly be overloaded for classes such as *Line*, *Transformer* etc.

2.2. Advantages of Object Orientation

It is now possible to elaborate on some of the advantages of object orientation for power system computation. They are as follows:

- Object oriented paradigm facilitates an implementation which is closer to reality. For example, we can have a line object only in the OOP language. Procedural languages like C have structures which can group data; however, it is not possible to associate methods with them.

- A strong coupling between a model and related methods can be established. For example, power flow in the `Line A` can be computed invoking method $ComputeFlow(v1, v2)$ on the object A (in

C_{++} syntax: as $A.ComputeFlow(v1, v2)$). Thus, using OOP, we are able to create syntax and models which closely matches the vocabulary in the problem domain. Further, an instance of another class say *Transformer* can also have a method by the name *ComputeFlow* to compute power flow in a transformer. It is even possible to have methods with same name in the same class but with varying signatures. Multiple constructors provide a trivial example of the same.

- Information hiding *i.e.*, details of models, data structure etc are hidden from the end user of the object. Information hiding facilitates encapsulation of data. A classic example is that of a sparse matrix. If it is implemented in a procedural language, the data structure is more or less bare. However, in a object oriented implementation, a user gets a feel of one cohesive entity. This cohesiveness makes a lot of difference in complexity of programming. For example, if a program could not model an object *car*, but instead treats it as an aggregation of an engine, a steering wheel etc, one can imagine the trouble in writing a simulator to train a driver. Class *SparseMatrix* will be discussed in more detail in the next two chapters.

While a class and its methods provide the basic framework for implementing an Object Oriented program, there is lot more to OOP. The next section reviews the concept of inheritance.

2.3. Inheritance

An important characteristic of complex systems is that they exhibit hierarchy. OOP languages provide an elegant way to capture hierarchy by *inheritance* or derivation of a subclass. A subclass inherits all the attributes and methods of the base class (also known as super class) and builds a bigger or more complex object on this foundation. Consider the example of a class *Bus* which has a name (identity), complex voltage, active and reactive power injections as its attributes. The behavior of PQ and PV buses can be captured by derivation from class *Bus*. Thus, inheritance facilitates reuse of classes. A method can be redesigned or redefined in a subclass, thus making an interesting example of code reuse. However, it introduces tight coupling in the program architecture. It is quite likely that with hindsight, which is sometimes far more revealing, one may disagree with this derivation. This is because a tight coupling makes the binding permanent. It is a bit early to illustrate this point as we have not yet developed sufficient number of classes. This aspect is illustrated by derivation of class *Graph* from class *SparseMatrix* later on in chapter 10.

2.4. Polymorphism and Dynamic Binding*

When a subclass is derived from a base class, it is possible to redefine the implementation of the methods is the subclass. Further, by dynamically binding the method, a single signature of the method in the program can handle multiple implementations of the function. C_{++} permits the same by use of *virtual* functions. Methods which are only declared as a part of the base class but implemented in the derived class are known as abstract methods or *pure virtual methods in* C_{++} . They are used to design a generalized interface for many specialized object usually implemented in subclasses.

Inheritance and polymorphism provide two very powerful tools that help in (1) programming to an interface and (2) programming in future. We illustrate the utility of these concepts by considering the example of a Linear System Solver (LSS) for various types of matrices.

2.4.1 Linear System Solver

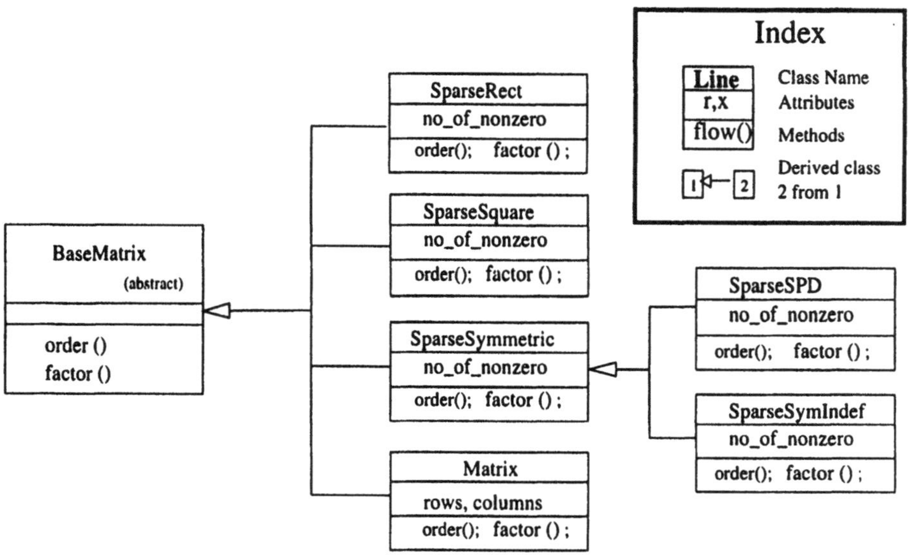

Figure 2.3. Derivation of Matrix subclasses in Unified Modeling Language

Consider the interface design of a Linear System Solver (LSS), which forms the backbone for many power system computations. Ideally, one would desire to have an interface which hides all the details of data structure for both, matrix and vector. Further, one expects implementation to be generic to many types of matrices. Thus, a generic interface and code for LSS should be as follows.

```
main() {
  int n=100;
  Matrix A(n,n);
  SparseSPD Y(n,n);
  vector b(n), v(n);
   ...
  LSS(A,b);
  LSS(Y,v);  //polymorphic code
  ..
  }
```

Development of a unique LSS interface can be achieved by defining an *abstract* class[5] *BaseMatrix*, from which specialized matrices can be derived as shown in figure 2.3. Given a program as shown above, specifications of actual implementation of LSS will vary from one matrix type to another. Using polymorphism with the same signatures, we can invoke specialized implementation. For example, if we pass matrix Y, which is an instance of derived class *SparseSPD*, corresponding implementation will be invoked. A large sparse LSS requires methods such as *order(), factor()*. Appropriate implementation depending upon the type of matrix has to be designed. For example, in case of both sparse symmetric and sparse Symmetric Positive Definite (*spd*) matrices, *order()* may be implemented by Minimum Degree Algorithm. While *factor()* in sparse *spd* matrices is implemented by Cholesky factorization; in sparse symmetric matrices, it can be implemented by incorporating 1×1, 2×2 diagonal pivots. As such, LSS implementation in turn would call relevant methods for ordering, factorization etc. as shown below:

```
 LSS(BaseMatrix & A, vector & b)
{
// invoke matrix type dependent implementation
     A.order();
     A.factor();
     A.solver(b);
                        //  vector x overwrites vector b
};
```

It should be noted that this *programming to interface*, coupled with hierarchical class implementation helps us to have a generic architecture. This single interface support is referred to as *polymorphism*[6]. Polymorphism promotes reuse of design/pattern/concepts.

[5] class having a common interface for its subclasses.
[6] polymorphism means having many forms.

The figure 2.3 also brings out another important concept which is representation of a design in a language understandable to the other OO literate. The design of the LSS is illustrated using a specific modeling language *viz.* Unified Modeling Language (UML)[Fowler and Scott, 2000]. UML has been accepted as the standard by Object Management Group (OMG). It is a language to express design; it does not specify analysis, design methodology or a developmental process. In fact, now it is possible to build a software with multiple programming languages. 'Component technology' which is now getting wide acceptance is based on the paradigm of connecting objects, whereby a previously developed software can be reused across platforms even in a distributed environment. For example, on a Windows system, executable components can be connected through a Dynamically Linked Library(DLL).

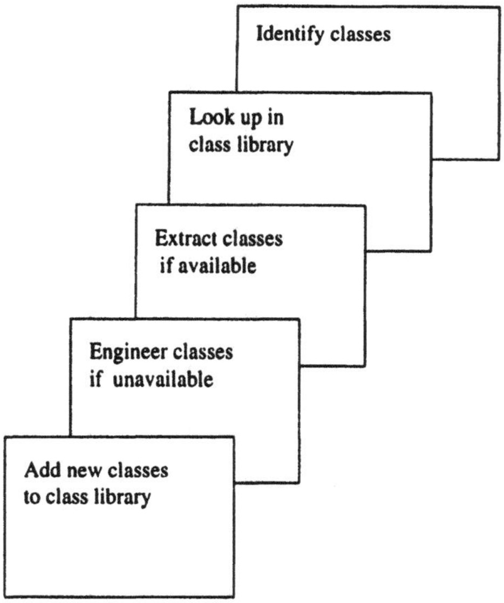

Figure 2.4. OO development in a nutshell

In the next sections, we view various power system applications in an OO perspective. Figure 2.4 illustrates important steps involved in developing an object oriented software. We begin by analyzing the power system analysis domain.

3. Analysis, Commonality & Classification of Objects

Analysis is an intensive activity in classification of a system into components that can be modeled as classes. Identification of classes is an iterative process. Usually, an elaborate and precise statement of a problem clearly brings out the important classes to be modeled for a given application. Also, details within the application statement influence this identification process. Let us consider an example.

Example 2.1 Identify the objects for developing a load flow program.

The first step is to define a semantic problem and solution statement.
Problem Statement: Given the network details like lines, transformers, loads, generator, shunts and connectivity, compute the complex voltages at all the busses for the system.
Solution Statement: The problem involves solving a nonlinear system of equations e.g., by Newton Raphson's approach. Note that an alternate solution statement could be to use Gauss Seidel or Fast Decoupled Load Flow method.

Analysis: From the problem statement, we realize the need to have classes for lines, transformers etc. The connectivity information requires further creation of a class *Network* that aggregates these network elements. The solution domain requires an iterative solver. The algorithm for load flow can be encapsulated by an algorithm object (which will be discussed soon). At every iteration of a load flow, a Linear System Solver (LSS) is required. Therefore, a class *SparseMatrix* with a method for LSS is required. Note that since the above statement is not detailed enough, we do not realize the need for class *Bus*, or class for PV and PQ nodes. An iterative refinement of the application statement will lead to identification of all such classes.

In general, description of an application involves (1) a definition of the problem and (2) a solution methodology. However, it is worth while to point out that, a *semantic* statement of any problem is not unique. A statement that encompasses problem definition and solution statement is referred to as *application statement.* Once, an application statement is formulated, it can be analyzed for identification of classes.

Solution methodology decides identification of computation classes. Elaboration of solution methodology leads to refinement in problem definition and vice versa. Usually, the two are not mutually exclusive, but have a relation between them as shown in figure 2.5. A very important question in analysis phase then becomes, as to *"how best to frame the application statement for an OO paradigm?"* Following rule can be formulated to reflect this:

`Rule:    capture commonality among multiple applications.`

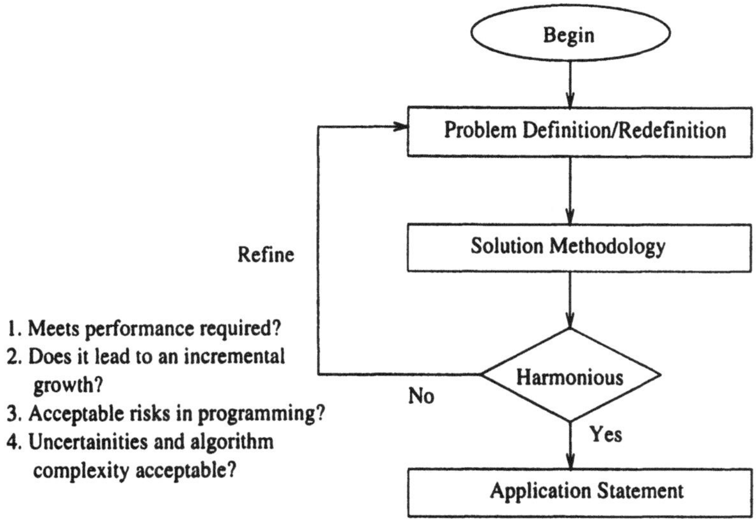

Figure 2.5. Cyclic Process for Application Statement

A good design requires foresight. A designer should identify and look for requirements of similar applications. The rule embodies the essence of integrating multiple applications, thereby promoting code reuse (classes). For example, applications like Load Flow Analysis, Short Circuit Analysis, State Estimation have large commonality in numerical computations. Applications like Observability Analysis, Network Topology Processing, Relay Coordination are all graph theoretic in nature. Extracting commonality requires study of similar applications, not necessarily from the point of view of the end user.

To organize commonality in computations and applications in the power system analysis domain, object classification (figure 2.6) is proposed. We classify objects broadly into following two types (1) data objects and (2) algorithm objects. Data objects usually correspond to physically identifiable objects like the lines, transformer etc. They encapsulate models of power system apparatus. The complexity of design is negligible as the models are well known, standard and easy to engineer. Data objects provide persistence of data and are generally stored in an off line data base. Primary function of data object is to store data, while that of algorithm object is to manipulate some data object(s) and return new information. Most of the data objects have high level of openness to avoid cumbersome programming. Their classes can be made to support simple methods like read and write, implemented by overloading >> and << operators in C_{++} .

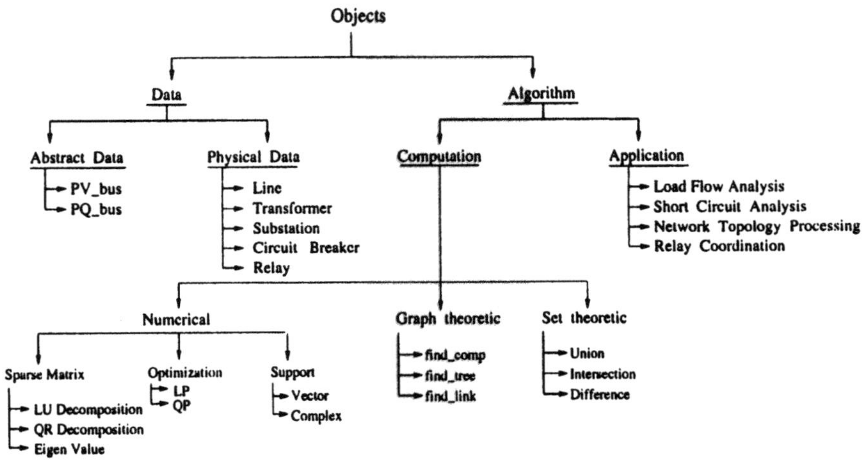

Figure 2.6. Classification & Glossary of Objects used in the Book

An algorithm object stresses non trivial implementation algorithms. It can further be classified into either a computation object or an application object. While computation objects may store and manipulate data such as matrices, an application object may store and manipulate data like a network. The following sections discuss modeling of each of these objects.

3.1. Modeling of Data Objects

Digital simulation to study interactions in power system is based on models of the power apparatus, lines, transformer, generator, reactors or loads. Modeling of apparatus is well established in literature [Hakavik and Holen, 1994; Zhu and Lubkeman, 1997]. Data objects usually abstract the behavior of power apparatus as per the requirement of an application. *A rule of thumb is to design a class for each apparatus that features in the vocabulary of the problem.* It is easy to model a data object, because of its close correlation with the physical object. On the other hand, some data objects arise because of the application philosophy, for example, a PV node and a PQ node in Load Flow Analysis.

A data object model can have two types of attributes: *primary* and *secondary* or derivable attribute. The primary attributes considered for an object line can be resistance (R), inductance (L), capacitance(C) per unit/length, connectivity details such as from node name, to node name in a substation, length, etc. Derivable attributes are modeled because either they feature significantly in the vocabulary of the problem domain or they provide simplicity in implementation. A model captures specific aspects of the system under consideration. Every model has its limitations. For example, a line object which implements a π model of a line cannot capture behavior at the intermediate points. For many applications like Load Flow Analysis, this level of modeling is sufficient. A distributed circuit model may be required while considering applications that have to capture fast transients.

Modeling is an application dependent activity. As such, in application like Network Topology Processing, it is required to model a circuit breaker & substation in detail, while in a Relay Coordination application, a wide range of relays may have to be modeled. Classes for each of these *supporting* power apparatus are identified and designed and put in a library for future use (figure 2.4). To read the parameters of an apparatus, input operator can be overloaded and made intelligent enough to identify nature of the data. At this point of time readers are encouraged to study `struct.h` file in the diskette supplied with this book for details of the data objects.

3.2. Designing Computation Objects

Typically, power system applications are computationally intensive. Computation object design focuses mainly on *data structure* and *algorithms.* The computation activity in power system analysis algorithms usually falls in any one of the following category:

1 Sparse matrix (numerical) computations;

2 Graph theoretic computations

3.2.1 Sparse Matrix Computation

It was realized in the early sixties that sparsity, if exploited successfully, can lead to drastic reduction in computation and storage requirements [Tinney and Walker, 1967]. Storage requirements of sparse matrices can be substantially reduced by storing only non-zero elements. Computational requirement can also be minimized by restricting computations to only non-zero elements. Following applications, involving sparse matrices, are typically encountered in power system computations.

- Solution of a linear system of equations with symmetric positive definite (*spd*), symmetric indefinite and complex symmetric diagonally dominant coefficient matrix. The application is developed in chapter 4 and used in load flow and short circuit analysis (chapters 8 and 9).

- Solution of linear and nonlinear optimization problems: The Jacobian and the Hessian matrices encountered in Optimum Power Flow problem are sparse. Only a subset of the numerical methods for optimization can exploit sparsity [Gill et al., 1984b]. This application is developed in chapter 6 and 7 and used for OPF in chapter 11.

- Numerical rank computation of the sparse matrix: Such problems arise in observability analysis required in state estimation and *static* voltage stability indices computations. Chapter 5 develops a sparse QR factorization solver and it is used in chapter 10 on power system state estimation.

- Eigenvalue computations in various stability problems. The chapter on power system dynamics (chapter 12) dwells into this application in more detail.

The field of sparse matrix computations including ordering is complex enough to justify a separate Object Oriented design and implementation. Libraries like *SPOOLES* [Ashcraft and Liu, 1999] and *SMOOTH* [Ashcraft and Liu, 1996] deal with the same.

3.2.2 Graph Theoretic Computation

The computational activity discussed so far has been numerically intensive. Applications such as network topology processing, observability analysis and relay coordination can best be solved by a graph theoretic approach. In addition to these applications, ordering methods like Minimum Degree Algorithm (MDA) used in sparse Linear System Solver (LSS) and QR decomposition of a sparse matrix are elegantly described by graph theoretic notations. This requires development of a class *Set* and class *Graph* for such applications. While considering a graph, vocabulary of the problem domain revolves around phrases like identifying links, finding a tree (forest), or finding components of a graph. This behavioral pattern does not have functional cohesiveness with sparse matrix, even though the raw data is similar. Class *Set* is developed in chapter 3, while class *Graph* is developed in chapter 10.

In addition to the classes listed in figure 2.6, many more additional classes are required to support development of application and computational objects. Of these, classes for complex numbers and vectors are most relevant for computational objects. Class for associative array, required for application objects, is developed in chapter 8. In terms of complexity, these objects can be visualized by a three tier hierarchy as shown in figure 2.7. Here the level of complexity grows from data object to application object, through computation object. At the topmost level,

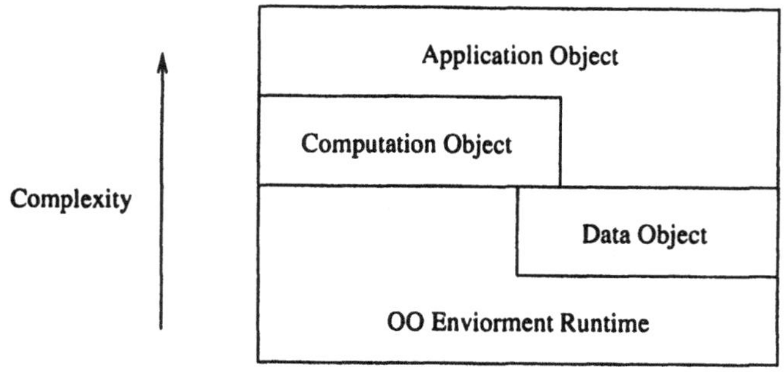

Figure 2.7. Three Tier Architecture for Power System Analysis

we have application objects which implement complex power system algorithms. These, in turn, use computation objects which by themselves are no less complex. Cumulative complexity of the computation is seen at this level. Usually, analysis and design of data object implies identification of attributes and methods to read/write the same. On the other hand, computation object design involves search for a suitable data structure, and computational methods.

The topmost design level, dealing with application object, has a dependency on both computation and data objects. Design at this level may simply use or lead to new computation class or strengthening of existing classes. Therefore, there is considerable interaction between application class development and computation class development, at least in the initial stages of development.

It can be observed that, analysis of the problem statement, typically, brings out the classes associated with physical objects; analysis of the solution statement identifies classes required for computations. Application statement tries to strike a harmony between problem and solution statement. This hierarchical object model provides an efficient way of handling complexity in power system computations. So far, we have developed the classification of objects encountered in power system analysis. Based upon the classification, it is now possible to develop a process diagram to design power system application functions. The authors developed this process diagram by reverse engineering. Many applications were initially designed and implemented without any formal paper design. However, over a period of time, a pattern in designing applications and grouping of objects emerged.

We now present the design process for power system applications, which is based upon the three tier architecture of figure 2.7. While the ideas used in this process are generic, the flow and vocabulary is suitable for the power system analysis domain experts.

4. Design Process

A *process* is an advice on what steps are to be taken while designing an application. Application objects are usually specific implementations of complex power system application algorithms such as Load Flow Analysis. These make use of both data and computation objects. In order to add a new application to the existing library, it is necessary to look at it from all the three perspectives. The activity begins by examining the data to be manipulated by the application and the algorithm that will be applied to accomplish the manipulation. The description of the application is also influenced by design specifications such as speed, stability consideration, robustness etc. Candidate classes are identified and the available class library (computational library and data library) is searched. If the required classes are available, they can be extracted and used. For example, most of the classes developed for Load Flow Analysis are reused in Short Circuit Analysis. However, if classes are not available, they are identified in the 'analyze' phase and engineered in the 'design' phase. Incremental strengthening of classes, typically by adding a few new methods is another important activity we have encountered. Such an incremental development promotes evolution leading to reuse of classes, without endangering basic design framework. It is this modular development of class library that makes the OO paradigm beautiful for power system applications.

An application design process is developed in figure 2.8 to capture the intricate dynamics of the overall development process. It consists of broadly three stages viz.

1 Data Object Design Process;

2 Computational Object Design Process;

3 Application Object Design Process.

While developing the process, following points are considered:

1 There are many ways to mathematically formulate a problem. Usually, these variations lead to usage of different solution (computation) techniques. Hence, different designs for the same problem evolve. Applications like Load Flow Analysis, State Estimation, Optimal Power Flow are some of the classical examples (refer figures 2.9a and 2.9b). While designing an application, development time and cost should be minimized, without compromising the performance criteria such

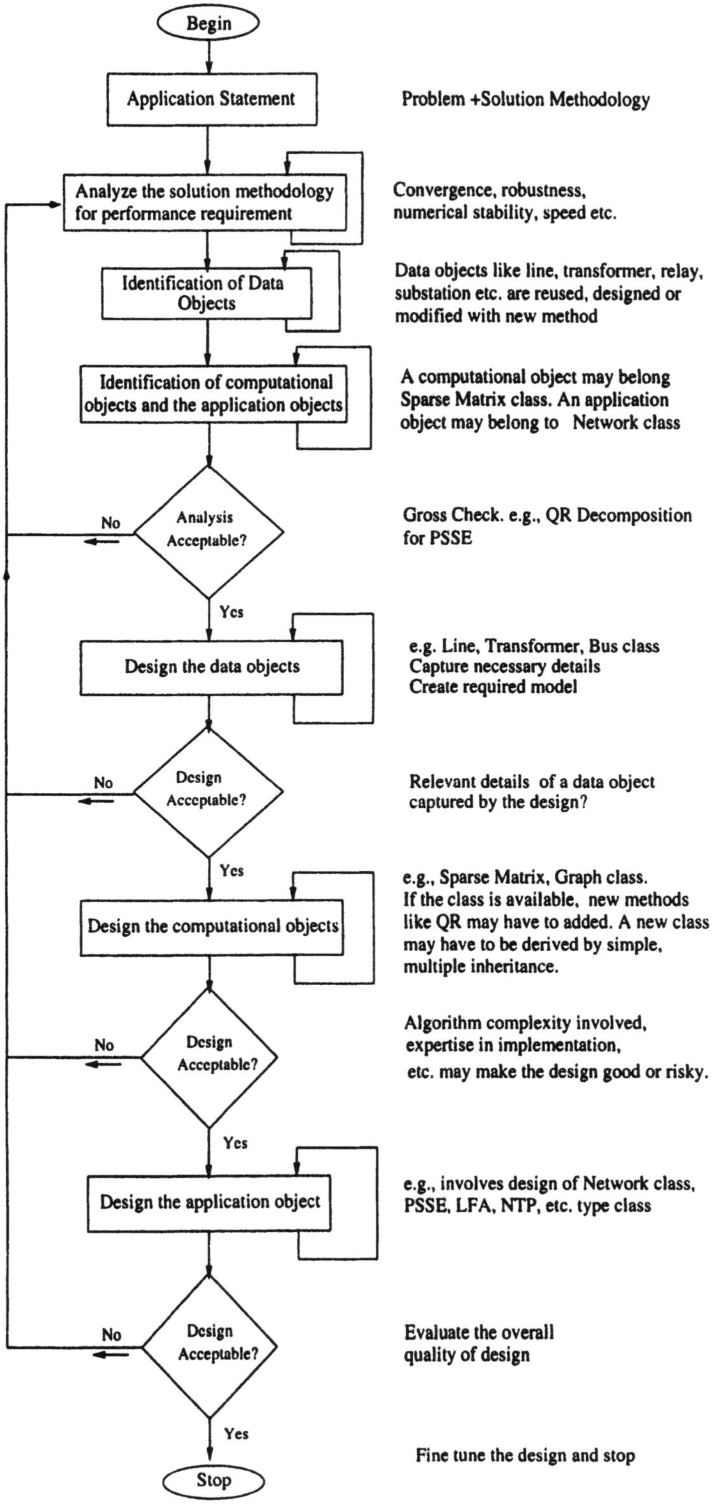

Figure 2.8. Data, Computation & Application Object Design Process

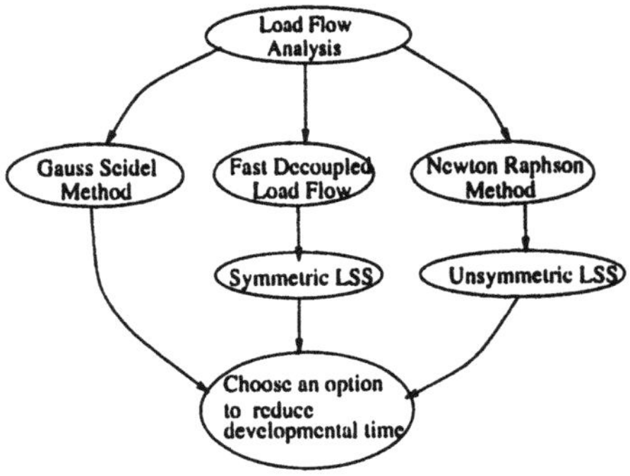

Figure 2.9a. Various Load Flow Solution Methodologies

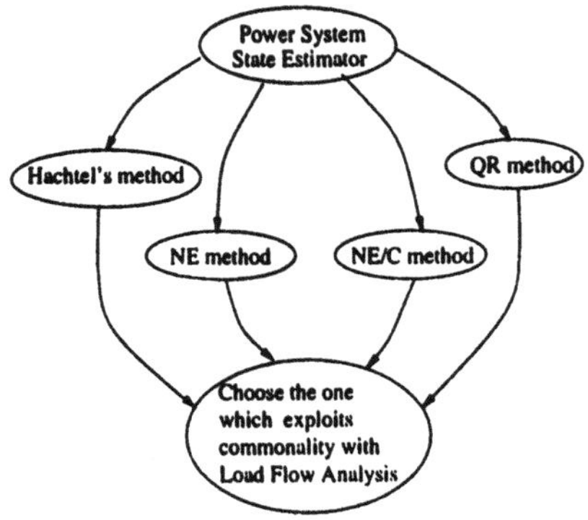

Figure 2.9b. Various State Estimation Methodologies

as speed, stability, etc. This can be achieved by optimal usage of existing classes.

2 The process should encourage evaluation of various design options at the initial stages of design. This may reduce the redesign and implementation effort later on, and cut down software development cost. Further, development using time tested classes helps in containment of risks.

3 The design should conform with the basic design metrics such as *coupling, cohesion, primitiveness, sufficiency and completeness* [Booch, 1994]. This issue is discussed in section 5.

4 Any new application development or variation of an existing application should be able to follow the same design process.

4.1. Data Object Design

Figure 2.8 begins with the data object design process. Most of the data objects are identified during the analysis of problem statement. If the data objects have been already designed then, we proceed to select solution approach; else, the corresponding objects have to be designed. The design of such an object is typically one time activity; it is reused in future applications development by maintaining a class library of data objects (refer to struct.h file). Data object design is usually quite simple. Example of design of class *Line* has already been discussed. During the design of data objects, following steps have to be undertaken:

1 Choose a model for power system apparatus depending upon the requirements of the application;

2 As per the model, identify the attributes of the class;

3 Decide upon the level of openness of the object; if the individual attributes have to be accessed by the application object, it may be worthwhile to promote openness even if, it means that details of the implementation are bare[7].

4 Identify the methods for initializing and printing the object details. This involves overloading operators >> and <<. Methods to compute the behavior or state of object when interfaced with other objects have to be identified and developed (*e.g.*, *ComputeFlow* in class *Line*). The design of methods may in turn lead to identification of new attributes *e.g.*, flow variables in class *Line*.

Design is an iterative process. Figure 2.8 illustrates this point by a self loop associated with each object design. Data objects are usually easy to engineer. The methods required by them are few but useful. They provide a mapping of real life objects to program objects.

4.2. Algorithm Object Design

The next step is to design the computational object(s). First the solution statement is analyzed for classes in the solution domain. At this point of time, if some classes for holding data objects (e.g., PV nodes or PQ nodes) are identified, then they have to be designed; if not, then the emphasis shifts to the computational objects. The class library of computational classes is checked for already engineered classes e.g., class for symmetric sparse matrix, matrix etc. If available, then a check is made on the implemented methods. If a suitable method (e.g., an algorithm from sparse matrix computations) is not available, then it has to be designed. The implementation complexity of the algorithm is analyzed in terms of primitiveness, sufficiency etc. If the complexity is significant, it may be worthwhile to derive a subclass and split an algorithm into set of smaller functionalities.

If the computational object is not available, it may have to be engineered from the beginning. Whenever either a new class for computation has to be implemented or a existing class has to be strengthened by addition of methods, it is worthwhile to evaluate the option of changing the solution strategy altogether. For example, one cannot implement Newton Raphson method for load flow analysis by using Choloskey decomposition. So if Choloskey decomposition is already implemented, it may be worthwhile to choose Fast Decoupled Load Flow option, as such

[7] Readers are encouraged to debate upon the pros and cons the of policy of openness.

a choice does not adversely effect the performance (speed, convergence) of a load flow application.

The design process at the application layer is similar to the computational layer. Here we deal with networks and the computational methods in power systems, treating the matrix computation or graph theoretic methods as already engineered tools. Fig 2.8 provides the details of this design process. The next section discusses the how to evaluate a design for *Quality*.

5. Quality of Design

Once, an application (or a class) has been designed, it is worth while to evaluate the design. Object Oriented design can be evaluated by following metrics[Booch, 1994]:

- cohesion;
- coupling;
- sufficiency;
- completeness;
- primitiveness.

A good design should have a high level of cohesion, loose coupling, and should have methods that are primitive, sufficient and complete. These metrics can be used to evaluate both, the overall architecture (figure 2.4) or design of a single class. For example, consider the design framework that we have developed in figure 2.8. It is but obvious that individual group of objects viz. data, computation, and applications are *cohesive.* On the other hand, there exists considerable decoupling in the three class of objects from modeling, design and implementation perspective. The methods that we use to model data object and the criteria that we use to gauge the quality of a model as good (e.g., should replicate the terminal behavior faithfully, capture dynamic behavior etc) has little in common with computation or application object. A computational object is evaluated typically, by its data structure, sparsity exploitation, numerical stability, computational effort etc. An application object is evaluated by its robustness, ability to capture system behavior, computational effort etc. Thus, the perspective in designing these objects is different and the design process reflects this decoupling.

The metrics of primitiveness, sufficiency and completeness usually reflect the concern of how big an object should be and how many methods should it support. Sometimes the number of methods of a class is known as its weight. One would like an implementation to grow incrementally. To begin with, we wish to implement a skeleton framework, relegating the development of details to a later date. This is the essence of *primitiveness*; such methods do require access to underlying representation of the abstraction for efficient implementation. For example, in matrix computation $+, -$ etc are primitive operations. On the other hand, *sufficiency* implies development of a class to such an extent that one can extract meaningful and efficient interaction from the objects. Usually, sufficiency implies a minimal interface while *completeness* refers to the ability to capture all the meaningful characteristics of the abstraction. In general, it takes a lot of effort to cater to the requirement of sufficiency; completeness can be met only over extended time horizons. As individual classes and then applications are developed in this text, readers are encouraged to evaluate them and check the quality of the abstraction.

6. Summary

This chapter presented a bird's view of OOP, analysis and design methods with a power system analysis and design perspective. Concepts of class, object along with templates, operator overloading inheritance and polymorphism were described. An outline of class *Line* and class *Matrix* was developed. It was shown that languages like C_{++} help us to map problem domain objects to program objects with ease. A classification of objects typically encountered in power system analysis was developed.

Classification requires insight and the way classification is done depends upon the perspective of the observer. In this text, the perspective is that of a programmer. Then, a process for designing an power system analysis applications was developed. OOP and design is meant to facilitate reuse of concepts, codes and design patterns. It is meant to design software that is flexible, resilient to change, and evolve in an incremental fashion. This is achieved by encapsulating the details of implementation and programming to an interface. The key reason for the success of OO paradigm is that it helps us to cope with the software complexity. It appeals to human cognition.

In power system computation, a few questions have been raised on the speed of OO implementation *vis-a-vis* procedural implementation. It is worthwhile to comment on the view that OO implementations are slower because of overheads in message passing, use of virtual functions etc does not hold much water. Usually OO codes are smaller in size; as will be seen later, virtual methods are not required often, nor are they curse by themselves. On the other hand, OO paradigm helps in more precise mapping of real life objects and concepts to program objects, thereby facilitating optimal and modular development.

7. Further Reading

One of the first references on application of OO paradigm to power system analysis is [Neyer et al., 1990]. [Booch, 1994] provides a lucid treatment on complexity in software and OO design. In nineties, various design methodologies [Booch, 1994; Rumbaugh et al., 1998] were developed and propagated. Subsequently, unified software development process was developed [Jacobson et al., 1999] and is now a standard. Similarly, the notations which express the design have been standardized through Unified Modeling Language (UML). [Fowler and Scott, 2000] makes an excellent reading on UML. Many OO languages exist but C_{++} and JAVA are the most popular in computing and power system analysis. [Prata, 1997] provides a good first introduction to C_{++} . [Stroustrup, 1997] provides a more comprehensive treatment on C_{++} . With the ever increasing complexity of the software, component technology is being used to develop dynamically linked libraries of run time executables written in various languages. [Osterlund, 2000] is a good tutorial introduction to component technology. [Hakavik and Holen, 1994; Zhu and Lubkeman, 1997] provide a good treatment on OO modeling in power system analysis. [Pandit et al., 2000a] has stressed the necessity of analyzing multiple applications to obtained a generic design. [Pandit, 2001] provides a comprehensive study of the OO power system analysis. [Zhu and Jossman, 1999] makes a case for use of design patterns in power systems. Utility of OO design and programming in ordering and sparse linear system solver is demonstrated by [Dongarra et al., 1994; Ashcraft and Liu, 1996; Ashcraft and Liu, 1999].

Chapter 3

DATA STRUCTURE FOR SPARSE MATRIX COMPUTATION

Complete disorder is impossible.
—Motzkin, D. B. West: Introduction to Graph Theory

1. Introduction

Power system computations can be classified into two types:

1 Sparse Matrix Computation;

2 Graph Theoretic Computation.

While doing a steady state analysis (e.g., load flow, optimal power flow etc), one typically needs to solve a system of linear equations, each equation having a few variables. Solution of differential and algebraic equations are required in time domain simulations like transient stability. Graph theoretic computations are required to evaluate the structure (connectivity) of a network for observability, islands, loops etc. Graph theoretic computations are also required to optimize computations in a sparse linear system solver. In this chapter, *data structures*, an important player in the sparse matrix and graph theoretic computations is discussed.

To understand how sparse matrices arise in power system, consider a network of n busses (or nodes). A square matrix (e.g., admittance matrix) will have n^2 elements. On an average, let each node be connected to k adjacent nodes. The total number of non-zeros in a square matrix will be equal to $(k+1) \times n$ where $+1$ is to account for the diagonal terms. If % sparsity is defined as ratio of number of zeros in the matrix to total number of elements $\times 100$, then percentage sparsity of such a matrix can be given by $(1 - \frac{k+1}{n}) \times 100$. When systems are modeled to more detail, n increases in much larger proportion to k. For example, in a system with 1000 nodes with an average connectivity of 4 per node, there are a total of $(4+1) \times 1000 = 5000$ non zero entries. A square, non-sparse matrix will have a provision for 10^6 non-zero entries. As such, the sparsity level of this matrix will be $\frac{10^6-5000}{10^6} \times 100 = 99.5\%$. In contrast, a 100×100 matrix with the same level of connectivity, will be only 95% sparse. Thus, the sparsity of a system usually increases with size. It is for this very reason that dividends of sparse matrix computation are noticeable only while dealing with large scale systems.

If this sparsity is not exploited, *i.e.*, accounted for explicitly, we will soon suffer *the curse of dimensionality.* Computational effort will scale up drastically and so will the storage requirements (main memory as well as the cache). *Therefore, it is essential that sparsity be exploited in large scale power system computing.* Technology of sparse matrix computations[1] [Sergio, 1984] is vast and has utility in many other fields like finite element methods. The above discussion leads us to draw the following two important conclusions about sparse matrix computations:

- storage should be minimized by storing only the non-zero elements;
- computations should be organized in such a way that only non-zeros have to be worked upon.

A third significant but not so obvious issue is that of controlling the growth of number of non-zeros as a consequence of computation. For example, the square of a sparse matrix A, having the following structure, is not at all sparse (figure 3.1). In fact, it is a bad idea to organize or

$$A = \begin{bmatrix} x & x & x & x \\ x & & & \\ x & & & \\ x & & & \end{bmatrix} \qquad A^2 = \begin{bmatrix} x & x & x & x \\ x & x & x & x \\ x & x & x & x \\ x & x & x & x \end{bmatrix}$$

Figure 3.1. Sparse Matrix Multiplication

design computations that end up with explicit multiplications and/or inversions of sparse matrices wherein not much can be done to contain the growth of non-zeros. On the other hand, a good design can lead to exploitation of sparsity. Sparsity exploitation is a theme that will repeat all through the book. In this chapter, we review the storage schemes and ways to manipulate (traverse) sparse matrices. The discussion begins with a simple storage scheme known as the *co-ordinate* storage scheme.

[1]The reader should notice use of word technology; it implies an applied science and involves design.

2. Co-ordinate Storage Scheme

An entry $a(i,j)$ in a matrix A can be identified by a three tuple $(i, j, value)$. The row and the column index are known as the co-ordinates. A simple scheme for sparse matrix storage is to store the co-ordinates and the values associated with the co-ordinates. All the row indices are stored in vector *row*, column indices in vector *col* and values in vector *value*. Let us consider an example.

Example 3.1: A sparse matrix A has a structure as shown in Figure 3.2. Incidentally, there are 12 non-zeros, *i.e.*, the matrix is $\frac{16-12}{16} = 25\%$ sparse. The co-ordinate storage scheme will require the three arrays as shown in the figure to store the matrix.

$$A = \begin{array}{c} \\ 0 \\ 1 \\ 2 \\ 3 \end{array} \begin{array}{c} \begin{array}{cccc} 0 & 1 & 2 & 3 \end{array} \\ \begin{bmatrix} a & d & & k \\ b & e & g & \\ & f & i & l \\ c & & j & m \end{bmatrix} \end{array}$$

$$\begin{aligned} \text{row} &= \begin{bmatrix} 0 & 3 & 2 & 1 & 2 & 1 & 0 & 2 & 1 & 0 & 3 & 3 \end{bmatrix} \\ \text{column} &= \begin{bmatrix} 3 & 2 & 3 & 0 & 1 & 1 & 1 & 2 & 2 & 0 & 0 & 3 \end{bmatrix} \\ \text{value} &= \begin{bmatrix} k & j & l & b & f & e & d & i & g & a & c & m \end{bmatrix} \end{aligned}$$

Figure 3.2. Matrix Representation

In general, the elements may not be organized in any particular order. Therefore, there is no simple way to locate an element (i, j), except by an explicit search over the complete array. The cost of accessing an element ($O(nz)$ where nz is the number of non-zeros) is quite high. Therefore, the coordinate scheme is seldom used in practice.

3. Sparse Matrix as Collection of Sparse Vectors

Co-ordinate storage scheme stores elements in any arbitrary order. A more efficient scheme can be designed by imposing a certain order during storage. The sparse matrix can be either packed by rows or by columns. This scheme then requires 3-vectors, one to store the beginning of each row (*row_begin*), one each to store the values (*value*) and the column index (*col*). For a $n \times n$ matrix, vector *row_begin* is of length n while the rest of the two vectors are each of length nz, the number of non-zeros in the matrix. Consider the matrix A of example 3.1. Using this scheme, (say, storage by rows), we can store the matrix A as shown in figure 3.3. Vector *row_begin* is used to store the index of the first element of each row, while vector *column* stores the column number for each non-

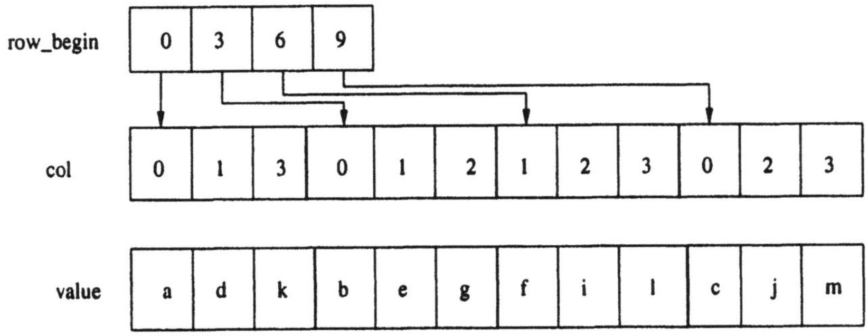

Figure 3.3. Sparse Matrix as Collection of Sparse Vectors

zero element. Not only is this scheme more compact than the previous scheme but it also reduces the cost of accessing an element (i, j). The row i begins at index $row\_begin[i]$ and ends at $row\_begin[i+1]-1$. The last row ends at $nz-1$. For example, row-0 begins at location 0 and ends at location 2, *i.e.*, $begin\_row[1] - 1$. This scheme reduces the amount of effort in accessing an element. However, it is fragile when addition of an element is required. Let us consider that an element has to be introduced in the row-0. Since, there is no space in the packed row-0, and the scheme requires elements of a row to be placed in an ascending order of column subscript, a complete data movement is required. Hence, the use of this scheme is also limited.

4. Linked List

This scheme is the most versatile storage scheme. The scheme is quite general and not meant for storing matrices alone. Figure 3.4a shows the representation of a linked list with five items. The first box contains the item and the next box has the pointer to the next element. The item itself may be an object e.g., an object *Line* discussed in chapter-2. A linked list can be implemented in two ways: (1) using arrays and (2) with pointers, *i.e.*, with dynamic memory allocation.

4.1. Array Implementation

Let the items of a linked list be physically placed in the following order (4,2,3,1,0). Then, figure 3.4b shows the way links need to be set up. In this implementation, a two dimensional array is constructed (table 3.1a), one for the list of objects and another for the link to the logical next item. There are five elements (objects) in the list (figure 3.4b). Incidentally, the first element is located in the last position. In this scheme, we do not sort the items (e.g., in ascending order) because

Index	Name	Next
0	-	5
1	item-4	NULL(-1)
2	item-2	3
3	item-3	1
4	item-1	2
5	item-0	4

Table 3.1a. Tabular Representation

Index	Name	Next
0	-	5
1	item-4	NULL(-1)
2	item-2	3
3	item-3	6
4	item-1	2
5	item-0	4
6	new-item	1

Table 3.1b. Appending new Element

of the obvious high cost involved in sorting. Instead, without physically dislocating the items, stored in a vector *Name*, an additional vector of integer called *Next* is used which stores the index of the next logical element. Vector *Next* implements links of figure 3.4b. The first entry in vector *Name*, *i.e.*, *Name*[0] does not contain any object; however, *Next*[0] is set to point to the first element. In this case *Next*[0] is set to 5. The *Next*[5] points to 4, which implies that the item-1 is stored at position number 4, *i.e.*, *Name*[4]. The last element is identified by the *Next* field set to -1. Figure 3.4b is a visual representation of links as per table 3.1a.

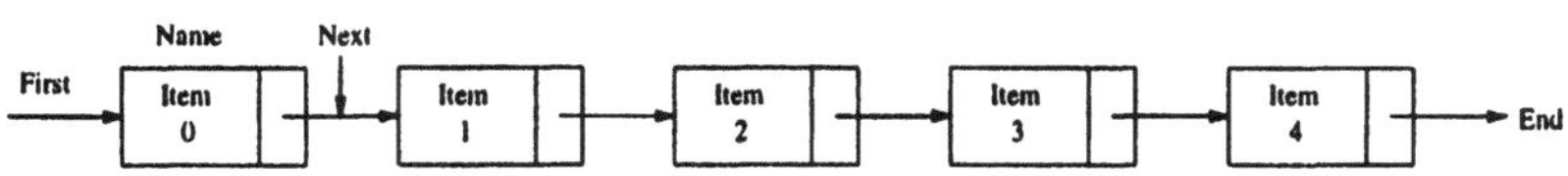

Figure 3.4a. Concept of linked list

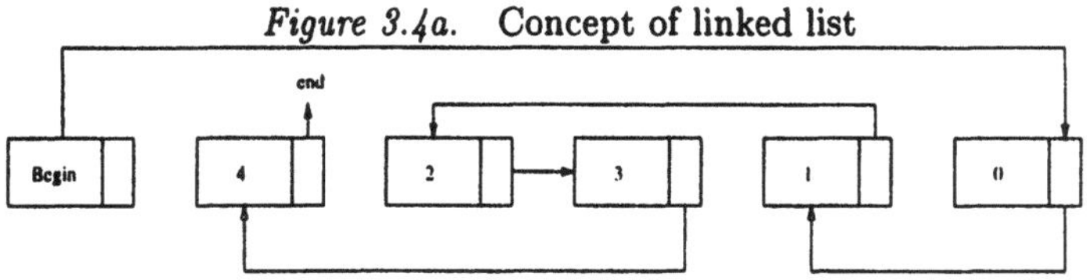

Figure 3.4b. Setting the links

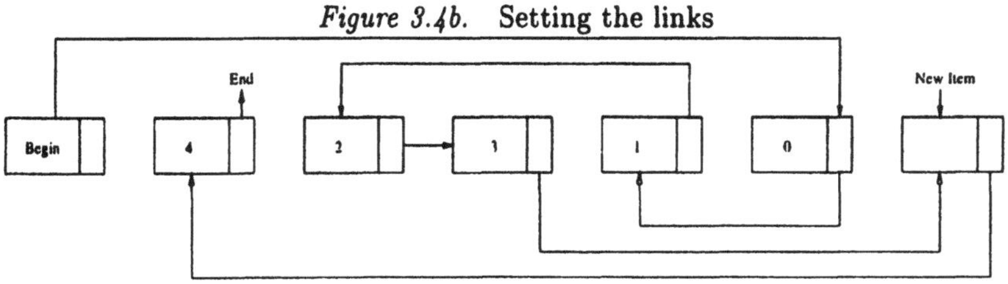

Figure 3.4c. Inserting an element to linked list

Now, suppose a new item has to be logically introduced between item 3 and 4. Logically, the element has to be placed between item 3 and 4. Physically, it will be appended in the last location, *i.e.*, location number 6 to minimize data movement. As such, the links of item 0,1,2 and 3 are not disturbed. What is required is to set up a link from item 3 to new item, *i.e.*, $Next[3]$ is set to 6, the location of the new item. The link from new item should point to item-4; hence, $Next[6]$ is set to 1 location of fourth item. Table 3.1b and figure 3.4c summarizes these results; figure 3.4c visualizes the links shown in table 3.1b. The ease with which an element can be logically inserted between two existing elements can be seen from this example. Similarly, deletion of an element can be achieved by modifying the link of the previous element such that the element to be deleted is bypassed without any physical movement of the data. It is for this very reason that linked list representation is preferred for sparse matrices (Example 3.2). The only drawback of an array implementation is that size of the array has to be fixed *apriori*. A more compact and scalable implementation of a linked list is discussed now.

4.2. A Pointer Implementation

In this scheme, each element of a linked list holds an object and a link to the next object. For example, if we are creating a linked list for a vector of floats, then it can be implemented by the following structure.

```
struct  Link(float value, Link *Next)
```

We discuss here a generic implementation using templates, which is not specific to any particular data type. First let us define a class *Link* as follows:

```
template<class T> class Link {
private:
T value;    // to hold the object
Link * ptrToNextLink; // the ptr to next link
friend class<T> List;
Link(T linkvalue, Link * nextptr); // a ctor
public:
 Link insert(T val); // add a link
   ......
};
```

With the class *Link* as a building block, a class *List* is defined as follows:

```
template<class T> class List {
protected:
Link<T> * ptrToFirstLink; // ptr to first link
public:
List() //default ctor;
List(const List & s);//copy ctor
virtual ~List(); // dtor will depend upon object type
virtual void add(T value);
};
```

One should notice that an instance of class *Link* actually holds the object and the pointer to the next link. It is declared as a friend class to *List* which means that friend class (in this case *List*) can access private and protected parts of the *Link*[2]. Now, the *Link* is defined in the *protected* part of the *List*. Therefore, it is a facilitator class for the *List*. A user of the class *List* is not even aware of the existence of *Link*.

This linked list structure though powerful, compact and elegant will not be used much in the computations we deal with. The reason is that the links may not be placed in a contiguous block of main memory. Consequently, a number of main memory fetches may be required to get the list in the cache which can adversely affect the overall speed. Therefore, usage of the array implementation of linked list is preferred, even if, it lacks the versatility of the pointer implementation. It will be referred as a *static linked list.*

Example 3.2: Design a static linked list representation for sparse matrices.

An element of a matrix can be described completely by minimum of three fields, *viz.* the row number (*row*), column number (*col*) and value (*val*). Depending upon the type of matrix, value *val* may be a *int, float, double or complex*[3]. In addition, to generate a linked list, we will require an integer variable to hold index (pointer) to the next element in the row. Similarly, to facilitate column access, an integer variable will

[2] In a way, friend declaration breaks encapsulation; thus, it is an alternative to public data objects designed in chapter-2.

[3] Data type complex is not available in C_{++} . It has to be defined as a parametrized class over *float or double.*

be required to hold index to the next element in a column. All these attributes can be grouped by a structure[4] in C_{++} .

```
template<class T> struct spmat<T> {
int row,col, link_row, link_col;
T val;
spmat():row(0),col(0),link_row(0),link_col(0),val(0){};
spmat(int & i, int & j, int & r_nxt, int & c_nxt, T & v)
 : row(i),col(j),link_row(r_nxt), link_col(c_nxt),val(v)
{} // initialization can be done in the function body
};
```

A fully developed version of *spmat* is available in the file Struct.h. Readers are encouraged to study the same. With this facilitator class *spmat*, we are now in a position to build a class for sparse matrix as follows.

5. Class *SparseMatrix*

```
template<class T> class SparseMatrix {
protected:
   int no_of_rows,no_of_cols,no_of_nz,size;
   int  * row_begin, * col_begin;
   spmat<T> * A;
   ..
public:
  SparseMatrix():no_of_rows(0),no_of_cols(0),A(NULL),
       row_begin(0),col_begin(0) {};// default ctor
  ~SparseMatrix();  // this has to be non-trivial dtor
  SparseMatrix(int m,int n,int nz1);
  //ctor for m x n matrix  having at most nz1 elements
  SparseMatrix(SparseMatrix &); //copy ctor
  void sort();  // establish links
  void permute();// renumber the matrix
  SparseMatrix<T> & operator+(const SparseMatrix <T>&A,
       const SparseMatrix <T> & B); //matrix addition
  int * friend mda(Sparse_Matrix &);
                                //return mda ordering
  void factor(); // LDLT factorization
  void LU_data_struct()// setting of data struct for L
  void QR_decomp(); // QR decomp of sparse matrix
};
```

[4]A structure in C_{++} is a public class having bare data associated with methods, *ctor* and *dtor*. The reader should debate use of openness albeit friend class implementation.

Class *SparseMatrix* is a generic class built upon the structure *spmat*. It sets up a static linked list representation for both the rows and columns of a sparse matrix. The beginning element of a linked list i for say a *row-i* is stored in *row_begin*[i]. Similarly, *col_begin*[i], stores the link to the first non-zero in the ith column. They facilitate access by row or column, a common requirement in sparse matrix computation. Thus, a $m \times n$ sparse matrix is implemented by $m + n$, static linked lists. Notice that attributes of the class are *protected* which means that only a subclass derived from it will have access to the attributes. The *public* interface lists various ways of construction of a matrix. Overloaded operators like $+, -, *, =$ permit the same syntax of matrix algebra to be used for addition, subtraction, multiplication and assignments of matrices[5].

Example 3.3: This example illustrates the use of a class *SparseMatrix* for storing a (8×7) sparse matrix A, shown in figure 3.5a. The storage of non-zero elements of sparse matrix A is as shown in figure 3.5b. The visual representation of linked list is shown in figure 3.5c. Any row (column) in the list can be traversed by starting with the *first* element given by vectors *row_begin* (*col_begin*). A scan through the *link_row* (*link_col*) helps in reaching an element. 0th row begins with (0,0) element which is stored in location 0 of the array. Therefore, *row_begin*[0]=0. The first element in the row-1 is (1,1) stored in location 4 (figure 3.5b). Therefore, *row_begin*[1]=4. Since, the first two columns also start with the diagonal element, *i.e.*, (0,0) and (1,1) (figure 3.5b), *col_begin*[0]=0; and *col_begin*[1]=4. The rest of the elements in *row_begin* and *col_begin* are as follows:

$$row\_begin = [0, 4, 7, 9, 11, 13, 16, 19]\,; col\_begin = [0, 4, 1, 5, 2, 3, 6]$$

If the *fourth* row has to be accessed, we begin at index *row_begin* [4]=11, *i.e.*, the starting element in row 4 is stored at the 11th position in the array of *spmat*. The row and column index of this element is (4,1) respectively. The next element in the row, pointed by *link_row* of element number 11, is at location number 10. This element is in the 3rd column of the 4th row. The *link_row* for element 10 points to NULL. This terminates the row. Any column can be accessed, in a similar manner, starting from *col_begin*.

If yet another non-zero element has now to be added to this list, then it will be placed at the 20th location in the list. Let the new element

[5] A piece of advice here; whenever, assignment operator is overloaded, it is a must to write a copy constructor. This is because, an overloaded assignment operator returns a copy of the object. In the absence of an explicit constructor, pointers may be equated leading to problems during operation of destructor.

$$A = \begin{bmatrix} x & & x & & x & & \\ & x & & & & x & \\ x & & & x & & & x \\ & & x & & & x & \\ & x & & x & & & \\ x & & & & & & x \\ & x & & x & & x & \\ & & x & & x & & x \end{bmatrix}$$

Figure 3.5a. Sparse matrix A

Posi-tion	structure spmat				
	row	col	val	link_row	link_col
0	0	0	x	1	7
1	0	2	x	2	9
2	0	4	x	NULL	17
3	1	5	x	NULL	8
4	1	1	x	3	11
5	2	3	x	6	10
6	2	6	x	NULL	12
7	2	0	x	5	13
8	3	5	x	NULL	14
9	3	2	x	8	19
10	4	3	x	NULL	15
11	4	1	x	10	16
12	5	6	x	NULL	18
13	5	0	x	12	NULL
14	6	5	x	NULL	NULL
15	6	3	x	14	NULL
16	6	1	x	15	NULL
17	7	4	x	18	NULL
18	7	6	x	NULL	NULL
19	7	2	x	17	NULL

Figure 3.5b. Static Linked List for A

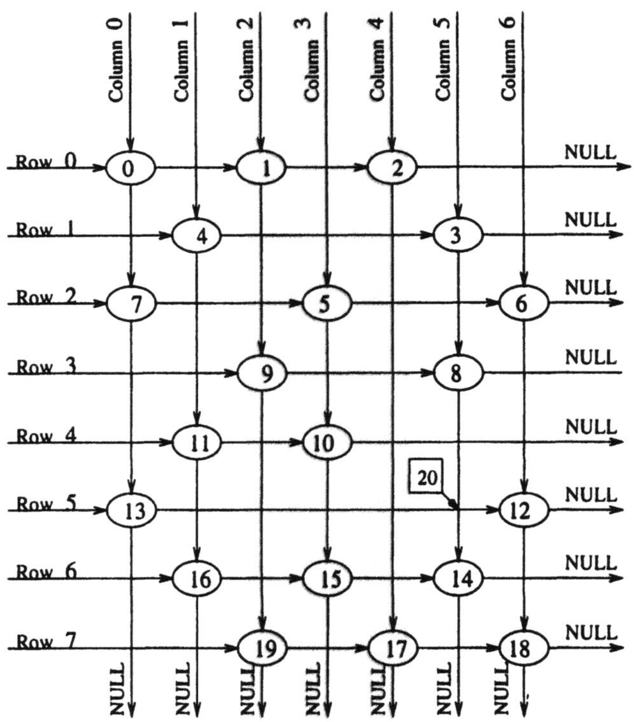

Figure 3.5c. The Linked list organization of sparse matrix A

$$A = \begin{bmatrix} x & & x & & x & & \\ & x & & & & x & \\ x & & & x & & & x \\ & & x & & & x & \\ & x & & x & & & \\ x & & & & & \mathrm{X} & x \\ & x & & x & & x & \\ & & x & & x & & x \end{bmatrix}$$

Figure 3.6a. Modified Matrix A

Position	structure spmat				
	row	col	val	link_row	link_col
8	3	5	x	NULL	**20**
13	5	0	x	**20**	NULL
20	5	5	**X**	**12**	**14**

Figure 3.6b. Modified Linked List

appear at the (5,5) position of matrix A (figure 3.6a). With this, some of the links of the linked list of figure 3.5b will get updated as shown in figure 3.6b.

The method *sort()* of the class *SparseMatrix* sets up the links (Figure 3.5c). Readers are referred to the implementation in spmat.h. Other methods of the class *SparseMatrix* are meant for designing linear system solvers. They will be discussed in the next chapter. The *sort* method is based on example 2.8 of [Duff et al., 1986]. It requires a linear effort (in term of numbers of non-zeros) to set up the links. This should be contrasted with usual sort procedures which would require $(nz\log(nz))$ effort.

6. Class *Vector*

In the previous section, a class *SparseMatrix* was developed. However, not much can be achieved by considering matrices in isolation. Most of the time, we have to work with Linear System Solvers. Thus, in addition to the development of a class *SparseMatrix*, development of a class *Vector* is also necessary. This section provides a brief introduction to how vectors can be handled in OOP languages like C_{++} . Class *Vector* is also available in the Standard Template Library of C_{++} . First, let us analyze the requirements of class *Vector*:

1 It should be able to model a n-tuple $[x_1, x_2, \ldots, x_n]^T$. An element x_i may be of type float, double, complex or even an unknown object (as of now) on which operations like + have been defined.

2 The methods should also support addition, subtraction, assignment of vectors in the same way as it is done in vector algebra.

These requirements are comparatively simple. From item 1, it can be concluded that class *Vector* should be implemented with templates (the same way as was done with sparse matrices). From item 2, it can be concluded that operators like +, -, * etc. should be overloaded, so that, a user of the class finds the interface clean and easy to work with. The design of the class is as follows:

```
template<class T> class Vector{
public :
  T * vec;
  int m;
  Vector();
  Vector(int sz);
  Vector(Vector<T> &v);
  ~Vector();
 void permute(int *old_order);  // used for LSS
 inline T & operator [ ] ( int i) { return vec[i];};
 Vector <T> & operator += (const Vector <T> & A) ;
 friend  istream & operator >> (istream & is, Vector<T> & A);
};
```

The implementation of class *Vector* is available in the file vector.h. Readers are encouraged to study it and write their own programs. We now proceed to discuss a data structure called *Set* which plays an important role in the graph theoretic manipulation.

7. Class *Set*

Motivation: The class *SparseMatrix* and *Vector* permit efficient storage and manipulation of sparse matrices e.g., in $\times$, $+$ etc. However, symmetric sparse matrix A can equally well be represented by an undirected graph. An edge between two nodes v_i and v_j of a graph exist if and only if $a_{ij} \neq 0$. The edge has a weight equal to the value of the (i,j) entry. Usually, the graph theoretic description is preferred, when the perspective in computing is manipulation of the structure of the embedded graph. For example, one may wish to evaluate whether a network described by an undirected graph is connected or not[6]. In matrix terminology it amounts to finding a block diagonal representation under identical row and column permutation. In a linear system solver, there is a need to evaluate the location of fills. All such manipulations have a perspective which is purely graph theoretic. Numerical values are irrelevant. Another interesting example is relay co-ordination of meshed system [Damborg and Venkata, 1984] wherein there is a need to evaluate all possible loops in the graph. For all such computations, the data

[6] If it is connected, only then, load flow study with a single slack bus can be performed.

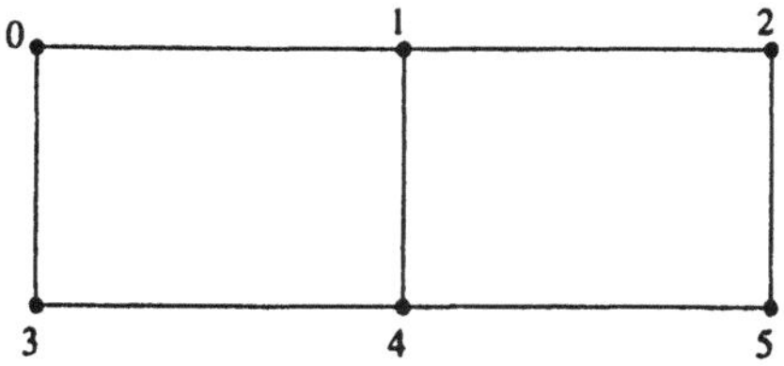

Figure 3.7a. Example 3.4 - graph

$$\begin{bmatrix} & 1 & & 1 & & \\ 1 & & 1 & & 1 & \\ & 1 & & & & 1 \\ 1 & & & & 1 & \\ & 1 & & 1 & & 1 \\ & & 1 & & 1 & \end{bmatrix}$$

Figure 3.7b. Example 3.4 - Incidence matrix

structures discussed so far, are not of much use. Usually, graphs can be described well in terms of adjacency. The following example illustrates this point.

Example 3.4: The graph shown in Figure 3.7a can be described by an incidence matrix (figure 3.7b). This representation can be stored compactly using the class *SparseMatrix* (Static linked list data structure). The graph can also be described by a vector of sets[7] $Adj[n]$ such that each set Adj[i] stores the adjacent vertices to the vertex i. Thus, in figure 3.7a, $Adj[0] = \{1,3\}$; $Adj[1] = \{0,2,4\}$; $Adj[2] = \{1,5\}$; $Adj[3] = \{0,4\}$; $Adj[4] = \{1,3,5\}$ and $Adj[5] = \{2,4\}$.

Design of class *Set* facilitates graph theoretic manipulations and hence is useful in analyzing sparsity patterns. Following attributes and methods define a class *Set*:

```
template<class T> class Set<T>{
protected :
  int cardin, status, vertex, MaxCardin;
   T *v;
public :
 Set();          //default constructor
 ~Set();         //default destructor
 Set(int size); //constructor for size number of members
 friend Set<T> operator-(const Set<T> &a,const Set<T> &b);
             // set difference operation
 Set<T> & operator = (const Set<T> &s1);
                //assignment operator
   };
```

[7] Another usage of template facility in class vector.

Class *Set* has been defined using the template facility, so that it can be instantiated by any data type. In this example, we instantiate the class *Set* with an integer data type. Attribute *cardin* gives the set cardinality, *i.e.*, number of members of the set; the set members are stored in an array *v*. Attribute *status* can be used as a flag, which may become useful in certain conditional operations. For example, it may be used to indicate active participation of a set in graph theoretic computations. Attribute *vertex* can alias the name for a set. In graph theoretic computations, while dealing with a set of adjacency nodes of a vertex v_i, it is associated with the corresponding node number-i. Maximum cardinality of a set, used for temporary allocation of the internal memory, is defined by *MaxCardin*. For a $n \times n$ matrix of a n- node graph, the maximum cardinality is n. However, large sparse matrices, in practice, require a very small value of *MaxCardin*. A constructor *set(size)* initializes a set so as to accommodate a user defined *size* number of members.

Set operations like union, intersection, difference are binary operations. By overloading the arithmetic operators like +, % and −, these operations can be performed on a set data type with great ease. The convenience of such a facility in simplification of programming, debugging and understanding of the code is simply beyond words !! List of overloaded operators supported by class *Set* are shown in table 3.2.

Table 3.2. Overloaded Operators: Class *Set*

Operator	Operation	Remark
=	A = B	Assign set B to set A.
−	A − B	Difference of two sets.
+	A ∪ B	Union of two sets.
%	A ∩ B	Intersection of two sets.
+=	A += B	$A \cup B$ returned in set A.
−=	A −= B	$A - B$ returned in set A.
%=	A %= B	$A \cap B$ returned in set A.
==	A == B	check if set A equal to B.
<=	A <= B	check if set A subset of B.

Example 3.5: Design a method to find the component to which a node *Node* belongs.

The following function *find_comp(Vector<Set<int> > & Adj, int Node)* implements a method to compute the component of node *Node* in a graph G. It returns a set that contains nodes in the component.

Set<int> & find_comp(Vector<Set<int> > & Adj, int Node)
{
1. Compute adjacency of each node i and store in set $Adj[i]$. This is easily done by traversing the static linked list representation of Example 3.3. Alternately, when *Adj* is passed, this information can also be passed.
2. Initialize set $comp = \{\phi\}$ to store the nodes that belong to the same component as *Node*.
3. Since, the set *comp* at least contains the adjacent nodes of *Node*, hence, update $comp+ = Adj[Node]$.
4. Mark *Node* as processed, *i.e.*, set status of $Adj[Node]$ as 1.
5. Check if each set member $v[i]$ of *comp* is marked *processed.*
6. If True, then the set *comp* contains the set of nodes that belong to the same component as *Node*. Return *comp*.
7. Else, add adjacency of the first unprocessed node $comp.v[i]$ to the set *comp*, *i.e.*, $comp+ = Adj[comp.v[i]]$. Mark node $comp.v[i]$ as processed, *i.e.*, $Adj[comp.v[i]].status = 1$. Go back to step-5.

}

Set theoretic manipulations are the backbone in which graph theoretic computations are performed. For example, Network Topology Processing (NTP) to be discussed later, use the *find_comp* function. Hence, it is necessary to discuss optimal implementation of sets.

7.1. Optimal Implementation of Class *Set**

A set by definition is a collection of distinct elements. From the point of view of implementation, a set can be classified as an *Unordered set*, a *Bit set* or an *Ordered set.*

7.1.1 Unordered Set

A set is said to be unordered if elements in the set do not appear in any particular order. Consider a representative union operation of two unordered sets, $s1$ and $s2$. Since, in the union set, a member can appear only once, each element of set $s1$ has to be checked if it belongs to the other set $s2$. $s2.cardin$ operations are required to check if an element of set $s1$ is in set $s2$. Union operation then requires an effort of $O\ (s1.cardin \times s2.cardin)$ which is much more than linear effort. Therefore, it is not advisable to use unordered set.

7.1.2 Bit Set

Let vertices of a graph G be represented by a set V; $V = \{1, 2, \ldots, n\}$. All sets involved in a graph theoretic implementation are subsets of V. Such sets can be represented by a collection of n bits. If vertex v is present in a set, then, bit v is set to 1, otherwise it is set to 0. The advantage of a bit set representation is that, set operations can be implemented by bit manipulations. For example, union operation corresponds to *OR* operation over n bits of the two sets. Similarly, intersection operation can be implemented by the *AND* operation on bit sets. C_{++} supports bit operations on a word length. The bit set implementation is very convenient when the order of n is small, typically $n < 64$. However, when n is large, say ≥ 10000, bit set implementation requires n integers to replicate n bits. Thus, memory requirement becomes prohibitive. However, the problem can be remedied by defining a $workvect[n]$ as a shared memory or work space to store n bits.

The illustrative union operation can be effectively implemented as follows:

- Initialize $workvect[0 : n - 1]$ to 0.
- Set $workvect[i]$ to 1, for all vertices i, of set $s1$.
- Set $workvect[i]$ to 1, for all vertices i, of set $s2$.
- Compress the $workvect$ to set S, by noting the positions of 1 in $workvect$. Thus, set S has the vertices of the union of set $s1$ and $s2$.

Observation: Consider a 1000 node system with set $A = \{1\}$ and set $B = \{999\}$. Then, step 2 and 3 would set workvect[1] and workvect[999] to 1. The total number of operations involved are just two. However, step 4 requires scanning of all the 1000 elements of the workvect. Thus, n operations are required.

As such, bit a set implementation is not very efficient for large scale systems with highly uneven sets.

7.1.3 Ordered Set

A set is said to be ordered, if all of its elements are arranged in ascending (or descending) order. Consider the following code for the union of two ordered sets $s1$ and $s2$ where the node numbers (or elements of the set) are arranged in an ascending order. A C_{++} like pseudo code for union operation implemented by overloaded operator + is as follows:

```
set s(s1.cardin + s2.cardin)// maximum possible
cardinality of union set
int i=j=k=0;
  while ((i<s1.cardin)&&(j<s2.cardin)) {
    if(s1.v[i] < s2.v[j])
     { s.v[k] = s1.v[i]; i++; k++;}
    else if(s1.v[i] > s2.v[j])
     { s.v[k] = s2.v[j]; j++;k++;}
    else if(s1.v[i] == s2.v[j])
     { s.v[k] = s1.v[i]; i++;j++;k++;}
     //useful in intersection operation
   }
  if(i==s1.cardin){copy rest of s2.v to s.v}
    else if(j==s2.cardin){copy rest of s1.v to s.v}
  s.cardin = k; return s;
```

The copy operation in the above code also increments k appropriately. It is clear that, union operation in this case can be performed in $O(s1.cardin + s2.cardin)$, which is typically much less than the $O(n)$ effort for bit set implementation $O(s1.cardin \times s2.cardin)$ effort for unordered set. This justifies the use of ordered sets. We now discuss the implementation of the intersection operation.

Intersection in ordered sets

Two efficient implementations for intersection are possible with ordered sets. One of it involves an algorithm similar to the union. The complexity of this operation will also be the same as the union operation. Alternately, a find operation, to search a single element in a set S, can be performed in $O(log(S.cardin))$ using the *divide and conquer* strategy. In this strategy, a set S is partitioned into two ordered subsets, each with cardinality of $\frac{S.cardin}{2}$. Only one of two subsets can have the element being searched. Recursive application of the find operation leads to the solution. At each step of recursion, the effort involved to solve the problem is halved; hence, the logarithmic order of complexity. Intersection of two sets $s1$ and $s2$ can be computed, using this technique, with complexity of $O(s1.cardin \times (logs2.cardin))$. Being a commutative operation, this implementation is of special importance when cardinality of any one of the sets, $s1$ or $s2$, is much smaller than the other. The reader is advised to refer to `set.h` file for more details.

Example 3.6: Show that using the divide and conquer strategy, el-

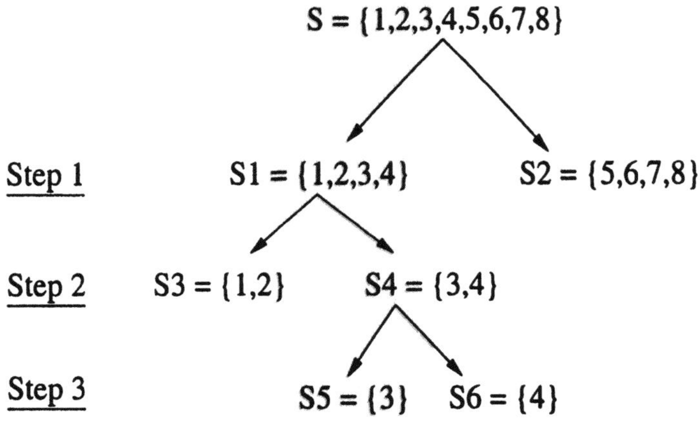

Figure 3.8. **Divide and Conquer Strategies**

ement 3 in set $S = \{1, 2, \ldots, 8\}$ can be found out in $log_2(2^3) = 3$ steps.

The figure 3.8 illustrates how the ordering in elements can be used to advantage. Since, *s.cardin* = 8, in the first step, set S is split into 2 sets $S1$ and $S2$. Since $3 < 5$, the ordering guarantees that if 3 exists in set S, it will have to be present in the set $S1$. Again, set $S1$ is subdivided in two sets $S3$ and $S4$, both of cardinality of 2. Application of similar logic leads us to the conclusion that if 3 exists in S, it must belong to $S4$. At step-3, each set is singleton and hence we locate 3 in $S5$. Thus, in $\log 2^3 = 3$ steps, we have ascertained that $3 \in$ S.

Operators $+ =$, $- =$, $\% =$ play a special role in efficient implementation of class *Set*. There are two ways of returning arguments in C_{++} [Meyers, 1996]. One is returning *by reference* and the other is returning *by value*. When operator + is used, e.g., $C = A + B$, the return has to be *by value*. In this case, a copy constructor, provided by the user, is invoked to obtain a hidden copy of instance $(A+B)$. This involves additional memory allocation for storing data of the hidden object. Contents of memory of $(A+B)$ have to be written to the memory of hidden object. The hidden object is subsequently passed to the assignment operator. It is then destructed, after the assignment is complete. All this work involves a non-trivial overhead.

In contrast, return *by reference* returns only the reference of the first object and hence does not involve use of copy constructor and the associated overheads. When unary operators like $+ =$ are involved, (e.g., $B+ = A$), then contents of addition are returned to set B itself. Thus, set B is modified by reference. Consequently, overheads in creation and deletion of hidden objects are avoided. The importance of *return by reference* vis-a-vis *return by value* should not be overlooked, even though it appears language specific. This is because, an inefficient memory management of set objects, that form the vocabulary of a problem domain, can lead to deterioration in performance of an otherwise efficient algorithm. Thus, it is advisable to use $+=$, $-=$ operators as compared to $+$ or $-$.

8. Summary

The chapter introduced data structure for sparse matrices. It is exploitation of sparsity that permits us to work with systems having thousands of nodes. Co-ordinate scheme, sparse vectors and linked list were identified as three possible data structures for storing and manipulating sparse matrices. Of the three, the linked list was identified as the most suitable data structure. It was shown that an array implementation of the same is desirable from the point of view of fast computing. However, it implies a small compromise on the scalability of the linked list. As will be seen in later chapters, this restriction is not very critical. A generic class *SparseMatrix* that can cater to *int, float, double* and *complex* data types was developed. The class development hides (or encapsulates) the data structure. It permits us to overload binary operators so that syntax of computing matches closely to the matrix algebra syntax. A brief introduction to class *Vector* was also provided. A class *Set* is developed which will be used for graph theoretic manipulations. An optimal implementation of class set is also discussed. The next chapter develops a linear system solver using the class *SparseMatrix* and class *Vector*.

9. Further Reading

The data structure for sparse matrices has been discussed in [Duff et al., 1986]. Class *Vector* is discussed in [Prata, 1997]. Class *Set* is implemented in the Standard Template Library (STL), as a generalized class map which serves various applications like dictionary, telephone directory, etc. [Stroustrup, 1997]. Set as a data structure is discussed in detail in reference [Aho et al., 1974]. Other useful books on data structures are [Weiss, 1999; Budd, 1994].

10. Review Questions

1 Rewrite the class *spmat* and *SparseMatrix* using the friend class declaration as discussed in linked list representation.

2 What is the worst case order of complexity to access a element in the co-ordinate scheme?

3 What are the advantages of sparse storage by vector scheme, over co-ordinate scheme?

4 Describe a scheme which stores a sparse matrix as a collection of sparse vectors where each vector compactly stores a column vector.

5 Let a matrix A (Figure 3.1) be stored as a collection of sparse row vectors. Set up the corresponding arrays *viz.* *begin_row*, *column* and *value*. Repeat the same for matrix A^2.

6 Write a MATLAB®program to compress a non-sparse representation of a sparse $(n \times n)$ matrix using a co-ordinate scheme and b) sparse-vector storage scheme. Then, using these data structure, write a program for matrix multiplication. Summarizes the experience.

7 Repeat problem 6, using the sparse-matrix class (use spmat.h); Summarize your experience. Use function *sort*() to set up the links.

8 Show that a vector V of length n of type set, *i.e.*, $Vector < Set >$ $V(n)$ can be used to store adjacency of a sparse $m \times n$ matrix. If two such vectors A and B for two $n \times n$ matrices are set up what would A+B represent.

9 List out the pros and cons of keeping the class *Vector* public. Do you agree with the authors choice. If not, implement your own vector class.

10 Design an interface for a Linear System Solver function (LSS) using the class *SparseMatrix* and class *Vector*. Evaluate the following design options (1)declaring it as a friend method in class and (2)deriving a class using multiple inheritance from class matrix and vector.

11 For simplicity, in this chapter the class *SparseMatrix* and *Vector* have been defined independently. Instead define an abstract class *matrix* and derive various classes like symmetric sparse matrix unsymmetric sparse matrix from the same. Show that using polymorphism how a single interface for function LSS $(matrix\&, vector\&)$ can be used in a code, irrespective of the type of matrix.

12 Generalize the algorithm of example 3.5 to compute all possible components of a network.

13 Apply the algorithm to example 3.4.

14 Let $S = \{1, 2, \cdots, 9\}$; execute $find(10)$ and $find(9)$ in a similar way to example-3.6. How many steps do you need?

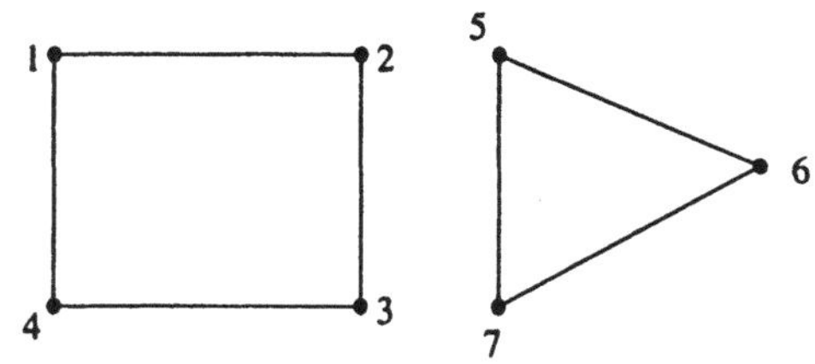

Figure 3.9. Problem-17

15 Show that worst case cost of accessing an element in co-ordinate scheme is $O(nz)$, in the *SparseVector* scheme $O(n)$. What is the average cost of accessing an element in the linked list.

16 Design a class graph by $Vector < set < int >> V(n)$ construction.

17 Consider a graph as shown in the figure 3.9. Apply $findComp$ method to compute the nodes in component for nodes 4 and 5 respectively. Suggest generalization to compute all possible components in a graph. Hence, describe a procedure to order a sparse matrix in a block diagonal form.

18 It was shown that if $n_1 << n_2$, find operation is efficiently implemented by divide and conquer strategy. Let $n_1 = 2^5$ and $n_2 = 2^5$, compare the performance of intersection algorithm implemented by 1)linear search like union and 2)divide and conquer algorithm. What conclusions do you draw?

Chapter 4

SPARSE SYMMETRIC LINEAR SYSTEM SOLVER

You see, there's a factor in this equation that neither of you nor anybody in Washington understands. The Shepherd Boy was removed from this earth, but the crown was passed. He's the key.

—Robert Ludlum, The Matarese Countdown

1. Introduction

Direct sparse Linear System Solvers (LSS) are used routinely in power system analysis functions. In fact, it is almost impossible to develop a program for a large scale system without being equipped with a linear system solver that can handle a large sparse coefficient matrix. An exception is Gauss Seidel method for Load Flow Analysis Program. The primary objective of this chapter is to introduce a direct sparse LSS and bring out the design issues. The emphasis is on specific requirements of power system applications like load flow analysis, OPF, short circuit analysis. We begin with Gaussian Elimination, the basic direct method for solving a linear system of equations. We presume that the coefficient matrix is both full rank and square (non-singular) and the diagonal pivots remain non-zero.

Consider the following system of equations:

$$Ax = b \text{ or } \begin{bmatrix} a_{11} & a_{12} & a_{13} & a_{14} \\ a_{21} & a_{22} & a_{23} & a_{24} \\ a_{31} & a_{32} & a_{33} & a_{34} \\ a_{41} & a_{42} & a_{43} & a_{44} \end{bmatrix} \begin{bmatrix} x_1 \\ x_2 \\ x_3 \\ x_4 \end{bmatrix} = \begin{bmatrix} b_1 \\ b_2 \\ b_3 \\ b_4 \end{bmatrix}$$

In Gaussian elimination, a set of elimination operations are performed on the augmented matrix $[A : b]$, such that matrix A is converted to an upper triangular matrix (U). Then, the vector x can be computed by simple back substitution. At each step of the elimination (total of n steps for a $n \times n$ matrix), the size of sub-matrix to be worked upon (also known as active sub-matrix) reduces by one row and one column. We illustrate the operations in Gaussian elimination on a LSS with 4 variables. Thus, A is a 4×4 matrix and b is a 4×1 vector. The active sub-matrix is shown by dotted lines.

$$\textbf{step 0} \quad \begin{bmatrix} a_{11} & a_{12} & a_{13} & a_{14} & : & b_1 \\ a_{21} & a_{22} & a_{23} & a_{24} & : & b_2 \\ a_{31} & a_{32} & a_{33} & a_{34} & : & b_3 \\ a_{41} & a_{42} & a_{43} & a_{44} & : & b_4 \end{bmatrix}$$

$$\textbf{step 1} \quad \begin{bmatrix} 1 & \vdots & u_{12} & u_{13} & u_{14} & : & \bar{b}_1 \\ \cdots & \cdots & \cdots & \cdots & \cdots & \cdots & \cdots \\ 0 & \vdots & {a_{22}}^1 & {a_{23}}^1 & {a_{24}}^1 & : & {b_2}^1 \\ 0 & \vdots & {a_{32}}^1 & {a_{33}}^1 & {a_{34}}^1 & : & {b_3}^1 \\ 0 & \vdots & {a_{42}}^1 & {a_{43}}^1 & {a_{44}}^1 & : & {b_4}^1 \end{bmatrix}$$

$$\textbf{step 2} \quad \begin{bmatrix} 1 & u_{12} & & u_{13} & u_{14} & : & \bar{b}_1 \\ & 1 & & u_{23} & u_{24} & : & \bar{b}_2 \\ & & \cdots & \cdots & \cdots & \cdots & \cdots \\ & 0 & \vdots & {a_{33}}^2 & {a_{34}}^2 & : & {b_3}^2 \\ & 0 & \vdots & {a_{43}}^2 & {a_{44}}^2 & : & {b_4}^2 \end{bmatrix}$$

$$\textbf{step 3} \quad \begin{bmatrix} 1 & u_{12} & u_{13} & & u_{14} & : & \bar{b}_1 \\ & 1 & u_{23} & & u_{24} & : & \bar{b}_2 \\ & & 1 & & u_{34} & : & \bar{b}_3 \\ & & & \cdots & \cdots & \cdots & \cdots \\ & & 0 & \vdots & {a_{44}}^3 & : & {b_4}^3 \end{bmatrix}$$

$$\textbf{step 4} \quad \begin{bmatrix} 1 & u_{12} & u_{13} & u_{14} & : & \bar{b}_1 \\ & 1 & u_{23} & u_{24} & : & \bar{b}_2 \\ & & 1 & u_{34} & : & \bar{b}_3 \\ & & & 1 & : & \bar{b}_4 \end{bmatrix}$$

At any step of Gaussian elimination the entries a_{ij} of the active sub-matrix are modified. The entries marked as u_{ij} and $\bar{b}_i$ do not change at later steps. In what follows, we use the following MATLAB® notations:

• A[i,:] denotes i^{th} row of matrix A;
• A[:,j] denotes j^{th} column of matrix A;
• A[i:n,j:m] denotes a submatrix of matrix A; from row i to row n and column j to column m;
• A[i,:] * B[:,j] denotes product of ith row of matrix A and j^{th} column of matrix B;
The algorithm for the Gaussian elimination can be described as follows:

1. **Begin:** Set the active sub-matrix $\hat{A} = A$ and index $k = 1$
2. **Scale:** Scale the row k of $\hat{A}$ and b_k by $\hat{a}_{k,k}$
   $\hat{A}[k, k:n] \leftarrow \hat{A}[k, k:n]/\hat{a}_{k,k}$; $\bar{b}_k = b_k/\hat{a}_{k,k}$
3. **Eliminate:** For each non-zero entry $\hat{a}_{i,k}$ below the diagonal, multiply the index row by $\hat{a}_{i,k}$ & subtract it from the row i.
   $\hat{A}[i, k:n] \longleftarrow \hat{A}[i, k:n] - \hat{a}_{i,k} * \hat{A}[k, k:n]$.
   Repeat the operation on the element b_i i.e.
   $b_i = b_i - \hat{a}_{i,k} \times \bar{b}_k$.
4. Size of the active submatrix is reduced by 1. Index k is incremented.
5. If $k \neq n+1$, loop back to the step 2; else proceed for backward substitution.

If the matrix A has a dimension n, it can be shown that Gaussian elimination requires $\mathcal{O}(n^3)$(read order of n^3) effort[1]. This implies that if we solve a LSS intially with a 100×100 matrix and then with a 1000×1000 matrix, the computational effort required in the second LSS is $\frac{10^9}{10^6} = 1000$ times more than the first one. Thus, when the system size grows by a factor of 10, the computational effort increases by a factor of 1000. *Hence, the basic Gaussian elimination procedure is not suitable for a large system.* In fact, not much can be achieved in large scale power system analysis without exploitation of sparsity.

When dealing with solution of a set of nonlinear equations (e.g., load flow analysis), one often comes across situations wherein the coefficient matrix A is fixed but the right hand side (rhs) vector changes at every

[1] $\mathcal{O}(n^3)$ implies that computation effort required to perform Gaussian Elimination can be expressed by a third degree polynomial. It is a Least Upper Bound; a problem whose computation effort cannot be bounded by a polynomial is known as a Non-deterministic Polynomial(NP) complete or a computationally intractable problem.

iteration. In such situations, applying Gaussian Elimination repeatedly is an overkill. The computational effort can be reduced by using a variant of Gaussian Elimination known as LU decomposition.

2. LU Decomposition

In this section, we show that LU decomposition records the process of elimination, thereby avoiding repetitive elimination. To begin with, consider the kth step of Gaussian Elimination, *i.e.*, elimination proceeding in the kth column with elements $a_{ik}(i > k)$ being eliminated. $\hat{A}[k:n, k:n]$ is the *active* sub-matrix. The row operation performed at this step is: $\hat{A}[i,k:n] = \hat{A}[i,k:n] - \frac{\hat{a}_{ik}}{\hat{a}_{kk}}\hat{A}[k,k:n]$ The above operation can be represented as follows:

$$M_{ik} \times M_{kk} \times \hat{A} = \tilde{A} \text{ where,}$$

$$M_{i,k}(i>k) = \begin{bmatrix} 1 & & & & & \\ & \ddots & & & & \\ & & 1 & & & \\ & & \vdots & \ddots & & \\ & & -\hat{a}_{ik} & & 1 & \\ & & & & & \ddots \\ & & \vdots & & & & 1 \end{bmatrix} \qquad M_{kk} = \begin{bmatrix} 1 & & & & & \\ & \ddots & & & & \\ & & \frac{1}{\hat{a}_{kk}} & & & \\ & & & \ddots & & \\ & & & & 1 & \\ & & & & & \ddots \\ & & & & & & 1 \end{bmatrix} \tag{4.1}$$

$$\hat{A} = \begin{bmatrix} 1 & * & * & * & * & * & * & * \\ & 1 & * & * & * & * & * & * \\ & & \hat{a}_{kk} & x & x & x & x & x \\ & & 0 & \ddots & x & x & x & x \\ & & \vdots & \vdots & \ddots & \vdots & \vdots & \vdots \\ & & \hat{a}_{ik} & x & x & \hat{a}_{ii} & x & x \\ & & \vdots & \vdots & \vdots & & \ddots & \vdots \\ & & x & x & x & x & x & \hat{a}_{nn} \end{bmatrix} \quad \tilde{A} = \begin{bmatrix} 1 & * & * & * & * & * & * & * \\ & 1 & * & * & * & * & * & * \\ & & 1 & * & * & * & * & * \\ & & 0 & \ddots & x & x & x & x \\ & & \vdots & \vdots & \ddots & \vdots & \vdots & \vdots \\ & & 0 & x & x & \hat{a}_{ii} & x & x \\ & & \vdots & \vdots & \vdots & & \ddots & \vdots \\ & & x & x & x & x & x & \hat{a}_{nn} \end{bmatrix} \tag{4.2}$$

Symbols $*$ and x represent an element of the matrix U and an element of the active sub-matrix respectively. Thus, an elimination operation involves premultiplying a partially factorized matrix with an *elementary* lower triangular matrix. When all non-zeros below the diagonal k are eliminated, products of elementary lower triangular matrices can be aggregated as follows:

$$\left[\; M_{n,k} \times M_{n-1,k} \; \ldots \; M_{k+1,k} \times M_{kk} \; \right] = \left[\; L_k \; \right] \tag{4.3}$$

where,

$$L_k = \begin{bmatrix} 1 & & & & & \\ & 1 & & & & \\ & & \frac{1}{\hat{a}_{kk}} & & & \\ & & \frac{-\hat{a}_{k+1,k}}{\hat{a}_{kk}} & 1 & & \\ & & \vdots & & \ddots & \\ & & \frac{-\hat{a}_{nk}}{\hat{a}_{kk}} & & & 1 \end{bmatrix} \tag{4.4}$$

The matrix L_k is called an elimination matrix. The matrix A is transformed into a unit upper triangular matrix U by sequence of premultiplications by lower triangular matrices i.e. $U = L_n L_{n-1} \cdots L_1[A]$. As each of the elementary lower triangular matrix L_k is non-singular, matrix A can be written as $A = L_1^{-1} L_2^{-1} \cdots L_n^{-1} U$. It can be verified that

$$L_k^{-1} = \begin{bmatrix} 1 & & & & & \\ & 1 & & & & \\ & & \hat{a}_{kk} & & & \\ & & \hat{a}_{k+1,k} & 1 & & \\ & & \vdots & & \ddots & \\ & & \hat{a}_{nk} & & & 1 \end{bmatrix}$$

Further, the product of two lower triangular matrices is also a lower triangular matrix. Therefore, in compact notations, we can write a non-singular matrix A as a product of a lower triangular matrix L and a upper triangular matrix U i.e. $A = LU$. If matrices L and U are known, then solution to the linear system of equations can be obtained as follows:

```
Solve [L][y] = [b] // forward substitution
      [U][x] =[y]  // backward substitution
```

The matrix L stores the pivoting information of Gaussian elimination. The key advantage of LU decomposition is that, it has to be computed only once and then solving multiple linear system of equations amounts to performing only repeated forward-backward substitutions. The choice of diagonals for matrices L and U is not unique. Matrix A has n^2 entries while L and U have respectively $n + (n-1) + \cdots = \frac{n(n+1)}{2}$ entries each.Therefore L and U have together $n^2 + n$ entries. So n entries have to be fixed apriori; then, n^2 unknowns are worked out from n^2 knowns. For example, we can design a decomposition such that L is a *unit* lower triangular matrix. The decomposition can be also expressed as a product of a unit lower triangular matrix, diagonal matrix and a unit upper triangular matrix ($A = LDU$).

2.1. Algorithm for LDU Decomposition

The LDU factorization can be computed in any one of the following ways: 1) by rows, 2) by columns or 3) combination of the two (frontal matrix approach). We describe an algorithm using the option-3 wherein the size of active matrix reduces by one row and column in each step.

Let L be represented columnwise and U rowwise as follows:

$$L = [l_1, l_2, \cdots, l_n] \qquad U = \begin{bmatrix} u_1^T \\ u_2^T \\ \vdots \\ u_n^T \end{bmatrix} \qquad D = \begin{bmatrix} d_1 & & & & \\ & d_2 & & & \\ & & \ddots & & \\ & & & \ddots & \\ & & & & d_n \end{bmatrix}$$

Then, $A = LDU = d_1 l_1 u_1^T + d_2 l_2 u_2^T + d_n l_n u_n^T$. Let the columnwise and rowwise representation of A be as follows:

$$A = \begin{bmatrix} c_1 & c_2 & \cdots & c_n \end{bmatrix} = \begin{bmatrix} r_1^T \\ r_2^T \\ \vdots \\ r_n^T \end{bmatrix}$$

If e_i represents the ith column vector of identity matrix I_n, where I_n is a $n \times n$ identity matrix, then it can be verified that:

$$Ae_1 = d_1 l_1 (u_1^T e_1) = d_1 l_1 \; (\mathrm{u}_{11} = 1).$$

Since,

$$l_{11} = 1; \qquad d_1 = a_{11}; \qquad l_1 \Rightarrow \frac{c_1}{d_1}.$$

Similarly, $e_1^T A = r_1^T = d_1 u_1^T$, & hence $u_1^T = \frac{r_1^T}{d_1}$. Now update matrix A to $A = (A - d_1 l_1 u_1^T)$. It can be verified that in the structure of A, $c_1 = r_1 = 0$.

$$A = \begin{bmatrix} 0 & & \cdots & & 0 \\ \vdots & \times & \times & \times & \times \\ 0 & \times & \times & \times & \times \\ \vdots & \times & \times & \cdots & \times \\ 0 & \times & \times & \cdots & \times \end{bmatrix}$$

Now, it can be seen that $Ae_2 = d_2 l_2 = c_2$; therefore $d_2 = a_{22}$ and $l_2 = \frac{c_2}{d_2}$. An algorithm for LDU is described below. Since a_{ij} is not used after

L_{ij}/U_{ij} has been computed, storage can be optimized by overwriting matrix A with L, U and D as follows:

$$A = \begin{bmatrix} d_1 & & & U \\ & d_2 & & \\ & & \ddots & \\ L & & & d_n \end{bmatrix}$$

```
begin L=[0];U=[0];
for i=1:n
 d[i]=A[i,i];
 L[i:n,i]=a[i:n,i]/a[i,i];
 U[i,i:n]=a[i:n,i]/a[i,i];
  for k=i+1:n;
   for j=i+1:n;
     a[j,k]=a[j,k]-d[i]L[j,i]U[i,k]
   end
  end
end
```

Algorithm for LU Decomposition

The LU decomposition algorithm as described above will fail if a zero pivot is encountered during the factorization of matrix. For example, $A = \begin{bmatrix} 0 & 1 \\ 1 & 0 \end{bmatrix}$ is non-singular but the very first pivot is zero. In such situations, row interchanges are performed to obtain primarily non-zero pivots. Also, row interchanges are used to obtain relatively large enough numerical values of pivots. Otherwise, division by small pivots may cause numerical grow and hence significant truncational errors. An interchange of rows (or columns) can be modeled by a permutation matrix (P) which has all the rows of an identity matrix but they are permuted to effect the row interchange. For example, it can be verified that by flipping 2nd and 3rd row of identity matrix, we obtain a permutation matrix which when premultiplied to a matrix A, flips row 2 to row 3.

$$\begin{bmatrix} 1 & 0 & 0 & 0 \\ 0 & 0 & 1 & 0 \\ 0 & 1 & 0 & 0 \\ 0 & 0 & 0 & 1 \end{bmatrix} \begin{bmatrix} a_1^T \\ a_2^T \\ a_3^T \\ a_4^T \end{bmatrix} = \begin{bmatrix} a_1^T \\ a_3^T \\ a_2^T \\ a_4^T \end{bmatrix}$$

Similarly, post multiplication by a permutation matrix results in column interchange. It can be shown that (1) a permutation matrix P is an orthogonal matrix (i.e., $P^TP = PP^T = I_n$) and (2) product of permutation matrices is also a permutation matrix. LU decomposition with row permutations for matrix A can be represented as $PA = LU$. There are many variants of LU decomposition. It can be shown that if the matrix is Symmetric Positive Definite(spd)[2], then a symmetric version of LU decomposition $A = LL^T$ exists. It is known as *Cholesky* decomposition. A square root free version of Cholesky decomposition is given by $A = LDL^T$.

2.2. Forward Substitution

In the forward substitution, following system of equation is solved:

$$\begin{bmatrix} l_{11} & & & & \\ l_{21} & l_{22} & & & \\ \vdots & l_{32} & \ddots & & \\ & \vdots & & l_{n-1,n-1} & \\ l_{n1} & l_{n2} & \cdots & & l_{nn} \end{bmatrix} \begin{bmatrix} y_1 \\ y_2 \\ y_3 \\ \vdots \\ y_{n-1} \\ y_n \end{bmatrix} = \begin{bmatrix} b_1 \\ b_2 \\ b_3 \\ \vdots \\ b_{n-1} \\ b_n \end{bmatrix}$$

It is easy to see that $y_1 = \frac{b_1}{l_{11}}$. To obtain y_2, subtract $y_1 \times [l_{21}, l_{31}, \cdots, l_{n1}]^T$ from sub-vector $[b_2, \cdots, b_n]^T$.

$$\therefore \quad \begin{bmatrix} l_{22} & & & \\ l_{32} & l_{33} & & \\ \vdots & l_{43} & \ddots & \\ & \vdots & & \\ l_{n2} & l_{n3} & \cdots & l_{nn} \end{bmatrix} \begin{bmatrix} y_2 \\ y_3 \\ y_4 \\ \vdots \\ y_n \end{bmatrix} = \begin{bmatrix} b_2^1 \\ b_3^1 \\ b_4^1 \\ \vdots \\ b_n^1 \end{bmatrix}$$

The sub-vector b^1 is obtained by updating the vector b as described above. At the ith step, the following linear system of equations is solved.

$$\begin{bmatrix} l_{ii} & & & & \\ l_{i+1,i} & l_{i+1,i+1} & & & \\ \vdots & l_{i+2,i+1} & \ddots & & \\ & \vdots & & l_{n-1,n-1} & \\ l_{n,i} & l_{n}, l_{n+1} & \cdots & & l_{nn} \end{bmatrix} \begin{bmatrix} y_i \\ y_{i+1} \\ \vdots \\ y_{n-1} \\ y_n \end{bmatrix} = \begin{bmatrix} b_i^{i-1} \\ b_{i+1}^{i-1} \\ \vdots \\ b_{n-1}^{i-1} \\ b_n^{i-1} \end{bmatrix}$$

[2]A matrix A is spd if A is symmetric and $x^TAx > 0$ ($\forall x \neq 0$).

where,

$$\begin{bmatrix} b_i^{i-1} \\ b_{i+1}^{i-1} \\ \vdots \\ b_n^{i-1} \end{bmatrix} = \begin{bmatrix} b_i^{i-2} \\ b_{i+1}^{i-2} \\ \vdots \\ b_n^{i-2} \end{bmatrix} - y_{i-1} \times \begin{bmatrix} l_{i,i-1} \\ l_{i+1,i-1} \\ \vdots \\ l_{n,i-1} \end{bmatrix}$$

$y_i = \frac{b_i^{i-1}}{l_{ii}}$ and fianlly $y_n = \frac{b_n^{n-1}}{l_{nn}}$.

2.3. Backward Substitution

In this phase, the following system of equations is solved.

$$\begin{bmatrix} u_{11} & u_{12} & \cdots & u_{1n} \\ & u_{22} & \cdots & u_{2n} \\ & & \ddots & \\ & & u_{n-1,n-1} & u_{n-1,n} \\ & & & u_{nn} \end{bmatrix} \begin{bmatrix} x_1 \\ x_2 \\ \vdots \\ x_{n-1} \\ x_n \end{bmatrix} = \begin{bmatrix} y_1 \\ y_2 \\ \vdots \\ y_{n-1} \\ y_n \end{bmatrix}$$

Hence, $x_n = \frac{y_n}{u_{nn}}$.

The sequence of operations to be performed in backward substitution are similar to forward substitution but the operations are executed in reverse order, *i.e.*, the sequence in which unknowns are computed is given by $x_n, x_{n-1}, \ldots, x_1$. Once x_n is known, for computation of x_{n-1}, sub vector $[y_1, \ldots, y_{n-1}]^T$ is updated. Hence,

$$\begin{bmatrix} y_1^1 \\ \vdots \\ y_{n-1}^1 \end{bmatrix} = \begin{bmatrix} y_1 \\ \vdots \\ y_{n-1} \end{bmatrix} - x_n \begin{bmatrix} u_{1n} \\ \vdots \\ u_{n-1,n} \end{bmatrix}$$

$$\therefore \quad \begin{bmatrix} u_{11} & \cdots & u_{1,n-1} \\ & \ddots & \vdots \\ & & u_{n-1,n-1} \end{bmatrix} \begin{bmatrix} x_1 \\ \vdots \\ x_{n-1} \end{bmatrix} = \begin{bmatrix} y_1^1 \\ \vdots \\ y_{n-1}^1 \end{bmatrix}$$

Hence, $x_{n-1} = \frac{y_{n-1}^1}{u_{n-1,n-1}}$.

Thus, while solving for x_{n-i} following system of equations is solved.

$$\begin{bmatrix} y_1^i \\ \vdots \\ y_i^i \end{bmatrix} = \begin{bmatrix} y_1^{i-1} \\ \vdots \\ y_{n-i}^{i-1} \end{bmatrix} - x_{n-i+1} \begin{bmatrix} u_{1,n-i+1} \\ \vdots \\ u_{n-i,n-i+1} \end{bmatrix}$$

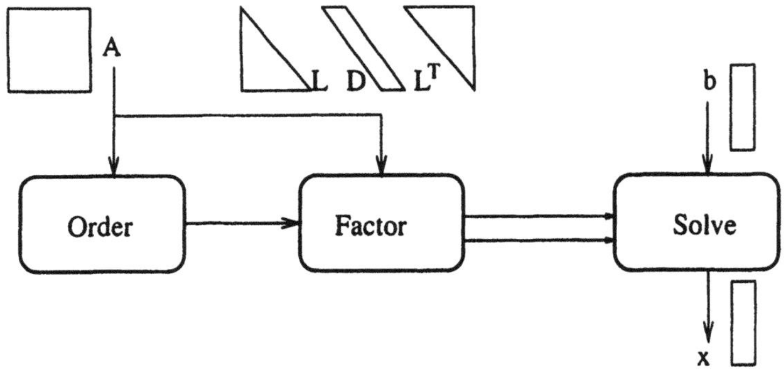

Figure 4.1. A "black box" formulation of a sparse direct solver

$$\begin{bmatrix} u_{11} & \cdots & u_{1,n-i} \\ & \ddots & \vdots \\ & & u_{n-i,n-i} \end{bmatrix} \begin{bmatrix} x_1 \\ x_2 \\ \vdots \\ x_{n-i} \end{bmatrix} = \begin{bmatrix} y_1^i \\ y_2^i \\ \vdots \\ y_{n-i}^i \end{bmatrix}$$

Hence, $x_{n-i} = \frac{y_{n-i}^i}{u_{n-i,n-i}}$ In some of the power system applications such as Load Flow and Short Circuit Analysis, matrices are symmetric and diagonally dominant. Therefore, instead of LU decomposition, LDL^T decomposition is used. Because of symmetry only the L and D matrices are stored. When the matrix A is square, large and sparse, it is obvious that a brute force implementation will not be able to exploit sparsity. In the next section, the sparsity issues involved in LU decomposition for spd matrices are considered.

3. LDL^T Decomposition for Sparse *spd* Matrices

Sparse spd matrix factorization involves following three phases: (1) Analyze (2) Factorize and (3) Solve (figure 4.1). If the matrix A is sparse, then it is likely that the matrix L is also sparse. The role of *analyze* phase is to (1) evaluate the structure of matrix L and (2) set up its data structure. It will be soon shown that in the analyze phase, an effort is made to keep the matrix L as sparse as possible. Once the data structure for matrix L is set up, the role of the *factorize* phase is to generate the numerical values for entries of L-matrix, *i.e.*, perform numerical factorization. The *solve* phase implements forward and backward substitution. A 3 step approach to LDL^T is usually used only with sparse spd matrices. It is not possible to apply this approach to non-symmetric indefinite sparse matrix.

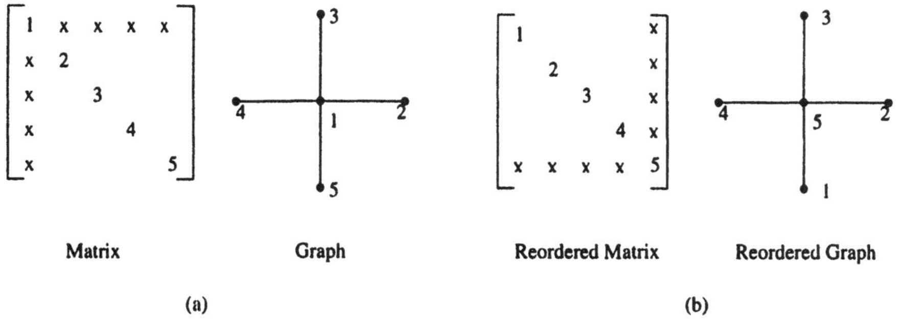

Figure 4.2. An Illustrative 5-node System

3.1. Analyze Phase

Example 4.1: Consider a symmetric matrix A as shown in figure 4.2(a). When the matrix is factorized the structure of matrices L and U is as follows:

$$L = \begin{bmatrix} \times & & & & \\ \times & \times & & & \\ \times & \bullet & \times & & \\ \times & \bullet & \bullet & \times & \\ \times & \bullet & \bullet & \bullet & \times \end{bmatrix} \qquad U = \begin{bmatrix} \times & \times & \times & \times & \times \\ & \times & \bullet & \bullet & \bullet \\ & & \times & \bullet & \bullet \\ & & & \times & \bullet \\ & & & & \times \end{bmatrix}$$

The $\times$ indicates a non-zero element in matrix A and $\bullet$ indicates a non-zero in matrix L and U which was zero in the original matrix A. A non-zero element in L and U ($L(i,j)$ and $U(i,j)$) which was zero in original matrix ($A(i,j) = 0$) is known as a *fill* or *fillin*. In this example, it can be observed that the matrices L and U have become completely dense during factorization. If possible, fills have to be avoided for the following reasons:

1 Fills propagate; if generated in early stages, they propagate and create a lot more fills in the later stages. Figures 4.3a and 4.3b show structure of a matrix A and a matrix U for 1000×1000 system. The black patch in the lower part of matrix U shows a high proportion of fills generated due to the propagation effect. More non-zeros in L or U imply correspondingly more effort in computation. If fills could be predicted and minimized, considerable computation during factorization can be saved.

2 A less sparse L or U implies correspondingly more effort in solve phase i.e. forward/backward substitution.

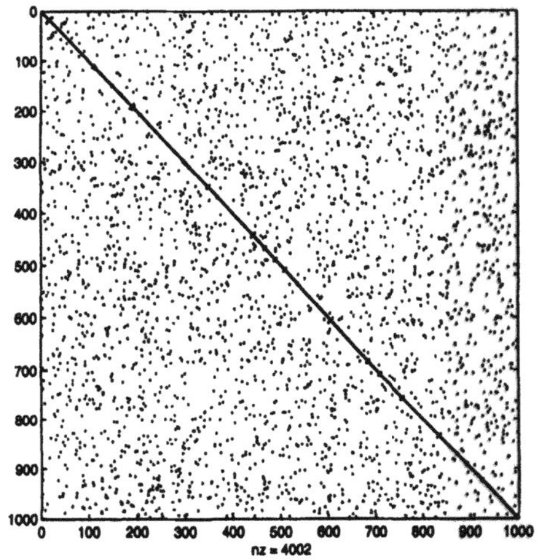

Figure 4.3a. Sparse Matrix (n=1000)

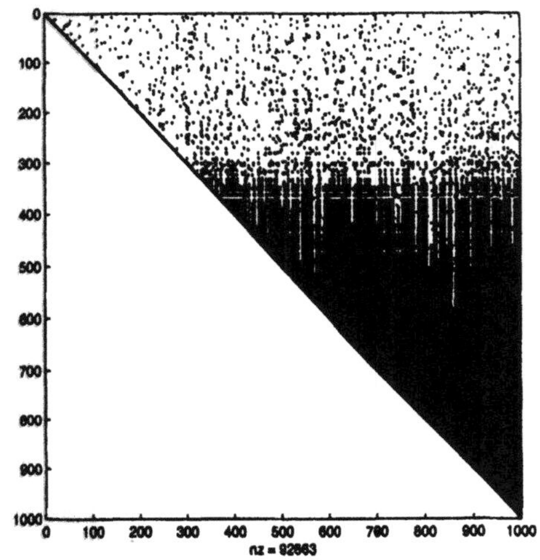

Figure 4.3b. The *U* Matrix

An important question then is: *Can a sparse spd matrix be factorized such that fills are reduced?* It so happens that fills can be reduced by re-ordering or permuting the matrix (flipping rows and columns appropriately). The next section discusses these issues.

3.1.1 Ordering

Suppose that matrix A in figure 4.2(a) is reordered by flipping rows 1 and 5 and columns 1 and 5 respectively. The structure of the permuted matrix is as shown in figure 4.2(b). It can be seen that such symmetric row and column exchanges preserve symmetry. It can be verified that the structure of new L/U matrix has no fills. In other words, no new non-zeros appear in L/U matrices. Hence, there are saving in computations, in both the L/U factorization and the F/B substitution. Now the question arises: *how to find out such an order?* To answer this question, we must first understand the mechanism of fill generation. The generation of fills can be explained by a graph theoretic approach discussed in the next section [Rose, 1972].

4. Symbolic Gaussian Elimination

The structure of a symmetric $n \times n$ sparse matrix can be captured by an undirected graph over n vertices. For every non-zero off diagonal entry $a_{ij}(i \neq j)$, an edge is connected between vertices i and j. From figures 4.2(a) and 4.2(b), it can be deduced that reordering a matrix is equivalent to renumbering the nodes of the graph. To develop a graph theoretic perspective, elimination process is carried out on the following example and the fills recorded on the graph.

Example 4.2: The structure of a matrix and its graph is as shown in figure 4.4. Record (1) structure of active submatrix and (2) elimination graph at each step of elimination.

Initially, in step 1, original matrix is the active sub-matrix and elimination operation is being performed by pivot $A(1,1)$. In the graph theoretic language, it is said that node 1 is being eliminated. Elements $A(3,1)$ and $A(4,1)$ are the only two non-zeros below the diagonal. When $A(3,1)$ is getting eliminated, the structure of the row 1 is superimposed on row 3. Similarly, elimination of $A(4,1)$ superimposes structure of row 1 on the row 4. This is recorded in the elimination graph as follows:

1 When elements $A(3,1)$ and $A(4,1)$ are zeroed, the size of active matrix reduces by one. Therefore, in the elimination graph for step 2, node 1 and its incident edges are deleted.

2 When the structure of row 1 is superimposed on row 3, three possibilities exist:

   (a) $A(1,j) = 0$; In such case, $A(3,j)$ is not affected.

   (b) For a non-zero $A(1,j)$ a non-zero $A(3,j)$ already exists in that location. In such case nothing extra has to be done. Correspondingly, edge between nodes 3 and j exists in elimination graph. This case does not happen in this particular example.

   (c) For a non-zero $A(1,j)$ the element $A(3,j)$ is zero in the original matrix. Corresponding edge is not present in the graph. However, the superposition converts a zero of matrix A to a non-zero, *i.e.*, a fill is created e.g., $A(3,4)$. Correspondingly, an edge is added between node 3 and node 4. Note that both nodes, 3 and 4 are the adjacent nodes of node 1, the node being eliminated.

3 Since elimination is carried out on all adjacent nodes of 1 (in this case nodes 3 and 4), all possible edges are created between adjacent nodes of 1, *i.e.*, the new elimination graph (in step 2) has a clique of adjacent nodes of 1. A clique is a completely connected graph; K_n indicates clique over n-nodes. Cliques upto K_4 are shown in figure 4.5. Since, in this example, there are only 2 adjacent nodes (3 & 4) to 1, clique K_2 (i.e. edge 3-4) is introduced. The rest of the symbolic elimination process is worked out in figure 4.4.

The example also suggests that whenever a node is eliminated, *i.e.*, used as a pivot, a fill is created if there exists a pair $(L_{ik}, U_{kj})(k <$

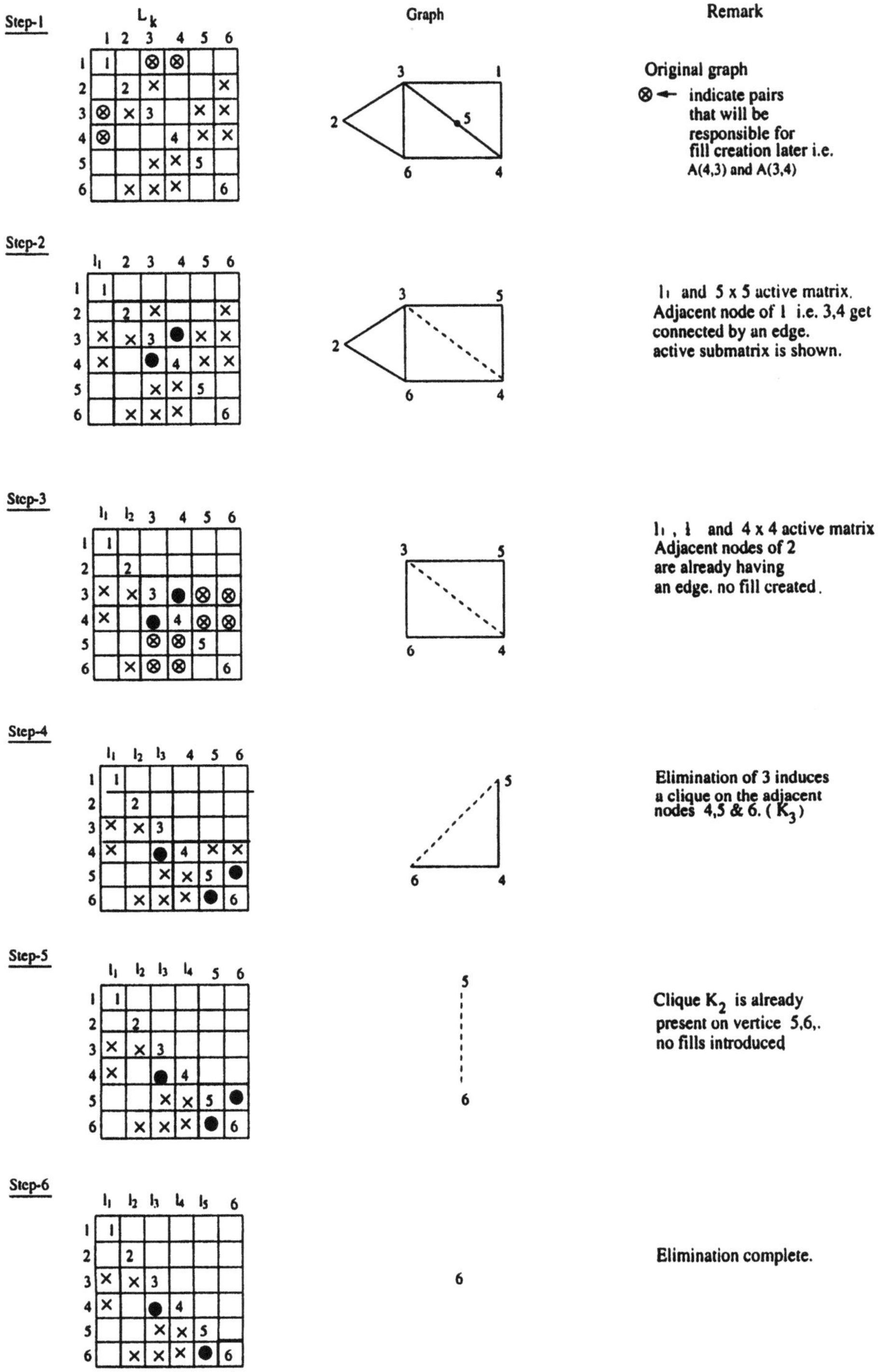

Figure 4.4. Example 4.2

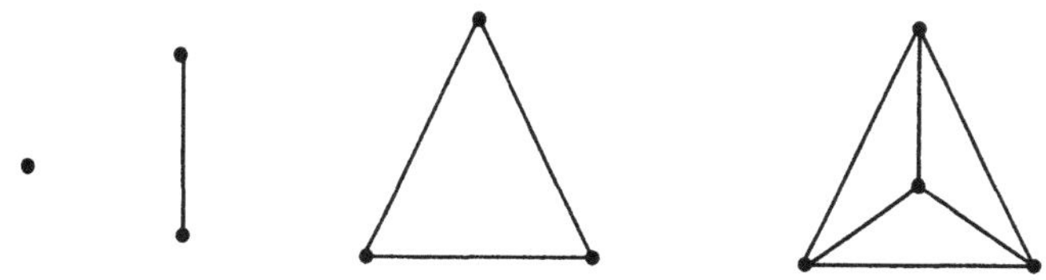

Figure 4.5.

min(*i*, *j*)) such that both L_{ik} & U_{ik} are non-zero. For example $L(3,1)$ $U(1,4)$ is responsible for creating fill at $U(3,4)$. In the elimination graph, a fill manifests as a new edge between two nodes, adjacent to the node being eliminated. Further, if an edge between two such nodes already exists, then it stays as it is.

Example 4.3: Explain the fill generation in figure 4.2(a) by elimination graph.

Since the first column and row of matrix A is non-sparse, the corresponding column and row of matrix L and U are also non-sparse. It can be seen from the matrix that for any a_{ij} which is zero in the original matrix, the corresponding $L(i,1)$ & $U(1,j)$ product is always non-zero, so a fill is inevitable. In terms of the elimination graph, this means that, the set of nodes $\{2,3,4,5\}$ adjacent to node 1 is completely connected thus, inducing a clique K_4. We conclude that when a node is eliminated, a clique is induced on the adjacent nodes. If an edge of clique is not present in the elimination graph, then it means that elimination induces a fill.

Example 4.4: Explain the zero fill condition in figure 4.2(b).

Since, nodes 1, 2, 3 and 4 are having only one adjacent node, no fills are created when they are eliminated in the aforesaid sequence. Alternatively, in the elimination sequence, it is easy to see that the first non zero in columns 1-4 is in the last row. Thus, pivoting affects only the last row when the node 5 is renumbered as 1 and vice versa.

5. Ordering Strategies for *spd* Matrices

We have now reached an important conclusion for sparse matrix factorization. *The number of fills depends upon the order or sequence in which the nodes are eliminated.* Reordering of nodes amounts to a symmetric permutation, *i.e.*, if rows i and j are interchanged then subsequently column i and j are also exchanged. In terms of permutation matrix (P), reordering can be represented as $\hat{A} = PAP^T$. Thus, reordering preserves symmetricity of matrix A. As both the row and column interchanges are used during symmetric permutation, the elements of b vector as well as the elements of x-vector have to be reshuffled in the same order. In matrix terminology, it can be written as follows:

$$Ax = b \quad \Rightarrow PAP^T Px = Pb \quad \Rightarrow \hat{A}\hat{x} = \hat{b}$$

where matrix P is used to indicate reshuffling operation or permutation. Note that, pivots are always the diagonal element of the matrix. A spd matrix has weighty diagonal. Consequently, the reordering of nodes does not threaten numerical stability of factorization.

With this background, it is now possible to tackle the important question of finding the best ordering scheme. An obvious brute force strategy would be to simulate the symbolic elimination using all possible permutations. However, there are $n!$ ways to order a graph. Table 4.1 indicates number of orders as a function of n.

Table 4.1. Maximum Number of Permutations in a Sparse Matrix

No of Nodes(n)	No of Permutations (n!)
1	1
2	2
3	6
4	24
5	120
6	720
10	3628800
20	2.4378×10^{18}

Clearly, enumeration strategy suffers from the *curse of dimensionality*. The question that arises now is, whether it is possible to develop a non enumerative optimal ordering strategy which can minimize fills? It has been shown that the problem of ordering is *NP complete* [Yannakakis, 1981]. Hence, it is only possible to devise efficient heuristics. A few important ordering strategies are now discussed.

5.1. Minimum Valency Ordering

Recall that a node with m adjacent vertices induces a clique of K_m on elimination; hence, *common sense suggests that elimination of nodes be carried out in the ascending order of their degree*[3]. For example, for the

[3]The degree of a node is the number of it's adjacent nodes.

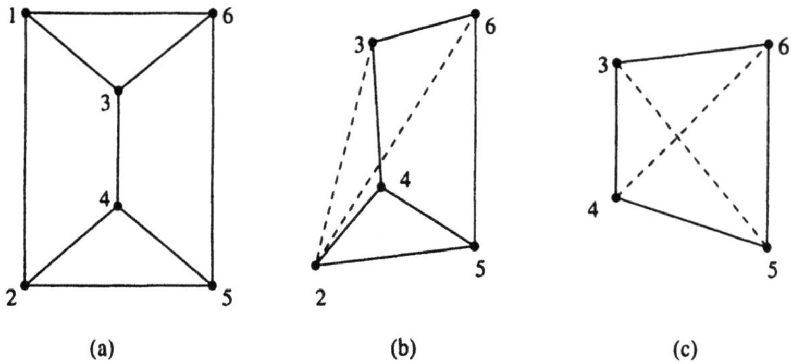

Figure 4.6. Example 4.5: Fills by Minimum Valency Ordering

5 node system (figure 4.2(a)) that we considered earlier, the minimum valency ordering (figure 4.2(b)) will give the best results. An obvious disadvantage of minimum valency ordering is that it cannot account for the generation of fills and their propagation. A node with low degree may have its degree inflated later on due to fills generated in elimination. Viewed from such a perspective, it may be wise to delay it's elimination. **Example 4.5:** All the nodes in the graph of figure 4.6(a) have a degree of 3. Thus, the ordering in figure 4.6 also corresponds to minimum valency ordering. Further, it is one among the 6! minimum valency ordering solutions. In the step 1, the elimination of node 1 leads to the elimination graph as shown in figure 4.6(b). Two fills are created when node 1 is eliminated. Notice that the degree of node 2 has now shot up to 4. As the order of nodes to be eliminated is decided apriori in case of minimum valency algorithm, it does not account for these dynamic changes in the degree of uneliminated nodes. As such, elimination of node 2 (step 2) leads to the elimination graph as shown in figure 4.6.

Notice that a clique K_4 is induced on rest of the nodes 3,4,5,6 with two additional fills (total fills 4). The matrix is now dense and hence, no further fills can be induced. Thus, the given minimum valency ordering induces a total of 4 fills. Since, the degree of an uneliminated node can change when its adjacent node is eliminated, it appears that after every elimination, the dynamic degree of uneliminated nodes should be recomputed. This leads us to the minimum degree algorithm.

6. Minimum Degree Algorithm(MDA)

This algorithm is implemented by the following steps 1) select a node of minimum degree for elimination 2) simulate elimination of that node; update the adjacency of its adjacent uneliminated nodes 3) select a new

node having minimum degree from the current elimination graph of un-eliminated nodes.
This process is repeated until all the nodes are eliminated. It can be seen that if such a strategy is employed in example 4.5, then node 3 would have been the next node to be eliminated in step 2. The elimination graph due to elimination of node 3 is shown in figure 4.7, which shows only one additional fill is generated after elimination of node 1. The new graph is K_4 and hence, no further fills can be induced. Therefore, total number of fills now is only 3.

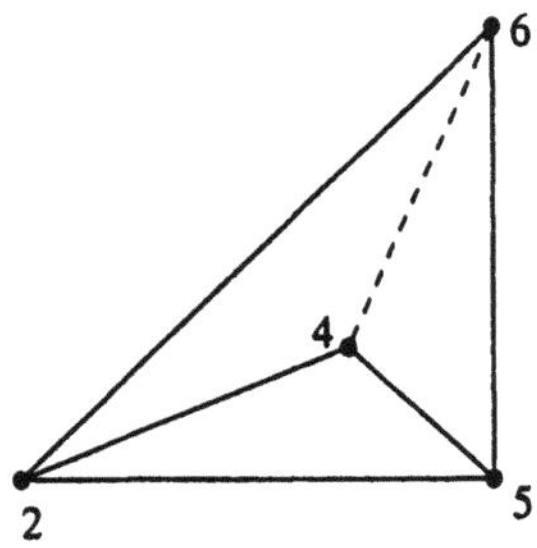

Figure 4.7. Fills by MDA Ordering for Example 4.5

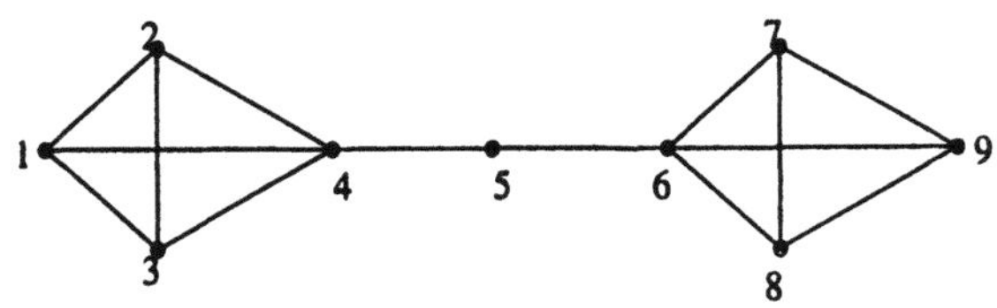

Figure 4.8. Example 4.6

Thus, deferring selection of the next minimum degree node until degree update has given better results. The next important question is, *Is MDA ordering optimal?* To answer the question let us consider an example.
Example 4.6: Find the number of fills for the graph in figure 4.8. For the ordering shown, it is very easy to verify that not a single fill will result in the elimination. Hence, ordering shown in figure 4.8 is optimal.
Example 4.7: Apply MDA to the graph of Example 4.6.

Node 5 has a minimum degree of 2. Hence, it will be eliminated first. Nodes 4 and 6 are adjacent to 5; as a consequence a fill will be created between 4 and 6. Thus, MDA need not give optimal ordering. After elimination of node 5, node 1 is the first node with minimum degree-3. Elimination of it will not create additional fill. It can be verified that no additional fills will result as elimination proceeds; the new ordering corresponds to $(5, 1, 2, 3, 4, 6, 7, 8, 9)$.

For the 1000 × 1000 node system discussed earlier, figures 4.9a-4.9b show the structure of matrix A and corresponding matrix U after MDA ordering. A few salient observations are as follows:

1 By and large, lower degree nodes are numbered first and higher degree nodes later; hence, MDA has a resemblance with minimum valency ordering.

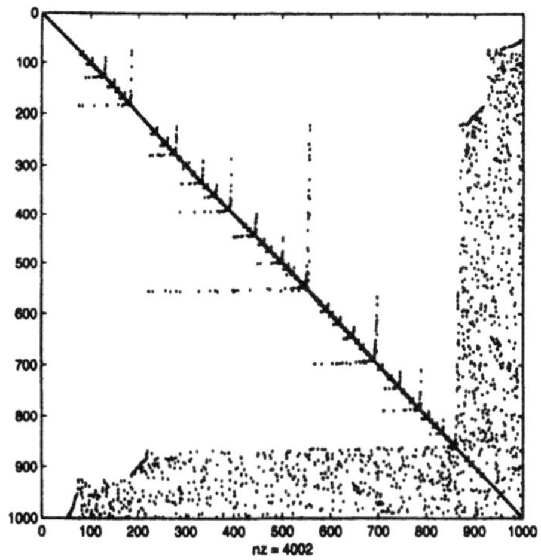

Figure 4.9a. A 1000 × 1000 MDA Ordered Matrix

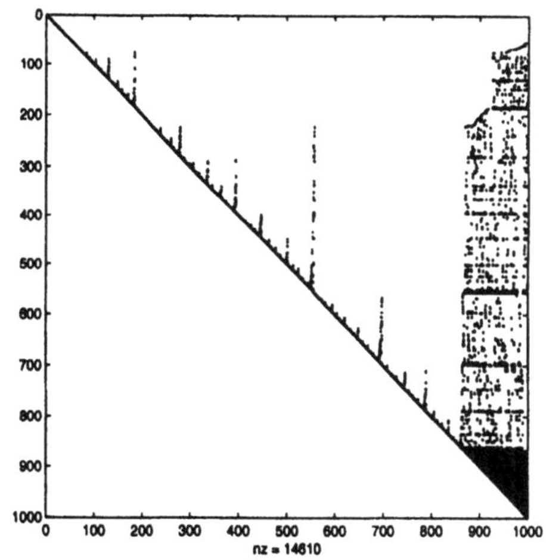

Figure 4.9b. U Matrix after MDA

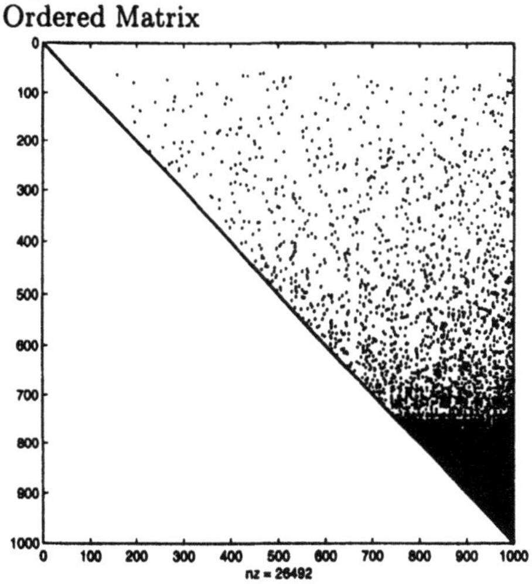

Figure 4.9c. U Matrix after minimum valency ordering

2 The number of fills after MDA ordering are far less than before ordering; while the original matrix has 4002 non-zeros, matrix U created without ordering of matrix A has 92683 non-zeros, and after MDA ordering has only 14610 non-zeros. Clearly, the black patch in figure 4.9b is considerably smaller than in figure 4.3b.

3 Performance of MDA is far superior to the minimum valency ordering. Figure 4.9c shows that, with minimum valency ordering, the matrix U has 26492 non-zeros which far exceed that due to MDA.

Before proceeding further, it is worthwhile to recapitulate the motivation for an ordering. A 'good' ordering reduces fills; it implies lesser computation effort in factorize and solve phases.

6.1. Graph Theoretic Algorithm for MDA

The MDA algorithm can be described elegantly by graph theoretic notations. Let G be undirected graph and v a node of G. Let $Adj_G(v)$ refer to the set of nodes adjacent to v in G. Let G_v represent the elimination graph obtained by deleting the node v and its incident edges and then, adding edges to create a clique of adjacent nodes of v in G, *i.e.* $K_{Adj_G(v)}$. The MDA algorithm can be described as follows:

```
Basic MDA {
  while  G ≠ {φ}
  {
      1.  Select a node y of minimum degree and
          order y  next;
      2.  G ← G_y; Initialize G to elimination
          graph G_y
  }
}
```

If a node does not belong to $Adj_G(y)$; its adjacency remains the same in the new elimination graph G_y. Hence, in step 2, adjacency of only those nodes that are adjacent to y have to be recomputed. It has been found out by researchers [George and Liu, 1989] that considerable effort (time) of MDA is spent in the degree update stage. In other words, suboptimal implementation of MDA may offset the benefits of ordering as weighed in computation time. Typically, benefits of ordering are pronounced with increasing size of the coefficient matrix. An important question now is *how to optimize computational effort in MDA?* The computational effort required in MDA depend upon (1) number of degree updates and (2) efficacy of the data structure used for storing and manipulating the graph. We now take a closer look at step 2 of basic MDA.

6.2. Degree Update*

The motivation for improving speed of MDA is that, it should take only a fraction of time of the factorize phase. Efforts in degree update can be cut down by reducing the number of updates (maximum n in the basic MDA). In other words, multiple number of nodes should be

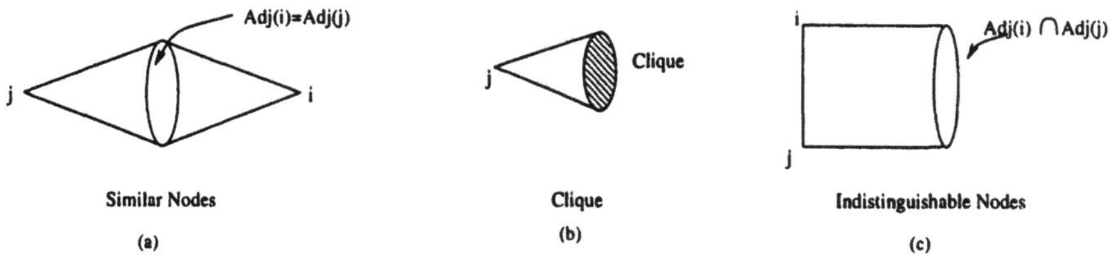

Figure 4.10. Similar and Indistinguishable nodes

eliminated before performing a degree update. For example, recall that in example 4.5, at step 3, the eliminated graph turns out to be a clique. The sequence of further eliminations can be recorded without any degree update. We now introduce the concept of disjoint, similar and indistinguishable nodes used to reduce the number of degree updates [Pandit, 2001].

6.2.1 Disjoint Nodes

Two nodes (i and j) having same degree (in fact minimum degree) in an elimination graph are said to be *disjoint* if and only if $Adj_G(i) \cap Adj_G(j) = \{\phi\}$ where $\{\phi\}$ indicates a null set. A set of nodes $\{i_1, i_2, \ldots i_k\}$ are said to be disjoint if every pair of nodes in the set is disjoint. All disjoint nodes can be eliminated at a time before a degree update. This concept is more commonly referred to as multiple elimination. In fact the concept of multiple elimination changes the MDA ordering and the fill pattern. In practice, it has been noticed that quality of ordering is not adversely affected by multiple elimination. Usually, in a large network, there are very many nodes with low degree which also implies a large proportion of disjoint nodes, at least, in the initial stages of elimination.

6.2.2 Similar Nodes

In an elimination graph G, two nodes i and j are are said to be similar (figure 4.10) if (1) they have same degree and (2)$Adj_G(i) = Adj_G(j)$. A set of nodes are similar if they are pairwise similar. If one of the similar nodes i is eliminated, then the rest can be eliminated, without creation of any additional fill. Thus, degree update for similar nodes can be cut down.

6.2.3 Indistinguishable Nodes

In an elimination graph G, two nodes i and j are said to be indistinguishable (figure 4.10) if (1)they have same degree and (2) $Adj_G(i) \cup \{i\} = Adj_G(j) \cup \{j\}$. A set of nodes are indistinguishable if they are pair wise indistinguishable. If a node i is selected for elimination, thereafter all its indistinguishable nodes can be eliminated without creation of a fill. It can also be verified that these indistinguishable nodes are also the later minimum degree choices. As far as MDA implementation is concerned, indistinguishable nodes can be merged together, *i.e.*, only a single degree update is required for a set of indistinguishable nodes. Hence, a representative node can be considered for elimination. In literature [George and Liu, 1989], such a node has also been referred to as a super node or a prototype node. The elimination of indistinguishable nodes in one step reduces degree updates, and therefore, speeds up MDA. As the elimination proceeds, sparsity may reduce and the percentage of indistinguishable nodes increase. In the example 4.5, at step 3 nodes 3, 4, 5 and 6 are indistinguishable. The black patches in the figure 4.9b and 4.9c for the 1000 × 1000 node systems, indicate a large number of indistinguishable nodes. The enhancements to MDA is an active area of research. More ideas like external degree, approximate minimum degree have also been developed to reduce the computation effort. While enhancements like similar and indistinguishable nodes reduce the number of degree updates, an efficient implementation will also require an efficient data structure. A pseudo code for set-theoretic implementation of MDA in an object oriented paradigm is given in figure 4.11.

6.3. Role of class *Set* in MDA

In the overall implementation of MDA, degree update is the most crucial step, which involves finding the new cardinality of all *affected* nodes. Various data structures such as quotient graph [George and Liu, 1980], storage by cliques have been used to implement MDA in literature [Liu, 1985]. The set data structure is both efficient and simple to comprehend. It can be seen that, the set theoretic operations like *union*, *intersection* and *difference* not only update cardinality of the *uneliminated* nodes appropriately, but use of overloaded arithmetic operators improve readability of the program. This results in an overall efficient implementation.

Having discussed MDA and its object oriented implementation, the other aspect of *analyze phase*, that needs to be addressed, is setting of data structure for L. For this, the coefficient matrix is reordered with the new order given by MDA. It is followed by symbolic factorization. The entire operation can be performed by an efficient algorithm described in [Duff et al., 1986], example 2.6.

```
mda(SparseMatrix & A) {
    Set Eliminated (A.NoOfNodes);
        //set to store eliminated nodes of A
    Set MatNode[A.NoOfNodes];
        //array of set for storing adjacency of each node of A
    A.adj(MatNode);
        //adjacency of each node of A returned in array MatNode
    while(Eliminated.cardin < A.NoOfNodes)
        //while all nodes have not been eliminated
    {
        1.Choose a node y such that
        it is an uneliminated node & it has a minimum degree
            Mark node y as eliminated.
        2.Find all disjoint nodes of node y.
            Store them in {DisjY}. Form {Z} = {y}U{DisjY}.
         Mark them as eliminated.
        3.For each member node z of {Z}
         {
            find all indistinguishable nodes {IndZ};
              mark them as eliminated.
            else
              find all similar nodes {SimZ};
              mark them as eliminated.
            Update the set of eliminated nodes
        4.For each uneliminated u node adjacent to z, perform
            degree update i.e.
            4.0 Adj{u} = Adj{u} ∪ Adj{z}
            4.1 Adj{u} = Adj{u} - {IndZ ∪  SimZ ∪ {z} ∪{u}}
        }
    }
}
```

Figure 4.11. Set Theoretic Pseudo Code for MDA

7. Summary & Additional Comments

LU decomposition of a sparse matrix is much more complicated than the corresponding non-sparse version because, the zeros of the original matrix A may be converted to non-zeros in matrix L/U. Such non-zeros in the L/U matrix are known as fillins. The benefits of sparse linear system of equations may be lost, if the fill generation is not controlled. Fills have a propagation effect. By ordering, *i.e.*, permuting the rows and columns of a matrix, growth of fills can be controlled. Thus, pivot selection in sparse matrix factorization has two dimensions (1) numerical stability and (2) sparsity exploitation, *i.e.*, keeping L/U matrices sparse. If the matrix A is spd then, diagonal pivots are preferred. MDA is a symmetric variant of the Markowitz ordering criteria [Markowitz, 1957]; it preserves symmetry and hence diagonal pivots. Therefore, analyze phase can be implemented without worrying about the numerical size of pivots. In such LSS, three phase implementation strategy is used. The three phases are (1) Analyze, (2) Factorize and (3) Solve.

Analyze phase finds an ordering to reduce fills and sets up a data structure for L/U matrices. In the factorize phase, actual numerical values of L/U matrices are computed. When the matrix A is not spd, it is possible to generalize LDU decomposition by considering 1×1 and 2×2 non-singular pivots. However, then the analyze phase and the factorize phase cannot be separated.

Sparse indefinite LSS implementations have a two phase approach viz, (1) Analyze+Factorize and (2) Solve. This distinction between 3-phase and 2-phase implementations is not just of academic interest. A three phase implementation implies that fills can be predicted *apriori*. Therefore, a static linked list data structure can be set up. On the other hand, in the two phase implementation, choice of pivot is dictated by the actual numerical values in the active submatrix. Hence, fills cannot be predicted apriori and therefore more generic linked list implementation is required.

This chapter introduced MDA as a popular and effective scheme for reducing fills in LU factorization of sparse spd matrices. It can be used to order matrices B' and B'' in fast decoupled load flow, admittance matrix ($YBUS$) for short circuit analysis, contingency analysis, OPF; GBUS for fast electromagnetic transient analysis as well as structurally symmetric Jacobian in the Newton Raphson load flow. It is also used for column ordering in sparse QR decomposition required in power system state estimator.

8. Further Reading

[Stewart, 1973] is one of the best text books on matrix computation. More recently, [Golub and Loan, 1989] presents a more exhaustive and up to date treatment of matrix computations. Chapters 1-6 of [Gill et al., 1991] provide easy to read, informative treatment on fundamentals of linear algebra and scientific computing, techniques for solving linear systems and linear least squares problems, and updating matrix factorizations. A specialized treatment on the direct methods for sparse matrices is available in [Duff et al., 1986]. The importance of positive definite systems and methods that are specific to them is brought out by [George and Liu, 1981]. Direct methods for sparse matrices are discussed in [Sergio, 1984; Wallach, 1986]. Much of the development in this vast area is reported in SIAM publications, ACM transactions on mathematical software and Linear Algebra and its applications. Treatment of MDA presented in [George and Liu, 1989] has both tutorial and research flavor. [Bunch and Parlett, 1971] introduces the concept of 1×1 and 2×2 pivots for symmetric but not spd linear system of equations. Recently, Object Oriented techniques have been applied for designing sparse matrix software. [Ashcraft and Liu, 1996; Ashcraft and Liu, 1999; George and Liu, 1997; Dobrian et al., 1999] is specifically recommended as a study material for learning applications of object oriented methods to direct sparse linear systems solvers. [George and Liu, 1989] is an excellent survey article on MDA. It introduces the concepts of multiple minimum degree. Role of elimination tree is discussed in detail in [Liu, 1990]. [Alvarado et al., 1991] makes a very interesting reading on sparsity issues in power system computation. The ideas developed in chapter are reported in [Pandit et al., 2001b]

9. Review Problems

1 Show that MDA is optimal in the minimal fill sense on tree graphs.

2 Why will Cholesky decomposition fail for matrices which are not spd?

3 Prove that in MDA once a node is selected for elimination, then all it's indistinguishable nodes can also be eliminated.

4 Prove that in MDA once a node is selected for elimination, then all it's similar nodes can also be eliminated.

5 Show that elimination of similar and indistinguishable nodes do not create additional fills.

6 A graph of a radial network is shown in figure 4.12 below. Show that MDA ordering will result in zero fills condition while minimum valency will not.

7 Explain why the percentage of indistinguishable nodes increase in the later stages of LU decomposition.

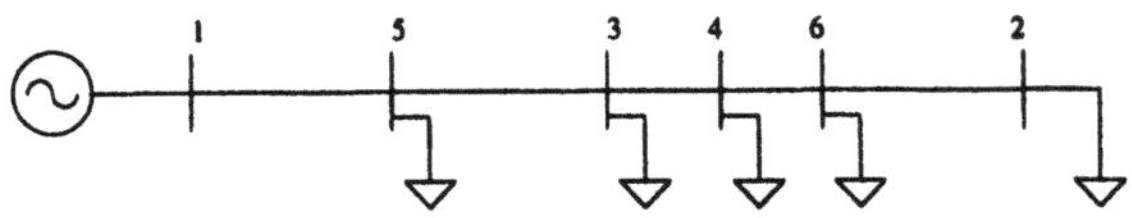

Figure 4.12. A Radial System

8 Consider the problem of LU decomposition of an admittance matrix in the following system of equations:

$$[I] = [YBUS][V]$$

Can the theory of LU decomposition developed in this chapter be applied by replacing the data type float with complex ? Describe a way of solving the complex linear system of equations using only real matrices.

9 Elaborate on the design issues for factorizing the admittance matrix.

10 Compute the inverse of the following matrix. Comment on the result.

$$\begin{bmatrix} 1 & & & \\ & 1 & & \\ & 2 & 1 & \\ & -3 & & 1 \end{bmatrix}$$

11 Show that product of two lower triangular matrices (upper triangular matrices) is also a lower triangular matrix (upper triangular matrix).

12 If a $m \times n$ matrix A has full column rank, show that the matrix A can be decomposed as $A = LU$, where, the $m \times n$ matrix L has a structure such that $L_{ij} = 0 \; \forall j > i$ and U is a unit upper triangular matrix.

13 Write a MATLAB® program to generate a sparse spd matrix, factorize and visualize it. Use *symmmd and colmmd* functions to order the matrix. Study the performance of LSS in terms of fills, solution time. What is the largest LSS on your PC that can be factorized without and with ordering ?

14 If $m \times n$ matrix A (with $m > n$) has full column rank, will the LSS of equations $Ax = b$ always have a solution. If not, what best can be done to solve the problem.

15 In the previous problem, if $m < n$ and the matrix has full row rank, comment on existence of (1) a LU decomposition and (2) solution of the LSS $Ax = b$.

Chapter 5

SPARSE QR DECOMPOSITION

It is not once nor twice but times without number that the same ideas make their appearance in the world.

—Aristotle: On the Heavens

1. Introduction

QR decomposition of a large sparse matrix is used to solve the linear least squares (LS) problem arising in power system state estimation. It can also be used to estimate rank of matrix (rank revealing QR decomposition). Its utility arises from the numerical stability offered by orthogonal transformations. In addition, a linear system of equations $Ax = b$, where matrix A can be indefinite and/or nonsymmetric[1] can be solved using this technique. In short, QR decomposition can be a workhorse for many types of computations. Since, the matrices that we encounter are large and sparse, in this chapter we introduce sparse QR decomposition.

Let $\{a_1, \cdots, a_n\}$ be a set of $m \times 1$ linearly independent vectors ($m \geq n$). Let us define a $m \times n$ matrix $A = [a_1, a_2, \cdots a_n]$; then $\text{rank}(A) = n$. The primary aim of QR decomposition is to express vectors $\{a_1, a_2, \cdots, a_n\}$ as a linear combination of n orthogonal vectors $\{q_1, \cdots, q_n\}$. The orthogonal set of vectors $\{q_1, \cdots, q_n\}$ satisfy the following properties:

$$||q_i||_2 = \sqrt{(q_i^T q_i)} = 1 \text{ (each vector is of unit length).}$$

[1]The fills in R of QR decomposition may be more than L/U matrices from LDU decomposition.

$$q_i^T q_j = 0 \ (i \neq j)$$

In a 2-dimensional case, we have $q_i^T q_j = ||q_i||_2 ||q_j||_2 \cos\theta$, where θ is the angle between the two vectors. Thus, if $q_i^T q_j = 0$, then $\theta = \pm\frac{\pi}{2}$, *i.e.*, the projection of q_i on to q_j is zero. We now describe a procedure known as the Gram-Schmidt orthogonalization procedure to extract the orthogonal basis $\{q_1, \cdots, q_n\}$ for column space[2]of A. It is appealing because of the geometrical construction that is used in extracting the orthogonal basis.

2. Gram-Schmidt Orthogonalization

Choose a vector q_1 in the same direction as vector a_1. Since, the vectors q_i $(i = 1, \cdots, n)$ are of unit length, vector q_1 is scaled to unit length, *i.e.*, $q_1 = \frac{1}{||a_1||_2} a_1$; let $R_{11} = ||a_1||_2$; then, $[a_1]_{m\times1} = [q_1]_{m\times1}[R_{11}]_{1\times1}$. Note that the vector q_1 spans the column space of $\{a_1\}$. The vector q_2 is chosen such that the set of vectors $\{q_1, q_2\}$ span the column space of $\{a_1, a_2\}$. Then,

$$a_2 = R_{12} q_1 + R_{22} q_2.$$

where R_{12} and R_{22} are scalars. Using the orthogonalization property, we can compute R_{12} as follows:

$$R_{12} = q_1^T a_2$$

$R_{22} q_2 = a_2 - R_{12} q_1 \Rightarrow R_{22} = ||a_2 - R_{12} q_1||_2$ and $q_2 = \frac{1}{R_{22}}[a_2 - R_{12} q_1]$
Since $\{a_1, a_2\}$ are linearly independent vector $[a_2 - R_{12} a_1] \neq 0$ and hence $R_{22} \neq 0$. In matrix notations, we can express the relationship as follows:

$$[a_1, a_2]_{m\times2} = [q_1, q_2] \begin{bmatrix} R_{11} & R_{12} \\ & R_{22} \end{bmatrix} \tag{5.1}$$

Proceeding in the similar manner, at the i-th step set of vectors $\{q_1, \cdots, q_{i-1}\}$, are known. The vectors span the column space of $\{a_1, \cdots, a_{i-1}\}$. Further, by the aforesaid construction, we have

$$[a_1, \cdots a_{i-1}] = [q_1, \cdots q_{i-1}] \begin{bmatrix} R_{11} & R_{12} & \cdots & R_{1,i-1} \\ & R_{22} & \cdots & R_{2,i-1} \\ & & \ddots & \vdots \\ & & & R_{i-1,i-1} \end{bmatrix} \tag{5.2}$$

[2]Column space of A denoted by *colspace*(A) is a set of all vector y such that for every $y \in$ *colspace*(A); there exist a vector x such that $Ax = y$.

The vector q_i can be computed from the following relationship:

$$a_i = R_{1i}q_1 + R_{2i}q_2 + \cdots + R_{ii}q_i$$

Using the orthogonal property of $\{q_1, \cdots, q_i\}$ we have

$$R_{ki} = q_k^T a_i \ (k < i);$$

Finally,

$$R_{ii}q_i = a_i - \sum_{k=1}^{i-1} R_{ki}q_k \Rightarrow R_{ii} = ||a_i - \sum_{k=1}^{i-1} R_{ki}q_k||_2$$

Diagonal $R_{ii} \neq 0$ because if $R_{ii} = 0$, it mean that a_i can be generated from set $\{q_1, \cdots q_{i-1}\}$. Since, $\{q_1, \cdots q_{i-1}\}$ exactly spans the column $\{a_1, \cdots a_{i-1}\}$, it means that $a_i \in$ colspace $\{a_1, \cdots, a_{i-1}\}$. This contradicts the condition that the rank(A)=n. In matrix notation, we can write $[a_1, \cdots, a_i]$ as follows.

$$[a_1, \cdots a_i] = [q_1, \cdots q_i] \begin{bmatrix} R_{11} & R_{12} & \cdots & R_{1i} \\ & R_{22} & \cdots & R_{2i} \\ & & \ddots & \vdots \\ & & & R_{ii} \end{bmatrix} \tag{5.3}$$

Proceeding in the similar manner, at the nth step, we have a complete decomposition of matrix A as a product of a matrix with orthogonal columns and a upper triangular matrix R.

$$[a_1, \cdots a_n] = [q_1, \cdots q_n] \begin{bmatrix} R_{11} & R_{12} & \cdots & R_{1n} \\ & R_{22} & \cdots & R_{2n} \\ & & \ddots & \vdots \\ & & & R_{nn} \end{bmatrix} = [q_1, \cdots, q_n][R] \tag{5.4}$$

Let $\{q_1, q_2, \cdots q_m\}$ be a orthogonal basis for the $\Re^m$. Then, equation (5.4) can also be written as follows:

$$A = [q_1, \cdots, q_n|, q_{n+1}, \cdots, q_m] \begin{bmatrix} R_{n \times n} \\ 0_{m-n \times n} \end{bmatrix} \tag{5.5}$$

Using the notation $Q = [q_1, \cdots q_m]$, we can write equation (5.5) in a compact form as follows.

$$A = Q \begin{bmatrix} R \\ -- \\ O \end{bmatrix} \tag{5.6}$$

where Q is $m \times m$ orthogonal matrix, *i.e.*, $Q^TQ = QQ^T = I_m$; R is $n \times n$ upper triangular matrix; and O is $(m - n \times n)$ zero matrix. Further, $R_{ii} \neq 0 \ \forall \ i = 1, \cdots, n$; if matrix A has full column rank. By proper choice of Q it can be ascertained that $R_{ii} > 0$. Conversely, if the matrix does not have full column rank, then at least one $R_{ii} = 0$ and hence R is singular.

Remark 1: It turns out that Gram-Schmidt decomposition is not a numerically stable implementation for QR decomposition [Golub and Loan, 1989]. Usually, QR decomposition is computed by either Householder reflections or Givens rotations. Both of these methods are now discussed.

3. Householder Reflection

Let v be a $m \times 1$ non-zero vector and H be a $m \times m$ matrix defined as follows:

$$H = I_m - \frac{2vv^T}{v^Tv} \tag{5.7}$$

It can be verified that:

$$Hv = -v \tag{5.8}$$

$$H^TH = HH^T = I_m \tag{5.9}$$

$$H = H^T \tag{5.10}$$

Thus, the matrix H is an orthogonal matrix. The vector v is known as a Householder vector. An important application of Householder matrix is the rotation of a vector x along the first column of an identity matrix e_1. It can be noticed that rotation preserves the distance of a vector, *i.e.*, for a vector $x \in R^m$, $||Hx||_2 = \sqrt{x^TH^THx} = \sqrt{x^Tx} = ||x||_2$. Let a vector $x \in R^m$ be rotated to the first column of identity matrix e_1 by Householders matrix H, *i.e.*,

$$Hx = ||x||_2e_1 \tag{5.11}$$

$$\Rightarrow \ (I_m - 2\frac{vv^T}{v^Tv})x = ||x||_2e_1 \tag{5.12}$$

The choice of v depends upon x and e_1. It can be verified that the following choice of v satisfies (5.12).

$$v = x \pm ||x||_2e_1 \tag{5.13}$$

The choice between + or − is based upon the requirement that for a numerically stable implementation, $||v||_2$ should not be too small. We illustrate the use of Householder matrix for computing QR decomposition by the following example.

Example 5.1: Obtain QR decomposition of the matrix A shown below:

$$A = \begin{bmatrix} 1 & 1 & 1 \\ 1 & 0 & 0 \\ 1 & 0 & 1 \\ 1 & 0 & 0 \\ 1 & 0 & 0 \end{bmatrix}$$

Note that QR decomposition can also be interpreted as $Q^T A = R$, *i.e.*, it reduces matrix A to $\begin{pmatrix} R \\ -- \\ O \end{pmatrix}$ by premultiplying it with an orthogonal matrix Q. This orthogonal matrix Q in turn can be a product of n orthogonal matrices, *i.e.*, $Q = Q_n Q_2, \cdots, Q_1$. When Householder matrix is used for QR decomposition, we evaluate orthogonal matrices $H_1, H_2, \cdots, H_n$ such that $H_n \cdots H_2 H_1 A = (\frac{R}{0})$.

Step 1: In the step-1, first column a_1 is rotated to e_1 axis of I_5 where I_5 is 5×5 identity matrix. Since $||a_1||_2 = \sqrt{5}$, by choosing $v_1 = a_1 + \sqrt{5}e_1 = (1 + \sqrt{5}, 1, 1, 1, 1)^T$; $H_1 = I_5 - \frac{2v_1 v_1^T}{v_1^T v_1}$ and premultiplying A by H_1 we obtain the following:

$$H_1 A = \begin{bmatrix} -\sqrt{5} & \vdots & -0.4472 & -0.8944 \\ \cdots & \cdots & \cdots & \cdots \\ 0 & \vdots & -0.4472 & -0.5854 \\ 0 & \vdots & -0.4472 & 0.4146 \\ 0 & \vdots & -0.4472 & -0.5854 \\ 0 & \vdots & -0.4472 & -0.5854 \end{bmatrix}$$

After the step 1, the 1st row and column are frozen. Therefore, the first row of $H_1 A$ corresponds to the first row of R-matrix. The first entry of R_{11} can be made positive by premultiplying the matrix A by $-H_1$.

Step 2: In this step, the matrix $[H_1 A]$ is premultiplied by orthogonal matrix $[H_2]$ so that it rotates the 1st column of the 4×2 submatrix (shown by dotted portion) to 1st column e_1 of identity matrix I_4. Vector v_2 is set as follows:

$$v_2 = -(0.4472, 0.4472, 0.4472, 0.4472)^T + \sqrt{4 \times 0.4472^2}(1, 0, 0, 0)^T$$

$= (-1.3416, -0.4472, -0.4472, -0.4472)^T$

$$\text{now, } H_2 = \begin{bmatrix} 1 & 0 \\ 0 & I_4 - \frac{2v_2v_2^T}{v_2^Tv_2} \end{bmatrix}$$

$$\text{Then, } [H_2][H_1][A] = \begin{bmatrix} -\sqrt{5} & -0.4472 & \vdots & -0.8944 \\ 0 & 0.8944 & \vdots & 0.6708 \\ \cdots & \cdots & \cdots & \cdots \\ 0 & 0 & \vdots & 0.8333 \\ 0 & 0 & \vdots & -0.1667 \\ 0 & 0 & \vdots & -0.1667 \end{bmatrix} \tag{5.14}$$

After this step the second row of matrix R is also frozen.

Step 3: In the step-3, the submatrix to be worked upon is shown by the dotted line. In fact, it is a 3×1 vector. It has to be rotated to the first column of I_3 matrix. The matrix H_3 has the following structure:

$$\begin{bmatrix} 1 & & \\ & 1 & \\ & & I_3 - \frac{2v_3v_3^T}{v_3^Tv_3} \end{bmatrix} \tag{5.15}$$

where $v_3 = (0.833, -0.1667, -0.1667)^T + (1,0,0)^T \times (0.8333^2 + 0.1667^2 + 0.1667^2)^{1/2} = (1.5833, -0.1667, 0.1667)^T$. Premultiplying H_2H_1A by H_3 it can be verified that:

$$H_3H_2H_1A = \begin{bmatrix} -\sqrt{5} & -0.4472 & \vdots & -0.8944 \\ 0 & 0.8944 & \vdots & 0.6708 \\ 0 & 0 & \vdots & -0.8660 \\ \cdots & \cdots & \cdots & \cdots \\ 0 & 0 & \vdots & 0 \\ 0 & 0 & \vdots & 0 \end{bmatrix} \tag{5.16}$$

At the end of step-3, complete QR decomposition is available. Product of matrices $H_3H_2H_1$ is a 5×5 orthogonal matrix and a matrix R is of size 3×3.

Remark 2: It can be seen that at the step-1 itself the sparsity of the matrix H_1A is lost. When the vector a_1 is rotated by a Householders matrix to e_1, the matrix A is modified as follows:

$$HA = [||a_1||e_1, Ha_2, \cdots, Ha_n]$$

Thus, the complete matrix is affected. Usually column Ha_i will not be sparse. Hence, Householders reflection is not preferred for QR decomposition of a sparse matrix A.

We now discuss Givens rotations and it's application to the QR decomposition.

4. Givens Rotation

Consider a 2×2 matrix of the following form:

$$Q = \begin{bmatrix} c & s \\ -s & c \end{bmatrix} \tag{5.17}$$

where $c = \cos\theta$ and $s = \sin\theta$. It can be verified that

$$Q^T Q = QQ^T = I_2,$$

Thus, Q is an orthogonal matrix. It is known as Givens rotation matrix. The role of Givens rotation is rotate a vector. Let $[x_1 \ x_2]^T$ be a 2 dimensional vector which has to be rotated to $y = [\sqrt{x_1^2 + x_2^2}, 0]^T$. Then,

$$Qx = y \Rightarrow \begin{bmatrix} c & s \\ -s & c \end{bmatrix} \begin{bmatrix} x_1 \\ x_2 \end{bmatrix} = \begin{bmatrix} \sqrt{x_1^2 + x_2^2} \\ 0 \end{bmatrix} \tag{5.18}$$

Using the orthogonal property of Q, it can be seen that c and s are given by the following equation.

$$c = \frac{x_1}{\sqrt{x_1^2 + x_2^2}} \text{ and } s = \frac{x_2}{\sqrt{x_1^2 + x_2^2}} \tag{5.19}$$

Givens rotations is suitable for sparse matrix computation. It can be applied selectively on any two rows of the matrix. A rotation between two rows x and y is illustrated below:

$$\begin{bmatrix} x_1' & x_2' & \dots & x_n' \\ 0 & y_2' & \dots & y_n' \end{bmatrix} = \begin{bmatrix} c & s \\ -s & c \end{bmatrix} \begin{bmatrix} x_1 & x_2 & \dots & x_n \\ y_1 & y_2 & \dots & y_n \end{bmatrix} \tag{5.20}$$

Both the rows x and y gets modified. When rows x and y are sparse then some of the zero elements becomes non-zero resulting in a *fill in.* Element x_1 is known as the *pivot.* Element y_1 is being *anhillated, eliminated* or *zeroed.* If the rotations have to be performed on a $m \times n$ matrix and an element $A(j, i)$ has to be eliminated using (i, i) pivot, then the rotations affect the ith and j^{th} rows. The matrix A is premultiplied by the rotation matrix of dimension $m \times m$. The structure of rotation matrix Q_{ij} is shown in figure (5.1).

$$\begin{array}{c} \\ \\ \\ i \\ \\ \\ j \\ \\ \\ \\ \end{array} \begin{array}{c} \begin{array}{ccccccccc} & & & i & & & j & & \end{array} \\ \left[\begin{array}{ccccccccc} 1 & . & . & 0 & . & . & 0 & . & . \; 0 \\ . & . & . & . & & & . & & . \\ . & . & . & . & & & . & & . \\ 0 & . & . & c & . & . & s & . & . \; 0 \\ . & & & . & . & . & . & & . \\ . & & & . & & . & . & & . \\ 0 & . & . & -s & . & . & c & . & . \; 0 \\ . & & & . & & & . & . & . \\ . & & & . & & & . & . & . \\ 0 & . & . & 0 & . & . & 0 & & . \; 1 \end{array}\right] \end{array}$$

Figure 5.1. The rotation matrix Q_{ij} for $m \times m$ matrix

The rotation matrix is same as the identity matrix except for the change in elements in locations (i,i), (i,j), (j,i), (j,j). The element (i,i) has a value c instead of 1, element (i,j) has value a s instead of 0, element (j,i) has a value $-s$ instead of 0 and element (j,j) has a value c instead of 1. When a vector $v \in \Re^m$ is premultiplied by Q(i,j), we obtain the following:

$$\begin{bmatrix} 1 & \cdots & 0 & \cdots & 0 & \cdots & 0 \\ \vdots & \ddots & \vdots & & \vdots & & \vdots \\ 0 & \cdots & c & \cdots & s & \cdots & 0 \\ \vdots & & \vdots & \ddots & \vdots & & \vdots \\ 0 & \cdots & -s & \cdots & c & \cdots & 0 \\ \vdots & & \vdots & & \vdots & \ddots & \vdots \\ 0 & \cdots & 0 & \cdots & 0 & \cdots & 1 \end{bmatrix} \times \begin{bmatrix} v_1 \\ \vdots \\ v_i \\ \vdots \\ v_j \\ \vdots \\ v_m \end{bmatrix} = \begin{bmatrix} v_1 \\ \vdots \\ cv_i + sv_j \\ \vdots \\ -sv_i + cv_j \\ \vdots \\ v_m \end{bmatrix} \tag{5.21}$$

Similarly, when Q_{ij} is multiplied by column vector a_i of a matrix A, only $a(i,i)$ and $a(j,i)$ are changed in column a_i. Thus, Givens rotations can be used for selectively anhillated the non-zeros in a matrix. The use of Givens rotation for QR decomposition is illustrated by the following example.

Example 5.2: Obtain QR decomposition for matrix A in previous example by Givens rotations. As in example 5.1, first the elements below diagonal (1,1) of A are zeroed by rotations. Then the elements below

diagonal (2,2) of partially modified A are zeroed. Finally, the elements below diagonal of column 3 are zeroed. The sequence of anhillation of elements is as follows:
Column 1: $\rightarrow A(2,1) \rightarrow A(3,1) \rightarrow A(4,1) \rightarrow A(5,1)$
Column 2: $\rightarrow A(3,2) \rightarrow A(4,2) \rightarrow A(5,2)$
Column 3: $\rightarrow A(4,3) \rightarrow A(5,3)$
In this example the diagonal elements of A, *i.e.*, the element having identical row and column index are used as the pivot. For clarity, at the end of each step the anhillated element and the pivot are marked in bold face. The steps are as follows:

Step 1: Anhillating A(2,1) with A(1,1) as pivot ($\theta = 45^0$), we have $x_1 = 1, x_2 = 1$

$$c = \frac{x_1}{\sqrt{x_1^2 + x_2^2}}; s = \frac{x_2}{\sqrt{x_1^2 + x_2^2}} \tag{5.22}$$

$$Q_{21} = \begin{bmatrix} \frac{1}{\sqrt{2}} & \frac{1}{\sqrt{2}} & & & \\ -\frac{1}{\sqrt{2}} & \frac{1}{\sqrt{2}} & & & \\ & & 1 & & \\ & & & 1 & \\ & & & & 1 \end{bmatrix} \tag{5.23}$$

$$A_1 = Q_{21}A = \begin{bmatrix} \mathbf{1.4142} & 0.7071 & 0.7071 \\ \mathbf{0} & -0.7071 & -0.7071 \\ 1.0 & 0 & 1.00 \\ 1.0 & 0 & 0 \\ 1.0 & 0 & 0 \end{bmatrix} \tag{5.24}$$

Step 2: Anhillate $A_1(3,1)$ with $A_1(1,1)$ as pivot; matrix to be worked upon is $Q_{21}A_1$.

$$Q_{31} = \begin{bmatrix} 0.8165 & & 0.5774 & & \\ & 1 & & & \\ -0.5774 & & 0.8165 & & \\ & & & 1 & \\ & & & & 1 \end{bmatrix} A_2 = \begin{bmatrix} \mathbf{1.7321} & 0.5774 & 1.1547 \\ 0 & -0.7071 & -0.7071 \\ \mathbf{0} & -0.4082 & 0.4082 \\ 1.0 & 0 & 0 \\ 1.0 & 0 & 0 \end{bmatrix} \tag{5.25}$$

Step 3: Anhillate $A_2(4,1)$ with $A_2(1,1)$ pivot

$$Q_{41} = \begin{bmatrix} 0.866 & & & 0.5 & \\ & 1.0 & & & \\ & & 1.0 & & \\ -0.5 & & & 0.866 & \\ & & & & 1.0 \end{bmatrix} A_3 = \begin{bmatrix} \mathbf{-2.00} & 0.5 & 1.0 \\ 0 & -0.7071 & -0.7071 \\ 0 & -0.4082 & 0.4082 \\ \mathbf{0} & -0.2887 & -0.5774 \\ 1.0 & 0 & 0 \end{bmatrix} \tag{5.26}$$

Step 4: Anhillate $A_3(5,1)$ with $A_3(1,1)$ pivot.

$$Q_{51} = \begin{bmatrix} 0.8944 & & & & 0.4472 \\ & 1 & & & \\ & & 1 & & \\ & & & 1 & \\ -0.4472 & & & & 0.8944 \end{bmatrix}; A_4 = \begin{bmatrix} \mathbf{2.2361} & 0.4472 & 0.8944 \\ 0 & -0.7071 & -0.7071 \\ 0 & -0.4082 & 0.4082 \\ 0 & -0.2887 & -0.5774 \\ \mathbf{0} & -0.2236 & -0.4472 \end{bmatrix} \tag{5.27}$$

At the end of step-4, the first column has been rotated in e_1 direction. The first row of R is available at this step. While anhillating the elements below the diagonal of column 2, element (2,2) is chosen as the pivot.

Step 5:

$$Q_{32} = \begin{bmatrix} 1 & & & & \\ & -0.866 & -0.5 & & \\ & 0.5 & 0.866 & & \\ & & & 1 & \\ & & & & 1 \end{bmatrix}; A_5 = \begin{bmatrix} 2.2361 & 0.4472 & 0.8944 \\ 0 & \mathbf{0.8165} & 0.4082 \\ 0 & \mathbf{0} & -0.7071 \\ 0 & -0.2887 & -0.5774 \\ 0 & -0.2236 & -0.4472 \end{bmatrix} \tag{5.28}$$

Step 6:

$$Q_{42} = \begin{bmatrix} 1 & & & & \\ & 0.9428 & & -0.3333 & \\ & & 1.0 & & \\ & 0.3333 & & 0.9428 & \\ & & & & 1 \end{bmatrix}; A_6 = \begin{bmatrix} 2.2361 & 0.4472 & 0.8944 \\ 0 & \mathbf{0.8660} & 0.5774 \\ 0 & 0 & -0.7071 \\ 0 & \mathbf{0} & -0.4082 \\ 0 & -0.2236 & -0.4472 \end{bmatrix} \tag{5.29}$$

Step 7:

$$Q_{52} = \begin{bmatrix} 1 & & & & \\ & 0.9628 & & & -0.2500 \\ & & 1 & & \\ & & & 1 & \\ & 0.2500 & & & 0.9682 \end{bmatrix}; A_7 = \begin{bmatrix} 2.2361 & 0.4472 & 0.8944 \\ 0 & \mathbf{0.8944} & 0.6708 \\ 0 & 0 & -0.7071 \\ 0 & 0 & -0.4082 \\ 0 & \mathbf{0} & -0.2887 \end{bmatrix} \tag{5.30}$$

Note that second row of R is obtained at the end of step 7.

Step 8:

$$Q_{43} = \begin{bmatrix} 1 & & & & \\ & 1 & & & \\ & & -0.866 & -0.5 & \\ & & 0.5 & -0.866 & \\ & & & & 1 \end{bmatrix}; A_8 = \begin{bmatrix} 2.2361 & 0.4472 & 0.8944 \\ 0 & 0.8944 & 0.6708 \\ 0 & 0 & \mathbf{0.8165} \\ 0 & 0 & \mathbf{0} \\ 0 & 0 & -0.2887 \end{bmatrix} \tag{5.31}$$

In the last step, we have
Step 9:

$$Q_{53} = \begin{bmatrix} 1 & & & & \\ & 1 & & & \\ & & 0.9428 & & -0.3333 \\ & & & 1.0 & \\ & & 0.3333 & & 0.9428 \end{bmatrix} ; A_9 = \begin{bmatrix} 2.2361 & 0.4472 & 0.8944 \\ 0 & 0.8994 & 0.6708 \\ 0 & 0 & \mathbf{0.866} \\ 0 & 0 & 0 \\ 0 & 0 & \mathbf{0} \end{bmatrix} \tag{5.32}$$

The matrix R obtained is identical to that obtained by Householders method except for signs appearing in the rows. The usual convention is to close the matrix Q in such a way that diagonal in R are positive.

5. Sparse QR decomposition

With this basic background on Givens rotations, we now focus on the sparsity issues like ascertaining sparsity in R, reducing intermediate fills and efficient organization of computations.

Remark 3: It can be verified that

$$A^T A = (R^T 0) Q^T Q \begin{pmatrix} R \\ 0 \end{pmatrix} = R^T R$$

where R is an upper triangular matrix. Since, matrix $A^T A$ is spd (A is full column rank), the upper triangular matrix R corresponds to the Cholesky decomposition of matrix $A^T A$. Therefore, for a large and sparse matrix A, the structure of R can be computed by the symbolic factorization of matrix $A^T A$. Note that the symbolic factorization does not require an explicit computation of matrix $A^T A$.

If columns of matrix A are permuted before QR decomposition, *i.e.*, $\hat{A} = AP$, P is a permutation matrix, then structure of R is given by symbolic factorization of $P^T A^T AP = \hat{R}^T \hat{R}$. To reduce fills in matrix R, a simple and efficient method is to permute of columns of A as per ordering obtained by applying MDA on structure of $A^T A$.

The new non-zeros arising in the sparse lower submatrix $(m - n) \times n$ of A (undergoing change) are finally anhillated (zeroed). Such elements which are zero in the original matrix, then get converted to non-zero in intermediate matrices and finally are anhillated known as *intermediate fills*. Intermediate fills are specific to QR decomposition; they do not arise in LDU decomposition of a square matrix. Intermediate fills increase the computational burden of QR decomposition. Givens rotations along with row ordering has been found to be very suitable for controlling intermediate fills.

5.1. Row Oriented Processing (ROP)

The concept of row oriented processing can be explained by considering the following augmented matrix:

$$A_{aug} = \begin{bmatrix} R^0 = 0 \\ -- \\ A \end{bmatrix} \tag{5.33}$$

In case of a linear system of equations or a LS solver, it simply means that we augment n extra equations of the form $\sum_{i=1}^{n} 0.x_i = 0$. Clearly, this does not affect the solution. *Now by Givens rotations, each row of A is sequentially anhillated with pivots from R matrix.* Finally, the augmented matrix is transformed to the following form:

$$\begin{bmatrix} R^m \\ -- \\ 0 \end{bmatrix} \tag{5.34}$$

where R^m is the final R matrix. It can be cross checked that

$$A_{aug}^T A_{aug} = A^T A = R^{m^T} R^m$$

By remark 3, R^m corresponds to R of $A^T A$, *i.e.*, R of QR decomposition of A. In an implementation of row oriented approach, it is not necessary to physically append matrix A with zeroes. The structure of R-matrix can be found out apriori by symbolic elimination of $A^T A$. The matrix R^0 is initialized to zero to start with. Then matrix R^1 is obtained by rotating the first row with matrix R^0. Similarly, the sequence of matrices $R^0, R^1, R^2, R^3, \ldots, R^m = R$ is obtained by rotating all the rows of the given matrix A. The algorithm for row oriented processing is given below:

1 Initialize R^0 with zeroes;

2 For $i = 1$ to m do steps 3, 4 and 5;

3 Fetch i^{th} row to be rotated;

4 Let α be the first non-zero column index in row i. If α^{th} row in matrix R^i is empty, then move the current row i to the αth row of matrix R^i (Review question 5), else rotate the row using the αth row of the matrix R^i and annihilate the first non-zero element (α^{th}) in the row;

5 If all the elements are not annihilated in i^{th} row, then repeat the previous step until the entire row becomes null.

Example 5.3: Row oriented QR decomposition is now illustrated for a sparse 4×3 matrix A. The initial matrix has a structure as shown below:

$$A = \begin{bmatrix} \times & & \times \\ \times & \times & \\ & \times & \\ \times & \times & \end{bmatrix} R^0 = \begin{bmatrix} 0 & 0 & 0 \\ & 0 & 0 \\ & & 0 \end{bmatrix} \qquad (5.35)$$

Let the ith row under rotation is denoted by v_i and $\_$ represent a zero element in v_i. Intermediate fills which are eliminated later are represented $\oslash$, whereas the fills in matrix R are represented by $\oplus$.

Step 1: The element to be annihilated is in column 1 and the diagonal in row 1 of R^0 (1,1) is zero, hence the row v_1 is copied to the first row of R^0.

$$v_1 = \begin{bmatrix} \times & \_ & \times \end{bmatrix} \; \& \; R^1 = R^{0'} = \begin{bmatrix} \times & & \times \\ & 0 & 0 \\ & & 0 \end{bmatrix} \qquad (5.36)$$

Step 2: The first element to be annihilated in the second row has column index 1. Since row-1 of R^1 is not zero and the rotation can be performed between first row of R^1 and the vector v_2. This leads to matrix $R^{1'}$ and vector v_2'.

$$v_2 = \begin{bmatrix} \times & \times & \_ \end{bmatrix} \; \& \; R^{1'} = \begin{bmatrix} \times & \oplus & \times \\ & 0 & 0 \\ & & 0 \end{bmatrix} \qquad (5.37)$$

The next element to be annihilated is in column 2 of vector v_2' and the row 2 of matrix $R^{1'}$ is zero. Hence, the vector v_2' is copied to row 2 of matrix $R^{1'}$; the matrix $R^{1'}$ is updated to matrix R^2.

$$v_2' = \begin{bmatrix} \_ & \times & \oslash \end{bmatrix} \; \& \; R^2 = R^{1'} = \begin{bmatrix} \times & \times & \times \\ & \times & \times \\ & & 0 \end{bmatrix} \qquad (5.38)$$

Step 3: The row v_3 has the first non zero element in column 2 and hence it gets rotated with respect to the row 2 of matrix R^2. This introduces an intermediate fill in column 3 of row v_3, *i.e.*, vector v_3'.

$$v_3 = \begin{bmatrix} \_ & \times & \_ \end{bmatrix} \; \& \; R^{2'} = \begin{bmatrix} \times & \times & \times \\ & \times & \times \\ & & 0 \end{bmatrix} \qquad (5.39)$$

The next element to be annihilated in row v_3' is in column 3 and row-3 of the of matrix $R^{2'}$ is zero. Hence, the vector v_3' is copied to row 3 of the matrix $R^{2'}$.

$$v_3' = \begin{bmatrix} - & - & \oslash \end{bmatrix} \ \& \ R^3 = R^{2'} = \begin{bmatrix} \times & \times & \times \\ & \times & \times \\ & & \times \end{bmatrix} \tag{5.40}$$

Step 4: The row v_4 has the first non zero element in column 1, hence it is annihilated by rotating with row 1 of matrix R^3. This introduces a intermediate fill in column 3 of vector v_4', which needs to be annihilated.

$$v_4 = \begin{bmatrix} \times & \times & - \end{bmatrix} \ \& \ R^{3'} = \begin{bmatrix} \times & \times & \times \\ & \times & \times \\ & & \times \end{bmatrix} \tag{5.41}$$

The next element which needs to be annihilated is in column 2, hence the row v_4' is rotated with row 2 of matrix $R^{3'}$.

$$v_4' = \begin{bmatrix} - & \times & \oslash \end{bmatrix} \ \& \ R^{3'} = \begin{bmatrix} \times & \times & \times \\ & \times & \times \\ & & \times \end{bmatrix} \tag{5.42}$$

The next element to be annihilated is in column 3 and hence it is rotated with row 3 of matrix $R^{3'}$. This leaves the row v_4 empty and matrix $R^{3'}$ becomes R^4.

$$R^4 = R^{3'} = \begin{bmatrix} \times & \times & \times \\ & \times & \times \\ & & \times \end{bmatrix} \tag{5.43}$$

The advantage of row oriented processing is that, the matrix to be rotated need not reside in main memory and only the row under rotation is required to be in the memory during factorization.

5.2. Column Oriented Processing (COP)

Column oriented QR decomposition processes columns of matrix A sequentially to compute the upper triangular matrix R. The algorithm produces a sequence of upper trapezoidal matrices $R(1\times n), R(2\times n), \cdots, R(n \times n)$ where $R(n \times n)$ corresponds to the matrix R obtained by QR decomposition of A. *At i-th step of the algorithm, all the non-zeros below the diagonal of the ith column are annihilated.* For example, in the first step of the algorithm, non-zero elements below (1,1) in first column of A are annihilated by selectively applying Givens rotations. In the second step non-zero elements of the second column below (2,2) are annihilated. Finally, at the nth step all the non-zeros below (n, n) in the nth column are annihilated and the matrix A is transformed to the upper triangular matrix R appended with $m - n$ null rows.

Example 5.4: Column oriented QR decomposition is now illustrated for a sparse 4×3 matrix A. The initial matrix has a structure as shown below:

$$A = \begin{bmatrix} \times & & \times \\ \times & \times & \\ & \times & \\ \times & \times & \end{bmatrix}$$

The matrix is transformed into a matrix R in three major steps as follows.

Step 1: Sequence of Givens rotations between the row pairs 1 and 2 and rows 1 and 4 are applied to annihilate elements (2,1) and (4,1) below the diagonal (1,1) of the first column. Note that rotations create three fill.

$$A_1 = \begin{bmatrix} \times & \otimes & \times \\ & \times & \otimes \\ & \times & \\ & \times & \oslash \end{bmatrix}$$

Step 2: In this step, the non-zeros below the diagonal (2,2) of the second column of matrix A_1 are annihilated. This involves Givens rotations operations on row pairs (2 and 3) and (2 and 4). Note that an additional fill is also created in this step.

$$A_2 = \begin{bmatrix} \times & \times & \times \\ & \times & \times \\ & & \otimes \\ & & \times \end{bmatrix}$$

Step 3: Finally, in this step the non-zero (4,3) in column 3 is annihilated by applying Givens rotations to row pair (3,4).

$$A_3 = \begin{bmatrix} \times & \times & \times \\ & \times & \times \\ & & \times \\ 0 & 0 & 0 \end{bmatrix} = \begin{bmatrix} R \\ 0 \end{bmatrix}$$

Now, if we look back at example 5.2, it can be seen that we had applied COP for obtaining QR decomposition. In the column oriented factorization, a column once processed is not required in later stages. At ith step, no intermediate fills can be created in the columns $1, \cdots, (i-1)$ which have been already processed. The elements to be annihilated in the column i can have row indices $i+1, \cdots m$; rows $1, \cdots, i-1$ are not used in steps $j \geq i$. Thus at every step size of the *active* submatrix to be processed is reduced by a row and a column. In fact the rows 1 to $i-1$ at ith step constitute the first $i-1$ rows of the matrix R. This can be contrasted with the row oriented processing where complete matrix R is updated at each step. Each step of ROP corresponds to annihilation of a row.

6. Row and Column Ordering

Just as in case of a sparse spd LDL^T discussed in chapter 4, ordering of rows and columns can influence the computations dramatically. Hence, to exploit sparsity and reduce computations, it is necessary to permute or order rows and columns of a sparse matrix A. As discussed in remark 3, columns of the matrix A should be permuted based upon application of MDA to structure of matrix $A^T A$. This results in reducing fills in matrix R. In case of sparse matrix A, additional issue to be considered is creation of intermediate fills and its relationship to the row order. Let us illustrate the importance of row ordering by an example.

Example 5.5: Let

$$A = \begin{bmatrix} 1 & 1 & 1 & 1 \\ 1 & 1 & & \\ 1 & & 1 & \\ 1 & & & 1 \\ 1 & & & \end{bmatrix} \tag{5.44}$$

Since $A^T A$ is non-sparse, R will also be nonsparse irrespective of the column order used. Thus, reordering of columns is not required. Let us apply COP for QR decomposition. It can be seen that when the elements below the diagonal of column 1 are anhillated, the structure of intermediate matrix is totally nonsparse.

$$\begin{bmatrix} \times & \times & \times & \times \\ & \times & \times & \times \\ & \times & \times & \times \\ & \times & \times & \times \\ & \times & \times & \times \end{bmatrix} \tag{5.45}$$

It appears that benefit of sparse matrix computation is not achievable. A total of 10 rotations will be required to obtain R.

Example 5.6 To evaluate the benefit of row ordering, let us now flip row 5 and row 1 before proceeding with the QR decomposition. Then

$$\hat{A} = \begin{bmatrix} 1 & & & \\ 1 & 1 & & \\ 1 & & 1 & \\ 1 & & & 1 \\ 1 & 1 & 1 & 1 \end{bmatrix} \tag{5.46}$$

The sequence of intermediate matrices after eliminating non-zeros below the diagonal are shown below.

$$A_1 = \begin{bmatrix} \times & \times & \times & \times \\ & \times & & \\ & & \times & \\ & & & \times \\ & \times & \times & \times \end{bmatrix} ; A_2 = \begin{bmatrix} \times & \times & \times & \times \\ & \times & \times & \times \\ & & \times & \\ & & & \times \\ & & \times & \times \end{bmatrix} \tag{5.47}$$

$$A_3 = \begin{bmatrix} \times & \times & \times & \times \\ & \times & \times & \times \\ & & \times & \times \\ & & & \times \\ & & & \times \end{bmatrix} ; A_4 = \begin{bmatrix} \times & \times & \times & \times \\ & \times & \times & \times \\ & & \times & \times \\ & & & \times \\ 0 & 0 & 0 & 0 \end{bmatrix} \tag{5.48}$$

Thus, it can be seen that only 7 rotations (30% saving) are required to compute R. We conclude that ordering of rows is equally important in sparse QR decomposition implementation irrespective of whether matrix R is sparse or not. So far, in the scheme for QR decomposition, that we have discussed, the pivot element has always be fixed as the diagonal element [$A(i,i)$ in COP or $R(i,i)$ in ROP]. This is known as a *fixed pivoting strategy.* An additional degree of freedom can be obtained in COP if *variable* pivots are permitted. The following example illustrates the same.

Example 5.7: Column oriented QR decomposition by variable pivot strategy is illustrated for the sparse 4×3 matrix A.

$$A = \begin{bmatrix} \times & & \times \\ \times & \times & \\ & \times & \\ \times & \times & \end{bmatrix} R^0 = \begin{bmatrix} 0 & 0 & 0 \\ & 0 & 0 \\ & & 0 \end{bmatrix} \tag{5.49}$$

The matrix is transformed into R in two major steps as follows.
Step 1: Sequence of Givens rotations between rows {(2 and 4) and (1 and 2)} are applied to annihilate elements (4,1) and (2,1) below the diagonal (1,1) of the first column. Thus, this step involves operations on the row pairs (2 and 4) and (1 and 2). Note that rotations create two fills represented by $\otimes$.

$$A_1 = \begin{bmatrix} \times & \otimes & \times \\ & \times & \otimes \\ & \times & \\ & \times & \end{bmatrix}$$

Step 2: In this step, the non-zeros below the diagonal (2,2) of the second column of matrix A_1 are annihilated. This involves Givens rotations operations on row pairs (3,4) and (2,3). Note that an additional fill is also created in this step at (3,3).

$$A_2 = \begin{bmatrix} \times & \times & \times \\ & \times & \times \\ & & \otimes \end{bmatrix}$$

The row 4 of the matrix A_2 is null and hence the rotation stops at step 2.

$$A_2 = \begin{bmatrix} \times & \times & \times \\ & \times & \times \\ & & \times \end{bmatrix} = \begin{bmatrix} R \\ 0 \end{bmatrix}$$

In the fixed pivot based implementation illustrated in Example 5.3, 5 rotations corresponding to following row pairs (1,2),(1,4),(2,3),(2,4) and (3,4) were required. Instead if the variable pivoting scheme is used with the following sequence of rotations (2,4),(1,2),(3,4) and (2,3) (shown below) only 4 rotations are required thereby saving a rotation. In such a small example problem, the degree of freedom for selecting matched row pairs is limited. However, for large problem, this additional degree of freedom can be quite rewarding in reducing the total number of rotations. Various ways of choosing the new pairs is now discussed. In practice, improvements in overall execution time will depend upon the saving in computations and additional overheads incurred implementing the variable pivot strategy.

7. VPAIR: On Line Row Ordering

[Gentleman, 1975] developed a variable pivot strategy for COP approach, wherein zeroing was first performed on sparser row pairs. [Robey and Sulsky, 1994] developed a variable pivot strategy VPAIR for sequencing of row pairs. In VPAIR, the rows are sequenced based on number of fillins that will be created during the rotation. This is measured by counting the number of mismatch in non zero pattern of row pair considered for rotation. A row pair which has least fillin creation (best matched pair) is chosen for rotation. As rotation operation on a row pair changes the structure of both rows, VPAIR ordering is best done online. Hence, ordering becomes dynamic. The tie breaking strategy recommended for VPAIR is based on choosing the sparset row pair. *In a detailed study of ordering for sparse QR decomposition in power system state estimation problem, [Pandian et al., 1999] conclude that best results are obtained by applying MDA for column ordering and online row ordering is done by VPAIR.* The COP is used in implementing QR decomposition. The following inference can be drawn from the previous discussion:

1 In row oriented processing, the row ordering is done apriori to the factorization and the ordering is followed strictly.

2 In row oriented processing when eliminating an element, if the intermediate fill gets generated, then it is eliminated in the same step. This should be contrasted with column oriented processing where in the elimination is delayed until the corresponding column is worked upon.

3 Column oriented processing supports online row ordering, hence the growth of fillins can be controlled.

8. A Large Scale Illustration

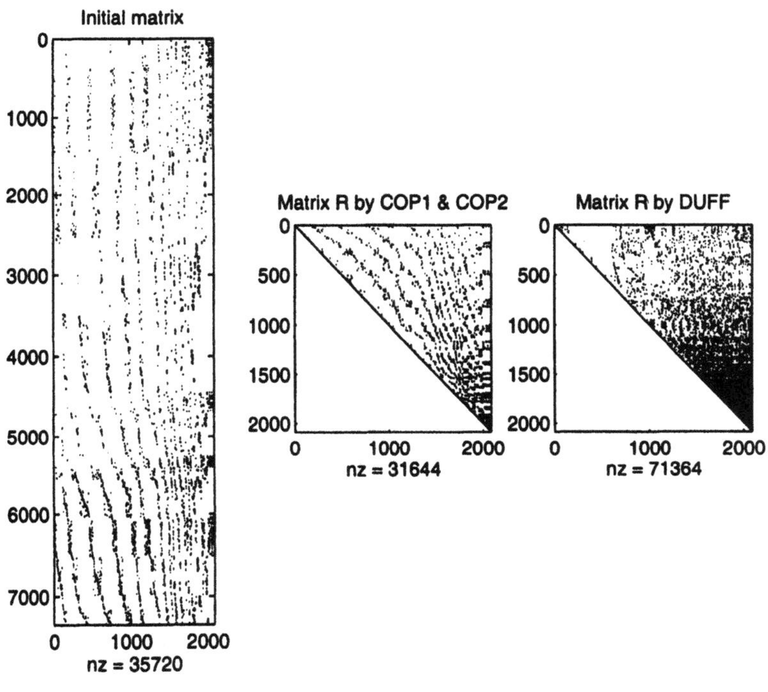

Figure 5.2. A 1044 node system: Initial matrix and structures of R

In this section, we illustrate the effect of choosing different ordering strategies as well computing procedures like ROP and COP on the efficiency of computing. For this illustration, the sparse matrix corresponds to the Jacobian of a 1044 node power system [Pandian et al., 1999] from power system state estimation. Following ordering strategies are considered.

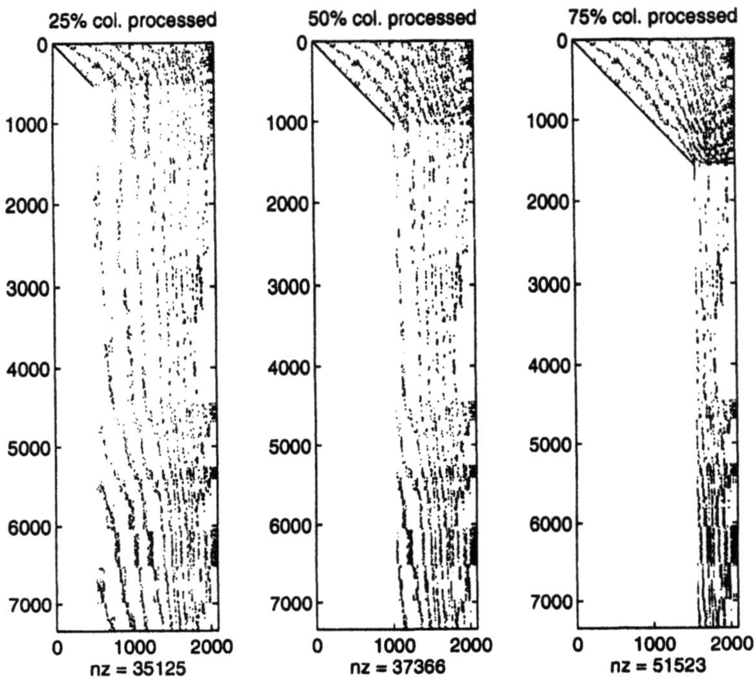

Figure 5.3. The Snapshots of Intermediate Steps for 1044 Node System by COP1

ROP: This is one of standard schemes used for QR decomposition in power system state estimation[Vempati et al., 1991]. ROP is used with apriori column and row ordering. MDA is used for column ordering and rows are ordered in an ascending order of the maximum column subscript with minimum sum of column subscripts as the tie breaking strategy.

COP1: The implementation is based on COP with apriori row and column ordering methods as used in ROP. While both the above schemes used *off line* ordering, the next two ordering schemes are based on *on line* ordering.

DUFF [Duff, 1974]: COP is used; column ordering is based upon minimum column degree implemented online, *i.e.*, degree of uneliminated columns are updated at each of elimination. A step in COP corresponds to elimination of all non-zero below a diagonal in a column. After elimination of a column, the column degree of the active submatrix is updated. Row ordering is by online VPAIR.

COP2: Again COP is used. Variable pivot VPAIR strategy is used for ordering rows online and column are ordered apriori by MDA applied to structure of $A^T A$. Figure 5.2 show the sparsity structure of the initial

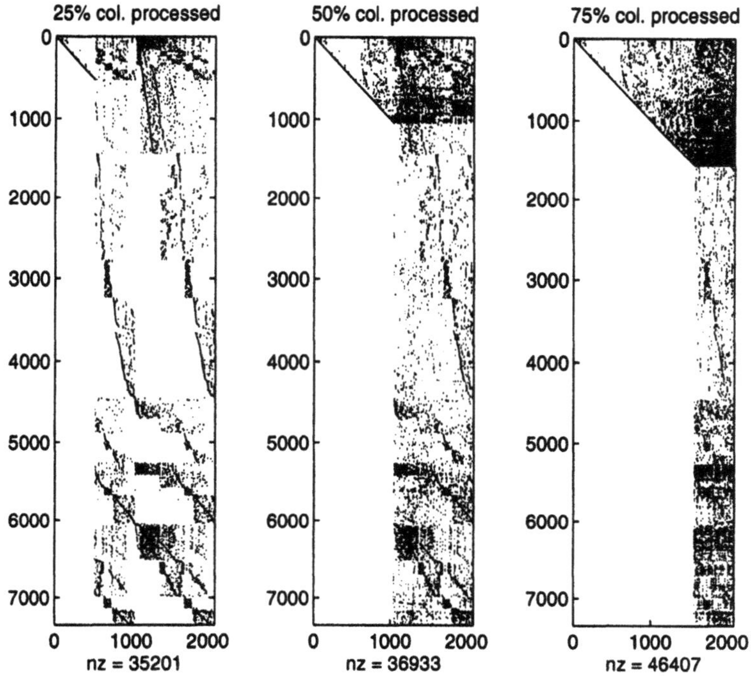

Figure 5.4. The Snapshot of Intermediate Steps for 1044 Node System by DUFF Scheme

matrix and R obtained by column ordering by MDA (ROP,COP1,COP2) and on-line minimum degree column ordering Duff (COP2) scheme. It can be seen that MDA scheme is preferable as it reduces fills drastically in comparison to DUFF scheme.

To gauge the effect of row ordering, snapshot of matrices after 25%, 50% and 75% elimination of columns is presented in figures 5.3,5.4 and 5.5. It can be seen by black patches that 1) DUFF schemes generates maximum fills, both in R and intermediate matrices while 2) Online VPAIR row ordering + MDA column ordering generates the least amount of intermediate fills as well as fills in R. Thus, it seems advisable to use VPAIR+MDA scheme for ordering. It is reported in [Pandian et al., 1999] that COP2 scheme also produces fastest implementation.

The data structure discussed in chapter 3 can be used for implementation. Since, intermediate fills can not be predicted beforehand, a heuristic allocating 4 times memory of the original matrix is used to account for intermediate fills.

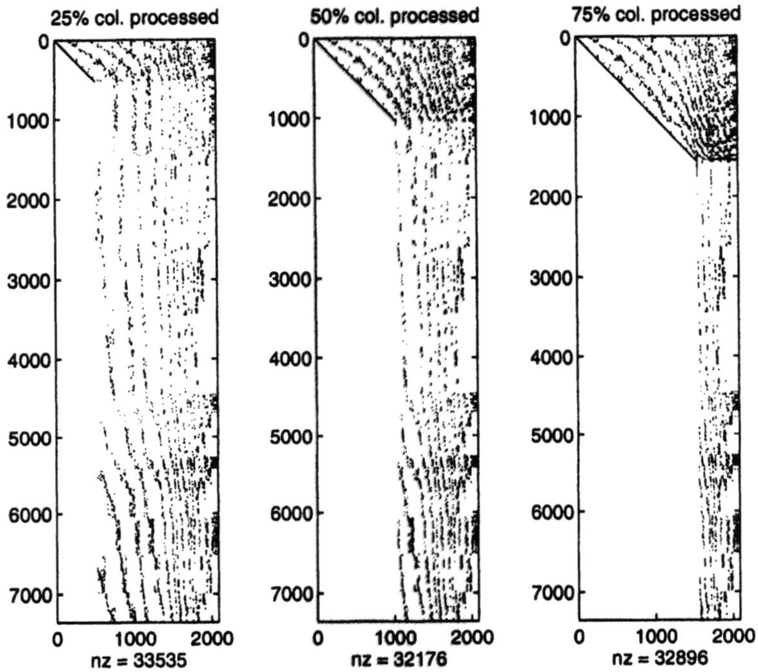

Figure 5.5. The Snapshot of Intermediate Steps for 1044 Node System by COP2

9. Summary

This chapter discussed QR decomposition for a large and sparse matrix. QR decomposition can be implemented through Gram-Schmidt orthogonalization, Householder reflections or Givens rotations. Givens rotations is the most suitable orthogonalization procedure for a sparse matrix because of its selective zeroing property. Computationally, sparse matrix QR decomposition can be implemented by a row oriented processing scheme or by a column oriented processing scheme.

The column oriented scheme is direct and more appealing as well as easy to implement. Earlier, it was considered unfavorable because the complete matrix had to be stored in core. In the present era of computations, this is not at all a problem. COP can be implemented using both fixed and variable pivot strategies. It is found that apriori ordering of columns by MDA and on-line row ordering using best match pairs (VPAIR strategy) produces the best results.

The QR decomposition is strongly recommended when dealing with numerically ill conditioned matrices. It can be used for both sparse linear system solver and sparse linear least squares problem. Because the matrix R in QR decomposition corresponds to Cholesky decomposition

of $A^T A$, in general, the R matrix is less sparse than the L and U matrix obtained by direct LDU decomposition of A. As such, its use in sparse LSS is restricted to only ill conditioned matrices. However, sparse QR decomposition is popularly used in solving sparse LS and LSE problem arising in power system state estimation.

10. Further Reading

For a first reading of QR decomposition, [Golub and Loan, 1989] is strongly recommended. [George and Heath, 1980] are responsible for providing an efficient implementation of QR decomposition using Givens rotations for sparse LS problem. They proposed ROP strategy and presented a static data structure implementation. [Duff, 1974] discuss the Duff strategy for row and column ordering. The strategy recommended by authors *viz.* VPAIR+MDA is due to [Robey and Sulsky, 1994; Hammarling, 1974]. [Anda and Park, 1994; Gentleman, 1973] discuss issues like dynamic scaling, square root free computations etc. For those interested in the first developments of Givens rotations, [Givens, 1958] is suggested for reading.

11. Review Questions

1 Show that the product of n orthogonal matrices is also an orthogonal matrix.

2 Why is Gram-Schmidt orthogonal procedure not used in numerical computations?

3 Why are Householder reflections not used in sparse QR decomposition?

4 Illustrate the selective anhillation property of Givens rotation.

5 Show that it is possible to achieve the following through Givens rotation.

$$\begin{bmatrix} c & s \\ -s & c \end{bmatrix} \begin{bmatrix} 0 & 0 & \cdots & 0 \\ x_1 & x_2 & \cdots & x_n \end{bmatrix} = \begin{bmatrix} x_1 & \cdots & x_2 & \cdots & x_n \\ 0 & \cdots & 0 & \cdots & 0 \end{bmatrix} \quad (5.50)$$

Find the scalars c and s. What is the corresponding value of θ?

6 What would be structure of R if the following linear system of equations is solved by (1) LDL^T decomposition and (2) QR decomposition.

Apply ordering if necessary before computing R.

$$\begin{bmatrix} 5 & 1 & 1 & 1 \\ 1 & 5 & & \\ 1 & & 5 & \\ 1 & & & 5 \end{bmatrix} \begin{bmatrix} x_1 \\ x_2 \\ x_3 \\ x_4 \end{bmatrix} = \begin{bmatrix} 1 \\ 1 \\ 1 \\ 1 \end{bmatrix} \tag{5.51}$$

7 Based on the results of question 6, can one conclude that 'for large sparse LSS, it is preferable to use LDL^T decomposition'.

8 Show that a $m \times m$ Householder matrix has $m-1$ unity eigen values and one eigen value -1. Identify the corresponding eigenvectors.

9 Show that in QR decomposition, ordering of rows has no effect on the structure of matrix R.

10 Evaluate the eigen values and eigen vector of a $m \times m$ Givens matrix used in QR decomposition.

11 Why is QR decomposition considered a numerically stable decomposition?

Chapter 6

OPTIMIZATION METHODS

Near a maximum the decrements on both sides are in the beginning only imperceptible.

—J. Kepler, V. M. Tikhomirov: Stories about Maxima and Minima

1. Introduction

Many power system analysis problems can be framed as unconstrained or constrained optimization problems. Typical examples are 1) Power System State Estimation(PSSE) which is framed as a sequential unconstrained or an equality constrained linear least squares problem, [Schweppe and Wildes, 1970] 2) Optimal Power Flow (OPF) which is framed as a nonlinear constrained optimization problem [Dommel and Tinney, 1968], 3) Optimal relay coordination which is framed as a Linear Programming (LP) problem [Chattopadhyay et al., 1996] and 4) Optimal scheduling for hydrothermal systems.

The field of numerical methods for optimization is rich with many methods for both constrained and unconstrained optimization. However, only a small subset of these methods are relevant to power system analysis because, methods that do not preserve sparsity are not of much use in large scale computations. In this chapter, we introduce fundamental principles of optimization and review methods that are used in power system analysis.

1.1. Unconstrained Optimization

Let $f(x_1, x_2 \cdots x_n)$ be a continuous at least twice differentiable function of variables $(x_1, x_2 \cdots, x_n)$, where $x_i \in \Re$. The necessary and sufficient optimality conditions for minimizing $f(x)$ are now summarized. Let x^* be the optimal point. Then, at the optimum x^*:

1 the gradient $\nabla f(x^*)$ should be zero;

2 The matrix of second derivative of $f(x)$, *i.e.*, $\nabla^2 f = \left[\frac{\partial^2 f}{\partial x^2}\right]$ should be positive definite. These conditions are summarized below.

min $f(x_1, x_2 \cdots x_n)$	max $f(x_1, x_2 \cdots x_n)$
$\nabla f^T = [\frac{\partial f_1}{\partial x_1}, \frac{\partial f}{\partial x_2}, \cdots, \frac{\partial f}{\partial x_n}]_{x=x^*} = [0]$	$\nabla f^T = [\frac{\partial f_1}{\partial x_1}, \frac{\partial f}{\partial x_2} \cdots, \frac{\partial f}{\partial x_n}]_{x=x^*} = [0]$
$x^T \begin{bmatrix} \frac{\partial^2 f_1}{\partial x_1^2} & \frac{\partial^2 f}{\partial x_1 \partial x_2} & \cdots & \frac{\partial^2 f}{\partial x_1 \partial x_n} \\ \frac{\partial^2 f}{\partial x_2 \partial x_1} & \frac{\partial^2 f}{\partial x_2^2} & & \vdots \\ & & \ddots & \vdots \\ \frac{\partial^2 f}{\partial x_n \partial x_1} & & & \frac{\partial^2 f}{\partial x_n^2} \end{bmatrix} x > 0 \ \forall x \neq 0$	$x^T \begin{bmatrix} \frac{\partial^2 f_1}{\partial x_1^2} & \frac{\partial^2 f}{\partial x_1 \partial x_2} & \cdots & \frac{\partial^2 f}{\partial x_1 \partial x_n} \\ \frac{\partial^2 f}{\partial x_2 \partial x_1} & \frac{\partial^2 f}{\partial x_2^2} & & \vdots \\ & & \ddots & \vdots \\ \frac{\partial^2 f}{\partial x_n \partial x_1} & & & \frac{\partial^2 f}{\partial x_n^2} \end{bmatrix} x < 0 \ \forall x \neq 0$

It should be pointed out that satisfying the zero derivative condition is only a necessary condition. The computed point may not be an optimum; it may be 1) a local minimum, 2) local maximum or 3) a saddle point. The matrix of second derivatives is known as the *Hessian*. It should be noted that the matrix of second derivatives is symmetric. The optimality conditions discussed pertain to a local optimum. The objective function $f(x)$ may have none, one or more than one local minima. Usually, a well formulated optimization problem will have at least one minimum. The problem of max $f(x)$ can be look upon as a min $-f(x)$ problem. Hence, in this chapter, we discuss only minimization problems. The emphasis is on the practical aspects, *i.e.*, algorithms and numerical methods.

1.2. How to Verify Positive Definiteness

The $n \times n$ symmetrical matrix H is positive definite if $x^T H x > 0$; $\forall x \neq 0; x \in \Re^n$. In a single variable problem, it means that $\frac{\partial^2 f}{\partial x^2}|_{x=x^*} > 0$, *i.e.*, the function has a cup shaped curvature as shown in figure 6.1. Thus,

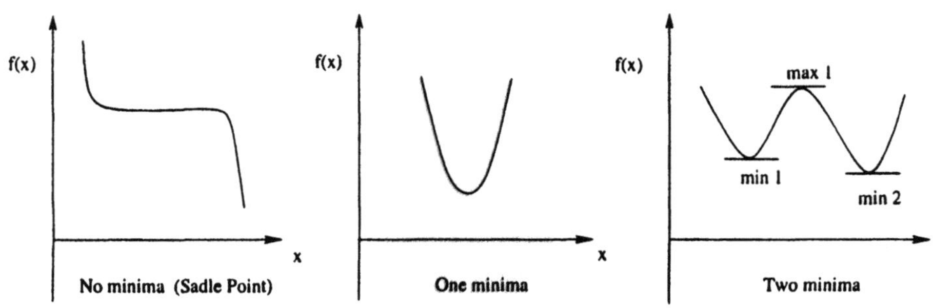

Figure 6.1. Cupshape Curvature

while this sufficiency condition is easy to verify on a single variable problem, it requires a nontrivial effort in case of multi-dimensional problems. One way to verify Symmetric Positive Definite (spd) nature of Hessian H is to attempt a Cholesky decomposition (LDL^T) of matrix H. If all diagonal elements of D are positive ($d_i > 0$), then, H is positive definite (review question 1). Note that the LDL^T decomposition requires ($O(n^3)$) effort in case the matrix H is non-sparse and an effort at least proportional to the number of non-zeros (nz), if the matrix H is sparse. In either case, the verification does involve a nontrivial effort.

The unconstrained optimization problem in power system analysis usually arises either as a 1) sequential linear least square problem or 2) as a line search minimization problem embedded in a nonlinear programming problem. Further, an equality constrained subproblem in nonlinear programming can be converted into a corresponding multi-dimensional unconstrained nonlinear optimization problem. We begin with a discussion on line search methods.

2. Line Search Methods

Line search forms the backbone of many algorithms for nonlinear programming problems. They are also known as one dimensional search methods. Typically, the problem is formulated as follows:
Given a point x_k and a direction d_k, find a suitable step size λ_k ($\lambda_k \in \Re$) to minimize $\theta(\lambda) = f(x_k + \lambda d_k)$. The updated optimal point is given by $x_{k+1} = x_k + \lambda_k d_k$. If the direction d_k is chosen to reduce the cost (objective function) then, $\lambda_k > 0$; the other possible constraint on the choice of λ for a constrained optimization problem is that $x_k + \lambda d_k$ is feasible. A point x_k is said to be *feasible* if it does not violate any equality or inequality constraint.

The optimality condition for minimizing $\theta(\lambda)$ requires that projection of the gradient along the descent direction should be zero or $\frac{\partial \theta}{\partial \lambda} = 0$, *i.e.*,

$$d_k^T \nabla f(x_k + \lambda_k d) = 0$$

It is presumed that the interval $[a, b]$ in which the line search is conducted is so chosen that only one minimum lies in it, *i.e.*, the function is quasi-convex. The objective function is said to be quasi-convex in the interval, if

$$f[\lambda a + (1 - \lambda) b] \leq \max\{f(a), f(b)\}, \lambda \in (0, 1).$$

A function $f(x)$ is quasiconcave if $-f(x)$ is quasiconvex. It may not be easy to cross check this condition in practice.

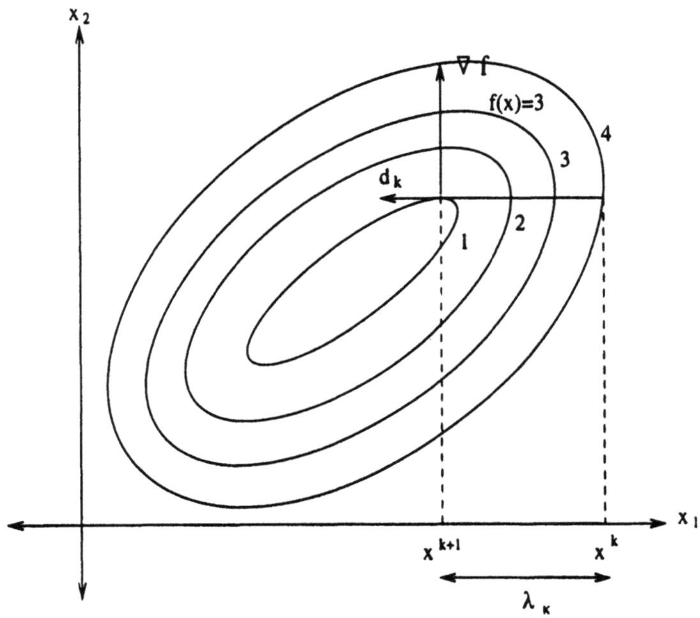

Figure 6.2.

Example[1] 6.1: Consider the objective function $f(x) = x^2 - x$. It is easy to verify that minimum of $f(x)$ lies at $x = 0.5$. The nature of the function is shown in figure 6.3. Now, let $x_0 = 0$, $d_0 = \nabla f(0) = -1$. If the next iterate is simply taken as $x_0 + d_0$, then $x_1 = 1$ and $f(1) = 0$; *i.e.*, no improvement in the function is observed. However, even if a uniform line search is made along d_0, say by dividing the segment $(x_0, x_0 + d_0)$ in 10 equal parts, we reach the minimum $x = 0.5$. This example illustrates the importance of line search along d_k in minimization process. For a non-linear problem, the complete step d_k may be unacceptable.

Primarily line search methods are of three types:

1 those which do not use derivative computation; they only require function evaluation;

2 those that do require derivative information and

3 inexact line search methods.

One illustrative method of each class is now discussed.

[1] We presume knowledge of steepest descent method here. Those unfamiliar with this method should check the example after reading section 3.1

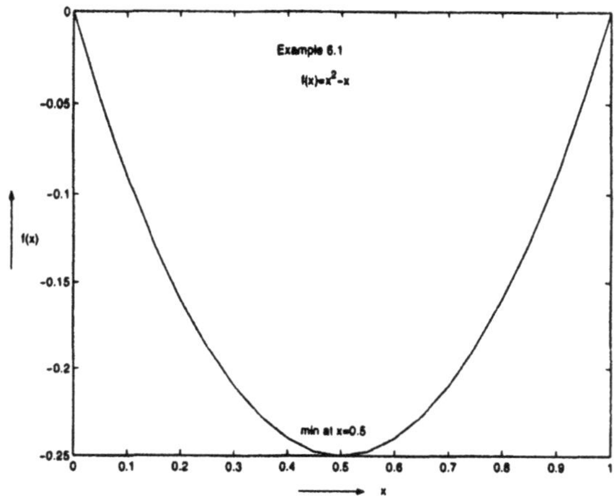

Figure 6.3. Example 6.1

2.1. Derivative Free Line Search Methods

In this class of methods, the interval in which a local minimum has to be computed is already known. The objective function is assumed to be quasi-convex. In a very simplistic approach, we may simply evaluate the objective function at n equidistant points (n being sufficiently large) and choose the point corresponding to the least cost as an approximation to the actual minimum. This is known as *uniform search* method. If the cost of evaluating objective function is low, *e.g.*, the economic dispatch problem, uniform search can be of use, but, in some other problems like MW loss minimization, this may involve a high cost because loss evaluation may require a load flow solution.

In an alternative class of methods known as the *sequential search* methods, the approach followed is to carry out objective function evaluations such that the interval in which the minimum lies is reduced successively. When the interval is reduced below a certain tolerance level, the search is terminated. Dichotomous search, Golden section search and Fibonacci search methods belong to this class. As an illustrative example, Fibonacci search method is now discussed.

2.1.1 Fibonacci Search Method

The method reduces this interval of uncertainty at every step $[a_k, b_k]$ by eliminating the intervals that do not contain the minimum. In general, the method finds a local minimum. The method is based on the Fibonacci sequence F_i defined as follows:

$$\text{Let} \quad F_0 = F_1 = 1$$

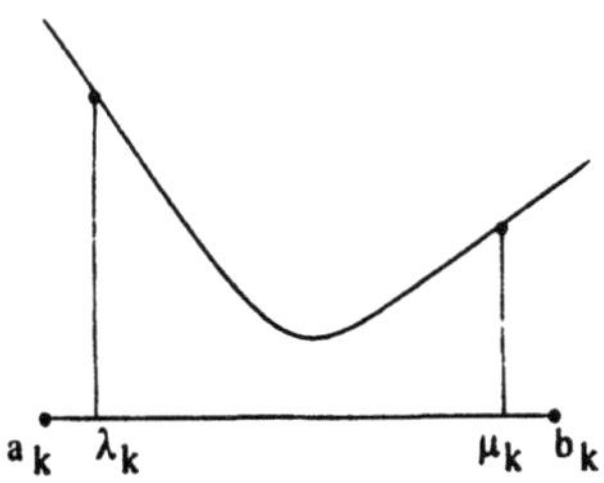

*Figure 6.4a.*Function behavior Case-1

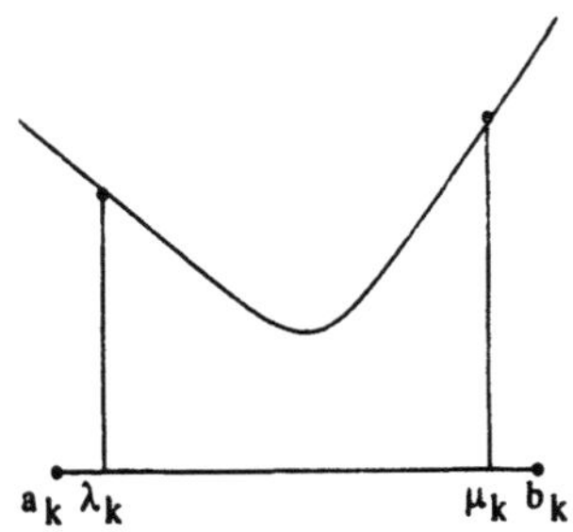

*Figure 6.4b.*Function behavior Case-2

$$\text{and } F_{i+1} = F_i + F_{i-1};\ i = 1, 2 \ldots$$

The sequence is given by $\{1, 1, 2, 3, 5, 8, 13, 21, \ldots\}$. First, by an appropriate choice of n, we choose the final length of uncertainty, $b_n - a_n = (b_1 - a_1)/F_n$. It reflects the accuracy level of the solution. Larger n implies identification of a smaller interval in which the minimum lies. Usually, the lesser of the function values $f(a_n)$ and $f(b_n)$ can be taken as a minimum. At a step k, the objective function behavior can be represented either by figures 6.4a or 6.4b. In this step, one of two intervals $[a_k, \lambda_k]$ or $[\mu_k, b_k]$ is eliminated, thereby, reducing the interval of uncertainty as follows:

Two new function evaluations are carried out at points λ_k and μ_k where:

$$\lambda_k = a_k + \frac{F_{n-k-1}}{F_{n-k+1}}(b_k - a_k)\ k = 1, 2, \ldots, n-1$$
$$\mu_k = a_k + \frac{F_{n-k}}{F_{n-k+1}}(b_k - a_k)\ k = 1, \ldots, n-1$$

Note that $b_k - \mu_k = \lambda_k - a_k$ (review question 4). The new reduced interval of uncertainty is computed as follows. If $\theta(\lambda_k) > \theta(\mu_k)$ (Figure 6.4a) then interval $[a_k, \lambda_k]$ can be eliminated and hence, a new reduced interval of uncertainty is given by $[\lambda_k, b_k]$. If $\theta(\lambda_k) \leq \theta(\mu_k)$ (figure 6.4b) then interval $[\mu_k, b_k]$ can be eliminated. The new interval is given by $[a_k, \mu_k]$. If $n \to \infty$, $\frac{F_{n-k}}{F_{n-k+1}} \to 0.612$ and, if the step size is fixed to this value, the method approaches the Golden section method.

2.2. Line search using derivatives

The other class of line search methods evaluate the derivative of $\theta(\lambda)$ and try to find a zero gradient point. The bisection search method and Newton method belong to this class. We now discuss the Newton method. In this method, a quadratic polynomial fit of the function $\theta(\lambda)$

is considered.

$$q(\lambda) = \theta(\lambda_k) + \theta'(\lambda_k)(\lambda - \lambda_k) + \frac{1}{2}\theta''(\lambda_k)(\lambda - \lambda_k)^2$$

The point λ_{k+1} is taken to be that point where the derivative of q with respect to λ is zero, *i.e.*,

$$\lambda_{k+1} = \lambda_k - \frac{\theta'(\lambda_k)}{\theta''(\lambda_k)}$$

Line search using quadratic approximation of the objective function is another popular method for constrained optimization problems.

Next we discuss an inexact line search method, that avoids 1) the expense of too many functional evaluations and 2) also ascertains sufficient degree of descent in the objective function value in a well defined sense.

2.3. Inexact line search: Armijo's Rule

Quite often, we encounter situations where, either the function evaluation is costly or the cost of evaluation of derivative and second derivatives is high. For example, in the problem of minimizing losses in a power system with reactive power control variables, every new function evaluation may require a load flow solution which is computationally costly. Also, in an OPF formulation, if all variables except the control variables are eliminated by computing sensitivities(reduced model formulation), then, the evaluation of the second derivatives is next to impossible. Thus, in practice, there is a need for line search methods that cut down on the number of objective function evaluations, and still reduce the objective function sufficiently, so as not to hamper the overall convergence of the nonlinear problem. Inexact line search methods try to achieve these two objectives.

Let $x_1 \in \Re^n$ be an initial point and $d \in \Re^n$ be a descent direction for $f(x)$ at x_1, *i.e.*, along d the functional value $f(x)$ reduces. Therefore, $\nabla f(x_1)^T d < 0$. We now wish to find a $\lambda \in \Re$ such that $\theta(\lambda) = f(x_1 + \lambda d)$, $\lambda > 0$ reduces sufficiently. The method works as follows:

Evaluate $\theta'(0) = d^T \nabla f(x_1)$ and choose two parameters $0 < \eta < 1$ and $\alpha > 1$ which respectively manage step length from being too large or small. The first order approximation of $\theta(\lambda)$ at $\lambda = 0$ is given by $\theta(0) + \lambda\theta'(0)$ (See figure 6.5). Let $\hat{\theta}(\lambda)$ be defined as $\hat{\theta}(\lambda) = \theta(0) + \lambda\eta\theta'(0)$ for$\lambda > 0$. Function $\hat{\theta}$ gives a more conservative estimate of function decrease than the first order approximation (figure 6.5). Let us Choose a fixed step length parameter $\bar{\lambda}$. Now, if $\theta(\bar{\lambda}) \leq \hat{\theta}\ (\bar{\lambda})$, then 1) either $\bar{\lambda}$ is selected as step size or 2) $\bar{\lambda}$ is sequentially doubled ($\alpha = 2$) until

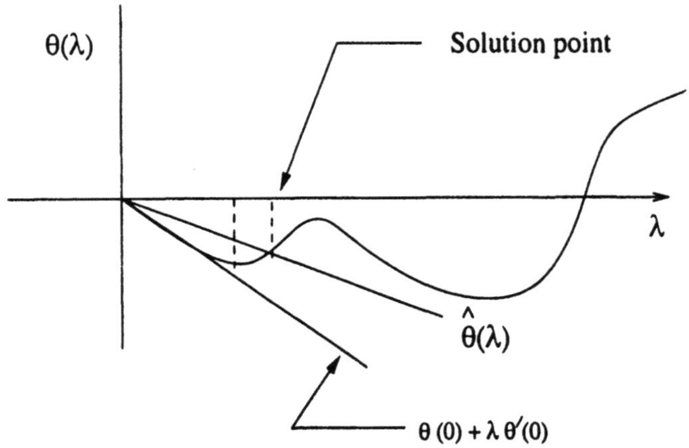

Figure 6.5. Armijo's Rule

$\theta(\lambda^{new}) < \hat{\theta}(\lambda^{new})$ condition is not violated. Conversely, if $\theta(\lambda) > \hat{\theta}(\lambda)$, then $\hat{\lambda}$ is sequentially halved until $\theta(\lambda^{new}) \leq \hat{\theta}(\lambda^{new})$. In short the method attempts to find the intersection point of nonlinear function $\theta(\lambda)$ and linear conservative approximation $\hat{\theta}(\lambda)$.

The line search methods are essentially a single dimensional search methods. Line search forms an important component of both unconstrained and constrained nonlinear programming algorithms. In this section we had presumed that descent direction is known. In the next section, we discuss the methods for finding descent direction for unconstrained optimization problems.

3. Multidimensional Search

Consider the problem of minimizing $f(x)$ where $x \in \Re^n$. In other words, the minimization is carried in a n-dimensional space. Further, we assume that $f(x)$ is at least twice differentiable.

In power system analysis, we seldom encounter truly unconstrained nonlinear optimization problem. Notable exceptions are least squares (LS) problem used either for curve fitting e.g., in a polynomial fit for overcurrent characteristic of a relay and eliminating white noise *e.g.*, in power system state estimation. However, a constrained optimization problem can be converted to an unconstrained optimization problem. In the following section, we discuss in brief two unconstrained minimization methods, which have a nice intuitive feel. They are:

1 Steepest descent method;

2 The method of Newton.

3.1. Steepest Descent Method

The direction of steepest descent is that direction along which the rate of decrease of function is maximum. If at point x_k, $\nabla f(x_k) \neq 0$, *i.e.*, point x_k is not an optimal point, then along the direction $d_k = -\nabla f(x_k)$,

$$d_k^T \nabla f(x_k) = -\nabla f(x_k)^T \nabla f(x_k) = -||\nabla f(x)||_2^2 < 0$$

where the second norm of a vector $x \in \Re^n$ is defined as follows:

$$||x||_2 = \sqrt{x_1^2 + x_2^2 \ldots + x_n^2}$$

By Taylor series expansion, we can express the function $f(x)$ around x_k as follows:

$$f(x_k+\epsilon d_k) = f(x_k)+\epsilon d_k^T \nabla f(x_k)+\frac{1}{2}\epsilon^2 d_k^T H(x) d_k \text{where } x \in (x_k, x_k+\epsilon d_k)$$

and ϵ is a scalar greater than zero. The matrix $H(x)$ is a $n \times n$ matrix of second order derivatives of objective function $f(x)$ and is known as Hessian. In the limit $\epsilon \to 0$, the Hessian matrix can be neglected leading to the following approximate relationship

$$f(x_k + \epsilon d_k) \approx f(x_k) + \epsilon d_k^T \nabla f(x_k)$$

Thus, the function reduces along d_k, *i.e.*, $d_k = -\nabla f(x_k)$ is a descent direction. In fact, it can be shown to be a direction of steepest descent (review question 5). In a steepest descent algorithm, the chosen direction of descent at a step k corresponds to $d_k = -\nabla f(x_k)$. Along the direction d_k, the function $f(x_k + \lambda d_k)$ can be minimized by using any one of the line search methods discussed in the previous sections. The updated point is given by:

$$x_{k+1} = x_k + \lambda_k d_k; \; k = k + 1;$$

The algorithm terminates when $||\nabla f(x_k)||_2 < \gamma$, where γ is the tolerance value. Though the method appears appears to be simple and attractive, it is well known that steepest descent method suffers from poor convergence.

3.2. Method of Newton

The method of Newton is based upon the quadratic approximation of the objective function at a given point x_k;

$$q(x) = f(x_k) + \nabla f(x_k)^T [x - x_k] + \frac{1}{2}[x - x_k]^T [H(x_k)][x - x_k] \quad (6.1)$$

where $H(x_k)$ is the matrix of second derivation of function $f(x)$ at x_k. A quadratic objective function captures the curvature information. Therefore, by using a quadratic approximation of the objective function, we expect to improve the convergence characteristic of the algorithm. If the initial point x_1 (a guess to the optimal solution x^*) is close to the optimal, then $H(x_k)$ will be non-singular. Therefore, the application of optimality conditions, *i.e.*, setting the gradient of $q(x)$ to zero leads to the following results:

$$[H(x_k)][x - x_k] = -\nabla f(x_k) \tag{6.2}$$

$$\text{Hence, } x_{k+1} = x_k - [H(x_k)]^{-1}\nabla f(x_k) \tag{6.3}$$

If the matrix $H(x_k)$ is sparse, then solving equation (6.2) using LDL^T is the right way to proceed with the implementation. Equation (6.3) is commonly used to represent the point x_{k+1}.

Example 6.2: Consider a function $f(x) = e^{-x}$ which is a monotonically decreasing function with $\min f(x) = 0$ in the limit $x \to \infty$. Now, $\frac{df}{dx} = -e^{-x}$. A minimization problem with minimum at finite x can be obtained by considering either of the following two functions $|e^{-x} - c|$ or $(e^{-x} - c)^2$ where $c > 0$. The optimal lies at $x = -\log c$. The steepest descent iterations starting from $x = 0$ point are given as follows:

$$x_{k+1} = x_k + e^{-x_k}$$

Also, $\frac{d^2 f}{dx^2} = e^{-x}$. Hence, the quadratic model of $f(x)$ at x_k is given by

$$q(x_k) = e^{-x_k} - e^{-x_k}(x_{k+1} - x_k) + \frac{1}{2}e^{-x_k}(x_{k+1} - x_k)^2 \tag{6.4}$$

It can be verified from equation (6.4) that the Newton method will have a constant step length of 1. Figure 6.6 show the result of application of steepest descent method and the method of Newton to minimization of $f(x)$. It can be seen that rate of convergence of steepest descent method is poorer than Newton method. The steepest gradient method after 1000 iterations reaches $x = 6.9322$ with objective function value of 0.000975 and gradient -0.000975. In contrast, in the Newton method, after 100 iterations ($x = 100$), the objective function is given $3.721e^{-44}$ and gradient $-3.721e^{-44}$, *i.e.*, the method has practically converged.

The rate of convergence of Newton method is quadratic near the solution. The method may not converge at all, if the initial guess is far away from the solution. It can be viewed as an application of Newton Raphson (NR) method[2] to solving the system of equations $\nabla f(x) = 0$.

[2] An application of NR method well known to the readers is the Newton Raphson Load Flow.

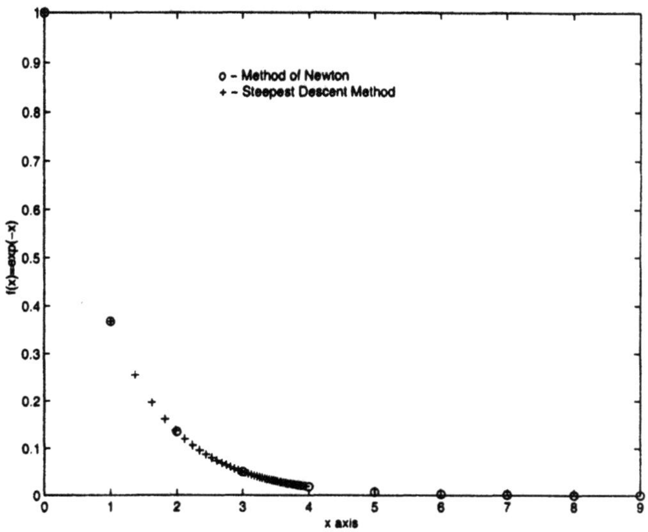

Figure 6.6. Example 6.2: Steepest Descent vs. Newton Method

The method converges to a solution when $||\nabla f(x_k)||_2 < \tau$, where τ is an user specified tolerance factor. One of the reasons why the Newton method may not converge, if started from a point far away from the optimal solution, is that matrix $H(x_k)$ may be singular. Even if it is nonsingular, it may not be spd in which case the direction as obtained from equation (6.2) may not be a descent direction, *i.e.*, $\nabla f(x_k)^T d_k = -\nabla f(x_k)^T H(x_k)^{-1} \nabla f(x_k) \geq 0$. However, instead of $H(x_k)$, if a spd approximation of $H(x_k)$ is used then clearly direction d_k will be a descent direction. The next method that we discuss, enlarges the region of convergence of the Newton method.

3.2.1 Levenberg-Marquardt Method

To guarantee descent at each step, it is preferable to use an approximation of $H(x_k)$ which is spd. In this method, this is achieved by adding a scaled identity matrix to $H(x_k)$ (review question 6). Thus, at each step, the following system of equations is solved:

$$[\xi_k I + H(x_k)][x_{k+1} - x_k] = -\nabla f(x_k)$$

where $\xi_k > 0$. Such a method is known as Levenberg-Marquardt type of method. Typical values of ξ_k are 0.25, 0.75, 2.5 etc. The method works as follows:
Choose a ξ_k and ascertain the *spd* nature of $[\xi_k I + H(x_k)]$ by attempting a LDL^T decomposition of $[\xi_k I + H(x_k)]$, such that $D_i > 0 \; \forall \; i$. If this is

unsuccessful, ξ_k is multiplied by a factor of 4. The process is repeated until such a factorization is possible. Then, solve the following system of equations using available factorization.

$$[\xi_k I + H(x_k)][x_{k+1} - x_k] = -\nabla f(x_k)$$

Because of spd nature of matrix $[\xi_k I + H(x_k)]$, direction $d_k = x_{k+1} - x_k$ as obtained from the above equation is a descent direction, *i.e.*,

$$d_k^T \nabla f(x_k) = -\nabla f(x_k)^T [\xi_k I + H(x_k)]^{-1} \nabla f(x_k) < 0$$

Now compute $R_k = \frac{f(x_k) - f(x_{k+1})}{q(x_k) - q(x_{k+1})}$, *i.e.*, the ratio of actual decrease to that predicted by a quadratic function. $q(x_k)$ is defined in Newton method (6.1). If R_k is close to unity, then quadratic approximation is acceptable and hence, ξ_k can be reduced. Conversely, if R_k is small, then a large value of ξ_k is required. With this motivation ξ_k is set as follows; if $R_k < 0.25$, put $\xi_{k+1} = 4\xi_k$; if $R_k > 0.75$ put $\xi_{k+1} < \xi_k/2$, else $\xi_{k+1} = \xi_k$.

Sometimes, it is either difficult or not possible to numerically evaluate the Hessian. In such situations, a quasi-newton approximation is used.

3.2.2 Quasi-Newton Methods

Consider the problem of minimizing $f(x) = c^T x + \frac{1}{2} x^T H x$, where H is spd. In a scheme belonging to this class, search direction is of the form $d_k = -D_k \nabla f(x)$, in lieu of $-H^{-1}\nabla f(x)$, as in Newton method. Matrix D_k is $n \times n$ spd matrix that approximates matrix H^{-1}. The spd nature of D_k ensures that d_k is a descent direction, *i.e.*,

$$\nabla f(x_k)^T d_k = -\nabla f(x_k)^T D_j \nabla f(x_k) < 0$$

In the next step, D_{k+1} is formed by addition to D_k, two symmetric matrices, each of rank-1[3]. For a quadratic function, it can be shown that within n steps such an update scheme will produce an exact representation of the actual inverse of Hessian [Bazaraa et al., 1993]. Even if the matrix H is sparse, it's inverse is likely to be non-sparse. Hence, the utility of Quasi-Newton methods to large scale optimization is limited except for projected gradient Sequential Quadratic Programming (SQP) methods. We conclude the discussion on unconstrained minimization with a brief description of a memoryless quasi-Newton update which is preferred when the matrix H is sparse.

[3]A symmetric $n \times n$ rank-1 matrix is of the form vv^T where v is a $n \times 1$ vector. If v is not a sparse vector, then it is non-sparse. Davidon-Fletcher-Powel method and Broyden-Fletcher-Goldfarb-Shanno (BFGS) update belong to the class of Quasi Newton methods. Local convergence of quasi-newton methods is usually superlinear.

3.2.3 Memoryless Quasi-Newton Updates

A memoryless version of quasi-Newton method along with inexact line search has been found useful [Bazaraa et al., 1993] for sparse-matrix computation where the updates are derived as follows:
Choose y_1 and $D_1 = I_n$, $d_1 = -D_1\nabla f(y_1)$. The first descent direction is simply the steepest descent direction. y_{k+1} is obtained by minimizing $f(y_k + \lambda d_k)$ with respect to λ.
Compute $p_k = [y_{k+1} - y_k]$
Compute $q_k = \nabla f(y_{k+1}) - \nabla f(y_k) = H(y_{k+1} - y_k) = Hp_k$
Update $D_{k+1} = I_n + \frac{p_k p_k^T}{p_k^T q_k}(1 + \frac{q_k^T q_k}{p_k^T q_k}) - \frac{q_k p_k^T + p_k q_k^T}{p_k^T q_k}$
Finally, $d_{k+1} = -D_{k+1}\nabla f(y_{k+1})$.

The update D_{k+1} forgets the previous updates and always restarting the inverse updates from an identity matrix approximation. The method requires storage of only 2-vectors and hence, is suitable for sparse computation.

It is possible to convert a constrained optimization problem to an unconstrained optimization problem by use of penalty functions. We now proceed to develop the methods for constrained optimization.

4. Nonlinear Constrained Optimization

The first task before us is to tackle the question of how to identify the optimal point in a presence of constraints. As can be seen in figure 6.7, the zero gradient point, *i.e.*, the unconstrained minimum solution will not be usually admissible in a constrained problem. In the next section we discuss Karush-Kuhn Tucker (KKT) conditions for optimality of constrained problems. These conditions were independently derived by Karush[1939] and by Kuhn and Tucker[1951].

4.1. Karush Kuhn Tucker Necessary Conditions

Consider the following minimization problem:

$$\text{Problem NLP}: \quad \min f(x),\ x \in \Re^n \tag{6.5}$$

$$\text{Such that} \quad h_i(x) = 0,\ i = 1, \cdots, m_1 \tag{6.6}$$

$$\text{and} \quad h_i(x) \leq 0,\ i = m_1 + 1, \cdots, m_2 \tag{6.7}$$

Let $\bar{x}$ be a feasible point. A point $\bar{x}$ is said to be feasible if it satisfies all the equality and inequality constraints listed above. Let us denote by I, the set of all constraints, satisfied by explicit equality relationship, *i.e.*,

$$\mathcal{A} = \{i : h_i(\bar{x}) = 0\}$$

Suppose that f and h_i for $i \in \mathcal{A}$ are differentiable and the rest of the constraints are continuous at $\bar{x}$. Further, suppose that $\nabla h_i(\bar{x})$ $i \in \mathcal{A}$

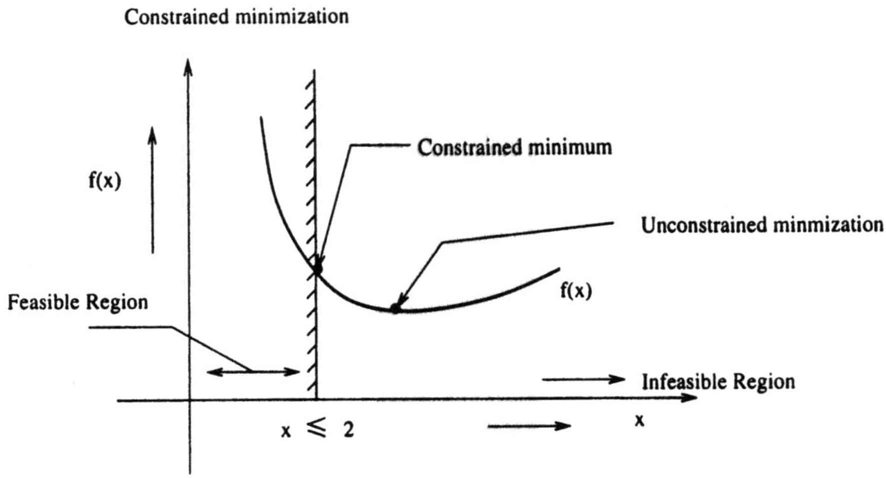

Figure 6.7. Constrained Problem

are linearly independent. This is also known as *regularity* condition. Then, the KKT conditions state the following *necessary* conditions for a feasible point $\bar{x}$ be a local optimal solution point.

$$\nabla f(\bar{x}) + \sum_{i=1}^{m_1} \mu_i \nabla h_i(\bar{x}) + \sum_{i=m_1+1}^{m_2} \mu_i \nabla h_i(\bar{x}) = 0 \tag{6.8}$$

$$\mu_i h_i(\bar{x}) = 0 \text{ for } i = 1, \cdots, m_2; \tag{6.9}$$

$$\mu_i \geq 0 \quad i = m_1 + 1, \cdots, m_2. \tag{6.10}$$

It is common to state KKT conditions in terms of a Lagrangian function defined as follows:

$$\mathcal{L}(x, \mu) = f(x) + \sum_{i=1}^{m_1} \mu_i h_i(x) + \sum_{i=m_1+1}^{m_2} \mu_i h_i(x)$$

Then, the necessary condition given by equation (6.8) can be restated in a compact fashion as $\nabla_x \mathcal{L}(\bar{x}, \bar{\mu}) = 0$. The μ_is are known as the Lagrangian multiplies. The constraints belonging to the set $\mathcal{A} = \{i; h_i(\bar{x}) = 0\}$, are known as the binding or active set at the optimal. Thus, all equality constraints belong to the active set. The subset $\mathcal{A}^+$ of set I corresponding to the inequality constraints for which $\mu_i > 0$ is known as the set of *strongly* active constraints and a corresponding subset I^- with $\mu_i = 0$ is referred to as set of *weakly* active constraints. The conditions $\mu_i h_i(x) = 0$ also known as complementary condition implies that if a constraint $h_i(x)$ is not active, *i.e.*, $h_i(\bar{x}) < 0$, then its Lagrangian

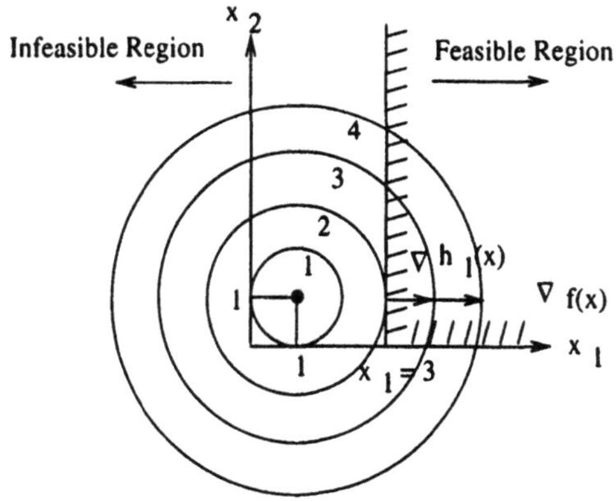

Figure 6.8. Example 6.3: A circular objective function

multiplier is zero. Equation (6.8) states that at the optimal point, the gradient of the objective function $\nabla f(\bar{x})$ is a linear combination of the gradient of the active or binding constraints. An unconstrained optimization problem can be viewed as a special case of a constrained optimization problem where none of the constraints are active. Hence, all the Lagrangian multipliers are zero and so is the gradient. Let us consider few simple examples to illustrate the KKT conditions.

Example 6.3: Consider the following minimization problem.

$$\min\ (x_1 - 1)^2 + (x_2 - 1)^2$$
$$\text{such that } x_1 \geq 3 \text{ and } x_2 \geq 0$$

The objective function corresponds to a circle centered at $(1, 1)$. The square of its radius (r) is the value of the objective function. It can easily be seen that at the center $(1, 1)$, $r = 0$, and hence, it is an unconstrained minimum. However, because of the constraint $x_1 \geq 3$, this point is inadmissible or infeasible. Then, it can be seen graphically (figure 6.8) that the optimal solution lies at $(3, 1)$; the constraint $x_1 \geq 3$ is active and the constraint $x_2 \geq 0$ is inactive. Further, $\nabla f(\bar{x})$ and $\nabla h_1(\bar{x})$ are in the same direction, *i.e.*, parallel to the x-axis.

$$\nabla f(\bar{x}) = 2 \begin{bmatrix} \bar{x}_1 - 1 \\ \bar{x}_2 - 1 \end{bmatrix} = \begin{bmatrix} 4 \\ 0 \end{bmatrix} \tag{6.11}$$

and

$$\nabla h_1(\bar{x}) = \begin{bmatrix} -1 \\ 0 \end{bmatrix} (x_1 \geq 3 \Rightarrow -x_1 \leq -3) \tag{6.12}$$

Clearly, $\nabla f(\bar{x}) + 4\nabla h_1(\bar{x}) = [0]$, *i.e.*, $\mu_1 = 4$. Since, $x_2 \geq 0$ is inactive at the solution ($x_2 = 1$), therefore $\mu_2 = 0$. Thus, the necessary KKT conditions are satisfied at the optimal point.

The above problem is a specific example of a class of optimization problems known as Quadratic Programming (QP) problem. The QP problem is defined as follows:

$$\min_{x} c + b^T x + \frac{1}{2} x^T Q x$$

$$\text{such that} \quad Ax \leq b$$

We discuss QP solvers in more detail in the next chapter. In a constrained optimization problem, The maximum number of linearly independent active constraints has to be less than or equal to n, the dimensionality of the problem. In the above problem, only one constraint is active in the two dimensional problem. In a Linear Programming (LP) problem to be introduced now, the optimal solution lies at a vertex, *i.e.*, n- linearly independent constraints are active.

A general n-dimensional linear programming problem is defined as:

$$\min_{x} \ c^T x$$

such that $A_i x = b_i$ for $i_1 = 1, \ldots, m_1$;
and $A_i x \leq b_i$ for $i = m_1 + 1, \ldots, m_2$.
The solution of a general LP problem is discussed in the next chapter. Note that both the objective function and the constraints are linear in LP.

Example 6.4: Consider the following two dimensional example problem:

$$\min \ x_2 - x_1 \text{ such that } x_1 \geq 1; \text{ and } x_2 \geq 2.$$

It can be seen that to minimize $x_2 - x_1$, x_1 should be maximized and x_2 minimized, without violating any constraint. Thus, optimal solution for this problem corresponds to $x_1 \to \infty$ and $x_2 = 2$; $c^T x \to -\infty$. This is usually referred to as an unbounded solution. In general, in a problem in engineering, unbounded solution implies an inadequate formulation. Let us now modify this problem so that it has a finite minima.

Example 6.5: In continuation of example 6.4, if additional constraints on variables are prescribed say, $x_1 \leq 3$ and $x_2 \leq 5$, then, it can be seen that optimum lies at (3,2) (Figure 6.9b). In this example, the feasible region is given by box and the solution lies at a vertex. Further,

$$\nabla f(x) = \begin{bmatrix} -1 \\ 1 \end{bmatrix}; \nabla h_2(x) = \begin{bmatrix} 0 \\ -1 \end{bmatrix} \text{ and } \nabla h_3(x) = \begin{bmatrix} 1 \\ 0 \end{bmatrix}$$

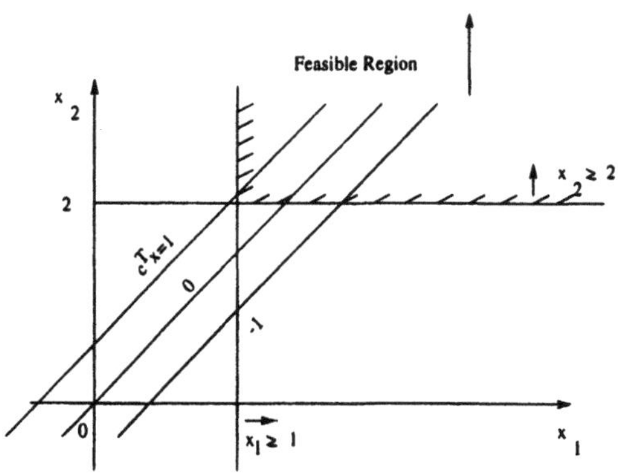

Figure 6.9a. Example 6.4

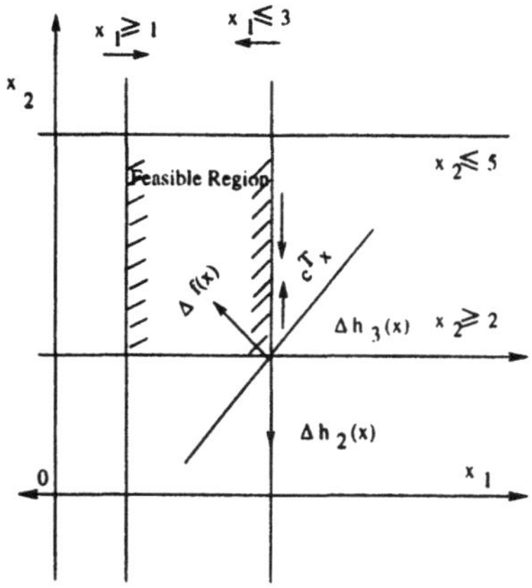

Figure 6.9b. Example 6.5

$$\begin{bmatrix} -1 \\ 1 \end{bmatrix} + 1 \times \begin{bmatrix} 1 \\ 0 \end{bmatrix} + 1 \times \begin{bmatrix} 0 \\ -1 \end{bmatrix} = \begin{bmatrix} 0 \\ 0 \end{bmatrix}$$

Thus, $\mu_2 = \mu_3 = 1$ and μ_1 and μ_4 are set to zero *i.e.*, again the KKT conditions are satisfied.

It can be shown that μ_i measures sensitivity of the objective function to constraint h_i. A $\mu_i > 0$ for an inequality constraint, implies that increasing $h_i(x)$ from 0 to a small value ϵ will reduce the objective function by $\epsilon\mu_i$. However, as this increase will create infeasibility, at the constrained optimal, $h_i(\bar{x})$ is constrained at its limit. A zero multiplier implies that increasing or decreasing the constrained value by ϵ will have no effect on the objective function. Inactive or weakly active constraints have zero Lagrangian multipliers.

4.1.1 Alternative form of KKT Conditions

An alternative form of the KKT condition is described by the following problem:

$$\min_x f(x)$$

$$\text{such that } h_i(x) = 0 \text{ for } i = 1, \ldots, m_1$$

$$h_i(x) \geq 0 \text{ for } i = m_1 + 1, \ldots, m_2 \text{ } i.e., \text{ } -h_i(x) \leq 0$$

Note that sign of inequality constraints has been reversed. The Lagrangian function is defined as follows:

$$\mathcal{L}(x, \mu) = f(x) + \sum_{i=1}^{m_1} \mu_i h_i(x) + \sum_{i=m_1+1}^{m_2} (-h_i(x))$$

$$i.e., \ \mathcal{L}(x,\mu) = f(x) + \sum_{i=1}^{m_1} \mu_i h_i(x) - \sum_{i=m_1+1}^{m_2} \mu_i h_i(x)$$

it can be seen that at the KKT point $(\bar{x}, \bar{\mu})$

$$\nabla_x \mathcal{L}(\bar{x},\bar{\mu}) = 0 \Rightarrow \nabla f(\bar{x}) - \sum_{i=1}^{m2} \bar{\mu}_i \nabla h_i(\bar{x}) = 0$$

$$\bar{\mu}_i h_i(\bar{x}) = 0 \text{ for } i = 1, \cdots, m_2$$

$$\text{and } \bar{\mu}_i \geq 0 \ \forall \ i = m_1 + 1, \cdots, m_2$$

Therefore both these forms of KKT conditions are equivalent.

The important point to note is that when the constraints are described by $h_i(x) \leq 0$, then it is incorporated with positive sign in the Lagrangian function. On the other hand, when nature of constraints is given by $h_i(x) \geq 0$, then it is incorporated in Lagrangian with negative sign. The optimality conditions remain the same *i.e.*, $\nabla_x \mathcal{L}(\bar{x}, \bar{\mu}) = 0$, $\mu_i h_i(x) = 0$ and $\mu_i \geq 0$ for all inequality constraints. Let us now consider an example of constrained optimization in power systems.

Example 6.6: Consider the classical Economic Dispatch (ED) problem framed as follows:

Minimize the total cost of generation subject to the constraint that the net load is met. The basic formulation neglects the network constraints and losses. Usually, the cost function of a generator $F_i(P_{G_i})(\$/MW)$ is a quadratic function, given by the following equation,

$$F_i(P_{G_i}) = c_{i_0} + c_{i_1} P_{G_i} + \frac{1}{2} c_{i_2} P_{G_i}^2, \text{for } i = 1, \cdots, g \tag{6.13}$$

where c_{i_1} and c_{i_2} are greater than zero. The ED problem can be formulated as follows:

$$\begin{gathered} \min \ f(P_{G_1}, P_{G_2}, \cdots, P_{G_g}) = \textstyle\sum_{i=1}^{g} F_i(P_{G_i}) \\ \text{such that } P_{G_1} + P_{G_2} + \cdots + P_{G_g} = P_L \\ P_{G_i}^{min} \leq P_{G_i} \leq P_{G_i}^{max} \end{gathered}$$

where P_L is the net load on the system and P_{G_i} refers to generation at bus i.

A constrained problem can be solved by two approaches *viz.*, penalty function approach, and Lagrangian multiplier approach. Both of these approach will be discussed in the rest of the chapter.

5. The Penalty Function Approach

In this approach, equality constraints $h_i(x) = 0$ are incorporated in the objective function. An equality constrained minimization problem defined as min $f(x)$ subject to $h_i(x) = 0$; $i = 1, 2 \cdots, m$, can be posed as an unconstrained minimization problem as follows:

$$\min\ f(x) + \rho \sum_{i=1}^{m} h_i^2(x) \tag{6.14}$$

where $\rho > 0$ and is a large number known as the penalty factor and $\sum_{i=1}^{m} h_i^2(x)$ is a penalty function.

It can be seen intuitively, that if ρ is large enough, then, minimization will force $\sum_{i=1}^{m} h_i^2(x)$ towards zero, *i.e.*, individually $h_i(x)$ will be forced to zero. Further, notice that the square terms in penalty function retain individual penalties as positive. So, there is no chance of a negative violation, say in $h_1(x)$, being canceled by a positive violation, say in $h_2(x)$. If there exist multiple solutions to set of equations $h_i(x) = 0; i = 1, \cdots, m$, (*e.g.*, when $m < n$ and $x \in \Re^n$) then, the problem of minimization will try to choose a x that minimizes $f(x)$ while keeping penalty zero. Such an approach has been used in equality constrained power system state estimation [Vempati et al., 1991]. The choice of penalty function is also not unique. For example, the following objective function can also be used in place of equation (6.14).

$$\min\ f(x) + \rho \sum_{i=1}^{m} |h_i(x)| \tag{6.15}$$

The only apparent drawback of objective function (6.15) is that it is not differentiable (review question 8 suggests how this difficulty can be circumvented).

To summarize, the role of exterior penalty function is to a transform a constrained problem into an unconstrained problem. If the original problem is nonlinear, we have to solve a sequence of unconstrained problems. Such an approach is sometimes referred to as a Sequential Unconstrained Minimization Technique (SUMT).

Let us now consider a problem with a single inequality constraint

$$\min\ f(x) \text{ such that } h(x) \leq 0$$

Clearly, the form $f(x) + \rho h^2(x)$ is inappropriate because both $h(x) < 0$ and $h(x) > 0$ points are penalized. Therefore, a more appropriate unconstrained problem is given by:

$$\min f(x) + \rho \max\{0, h(x)\}\ x \in \Re^n \tag{6.16}$$

If $h(x) \leq 0$, then $\max\{0, h(x)\} = 0$ and no penalty is incurred. However, if $h(x) > 0$, an infeasible point then, $\max\{0, h(x)\} = h(x) > 0$ and penalty $\rho h(x)$ is realized. The above function is not differentiable. To obtain differentiability, the unconstrained problem can be modified as follows:

$$\min \ f(x) + \rho[\max\{0, h(x)\}]^2 \tag{6.17}$$

For a problem that has both equality and inequality constraints, the following unconstrained problem can be framed.

$$\min \ f(x) + \rho \sum_{i=1}^{m_1} h_i^2(x) + \rho \sum_{i=m_1+1}^{m_2} [\max\{0, h_i(x)\}]^2 \tag{6.18}$$

This function is referred to as the *auxiliary function.*

5.1. SUMT: Algorithm for Penalty Function Method

Due to the difficulties associated with large weights, most algorithms use a sequence of increasing penalty parameters leading to a Sequential Unconstrained Minimization Technique(SUMT). A conceptual algorithm is as follows [Bazaraa et al., 1993].

Initialization step: Choose a scalar $\epsilon > 0$, the initial point x_1, a penalty parameter $\rho_1 > 0$, and a scalar $\beta > 1$. The scalar β is used to control penalty and ϵ is used for termination criteria. Let $k = 1$, and go to main step.

Main step:

1 Start with x_k and solve the problem

$$\min_x \ f(x) + \rho_k \sum_{i=1}^{m_1} h_i^2(x) + \rho_k \sum_{i=m_1+1}^{m_2} [\max\{0, h_i(x)\}]^2$$

Let x_{k+1} be the optimal solution.

2 If $\rho_k[\sum_{i=1}^{m_1} h_i^2(x_{k+1}) + \sum_{i=m_1+1}^{m_2} \max[0, h_i(x_{k+1})^2] \leq \epsilon$ stop; else set $\rho_{k+1} = \beta\rho_k$ and $k = k + 1$. Repeat the main step

Power System State Estimation (PSSE) using QR decomposition is a successful application of the penalty function approach in power system analysis. To simplify the algorithm and cut down on computation, a fixed large penalty ρ and a stable LSS (QR decomposition approach) is used. With such a modification, the step 2 becomes redundant.

Example 6.7: The classical ED problem can also be formulated as an unconstrained optimization problem as follows:

$$\min \sum_{i=1}^{g} F_i(P_{g_i}) + \frac{\rho}{2}(P_{G_1} + P_{G_2} + \cdots + P_{G_g} - P_L)^2$$

Here ρ refers to penalty and not the Lagrangian multiplier. Setting the gradient of an auxiliary function to zero, the following system of equations result.

$$\begin{bmatrix} c_{12}+\rho & \rho & \rho & \cdots & \rho \\ \rho & c_{22}+\rho & \rho & \cdots & \rho \\ \vdots & \vdots & \ddots & \vdots & \vdots \\ \vdots & \vdots & \vdots & \ddots & \vdots \\ \rho & \rho & \rho & \cdots & c_{g2}+\rho \end{bmatrix} \begin{bmatrix} P_{G_1} \\ P_{G_2} \\ \vdots \\ \vdots \\ P_{G_g} \end{bmatrix} = \begin{bmatrix} -c_{11}+\rho P_L \\ -c_{21}+\rho P_L \\ \vdots \\ \vdots \\ -c_{g1}+\rho P_L \end{bmatrix} \quad (6.19)$$

It can be shown that coefficient matrix in equation (6.19) is the Hessian of the auxiliary function (review question 10). Dividing each equation by ρ, we have

$$\begin{bmatrix} 1+c_{12}/\rho & 1 & 1 & \cdots & 1 \\ 1 & 1+c_{22}/\rho & 1 & \cdots & 1 \\ \vdots & \vdots & \ddots & \vdots & \vdots \\ \vdots & \vdots & \vdots & \ddots & \vdots \\ 1 & 1 & 1 & \cdots & 1+c_{g2}/\rho \end{bmatrix} \begin{bmatrix} P_{G_1} \\ P_{G_2} \\ \vdots \\ \vdots \\ P_{G_g} \end{bmatrix} = \begin{bmatrix} P_L - c_{11}/\rho \\ P_L - c_{21}/\rho \\ \vdots \\ \vdots \\ P_L - c_{g1}/\rho \end{bmatrix} \quad (6.20)$$

As $\rho \to \infty$, it is clear that coefficient matrix tends to be singular. Further, all the row equations approach $\sum P_{G_i} = P_L$, *i.e.*, only feasibility is emphasized. The effect of cost coefficients is lost. The right choice of ρ that gives appropriate penalty without sacrificing numerical stability is important.

6. Lagrangian Multiplier Approach

In this section, we discuss methods that compute a KKT point. We introduce the approach through ED problem.

Example 6.8: We solve the ED problem of Example 6.6 by applying KKT condition.
Let us define a Lagrangian function $\mathcal{L}(P_{G_1}, P_{G_2}, \cdots, P_{G_g}, \mu)$ as follows:

$$\mathcal{L}(P_{G_1}, P_{G_2}, \cdots, P_{G_g}, \mu) = \sum_{i=1}^{g} F_i(P_{G_i}) + \mu[P_{G_1} + P_{G_2} + \cdots + P_{G_g} - P_L]$$

KKT condition requires that at the optimal

$$\frac{\partial \mathcal{L}}{\partial P_{G_i}} = 0 \Rightarrow c_{i_1} + c_{i_2} P_{G_i} + \mu = 0 \text{ for } i = 1, \cdots, g$$

and $P_{G_1} + P_{G_2} + \cdots P_{G_g} - P_L = 0$ This leads us to the following $g+1$ equations in $g+1$ variables.

$$\begin{bmatrix} c_{1_2} & & & & 1 \\ & c_{2_2} & & & 1 \\ & & \ddots & & \vdots \\ & & & c_{g_2} & 1 \\ 1 & 1 & \cdots & 1 & 0 \end{bmatrix} \begin{bmatrix} P_{G_1} \\ P_{G_2} \\ \vdots \\ P_{G_g} \\ \mu \end{bmatrix} = \begin{bmatrix} -c_{1_1} \\ -c_{2_1} \\ \vdots \\ -c_{g_1} \\ P_L \end{bmatrix} \tag{6.21}$$

The above set of equations can be solved efficiently by Gaussian elimination with zero fillins. A little contemplation shows that since $c_{i_2} > 0$, no zero pivot is involved during elimination and therefore the method is numerically stable. Further, in this particular example, it can be seen that at the optimal point (refer review question 11 for the value of μ):

$$\left[\frac{\partial \mathcal{L}}{\partial P_G}\right] = [0] \Rightarrow \frac{\partial F_i}{\partial P_{G_i}} + \mu = 0 \Rightarrow \frac{\partial F_1}{\partial P_{G_1}} = \frac{\partial F_2}{\partial P_{G_2}} = \cdots = \frac{\partial F_{G_g}}{\partial P_{G_g}} = -\mu$$

Refer to review question 11 for exact expression of μ. This condition is described in power system textbooks [Wood and Wollenberg, 1996] by the following statement. *At the optimal solution the incremental cost of generation is identical.* It can be easily seen that for this particular problem, the Lagrangian approach is much more attractive than the penalty function approach. This brings us to an important conclusion that "*The solution approach has to strike harmony with the problem*". We now generalize the above approach of finding a KKT solution for a nonlinear equality constrained problem, *viz.*

$\min_x \ f(x)$ subject to constraint $h_i(x) = 0;\ i = 1, \cdots, m$

6.1. Method of Newton

The algorithm computes the solution to the KKT equations by applying Newton Raphson method to the KKT equations. The steps are as follows:

1. Define $\mathcal{L}(x, \mu) = f(x) + \sum_{i=1}^{m} \mu_i h_i(x)$.
2. Solve the following $n+m$ equations in $n+m$ variables:

$$\nabla_x \mathcal{L}(x,\mu) = \nabla f(x) + \sum_{i=1}^{m} \mu_i \nabla h_i(x) = 0 \quad \text{and} \quad h_i(x) = 0 \qquad (6.22)$$

In the ED problem, these equations are linear. However, when equations (6.22) are nonlinear, *e.g.*, cost function contains a cubic term or network power flow equations are modeled, then they can be solved iteratively by Newton Raphson method as follows:

Start with a guess (x_1, μ_1)
While $k = 1$ or $||\nabla \mathcal{L}||_\infty$ or $||h(x_k)||_\infty > \epsilon$, do {

i) Solve the following augmented linear system of equations.

$$\begin{bmatrix} \left[\frac{\partial^2 f}{\partial x^2} + \sum \mu_i \frac{\partial^2 h_i}{\partial x^2}\right] & \left[\frac{\partial h}{\partial x}\right] \\ \left[\frac{\partial h}{\partial x}\right]^T & 0 \end{bmatrix} \begin{bmatrix} [x_{k+1} - x_k] \\ [\mu_{k+1} - \mu_k] \end{bmatrix} = \begin{bmatrix} [-\nabla \mathcal{L}(x_k, \mu_k)] \\ [-h(x_k)] \end{bmatrix} \qquad (6.23)$$

ii) $k = k + 1$ }

The review question 17 shows that the equations (6.23) can be further simplified as follows:

$$\begin{bmatrix} \nabla^2_{xx} \mathcal{L}(x_k, \mu_k) & \left[\frac{\partial h}{\partial x}\right] \\ \left[\frac{\partial h}{\partial x}\right]^T & 0 \end{bmatrix} \begin{bmatrix} x_{k+1} - x_k \\ [\mu_{k+1}] \end{bmatrix} = \begin{bmatrix} -\nabla f(x_k) \\ [-h(x_k)] \end{bmatrix} \qquad (6.24)$$

where the Hessian $\nabla^2_{xx}\mathcal{L}(\bar{x}, \bar{\mu})$ is defined as follows:

$$\nabla^2_{xx}\mathcal{L}(\bar{x}, \bar{\mu}) = \nabla^2_{xx} f(\bar{x}) + \sum_i \bar{\mu}_i \nabla^2_{xx} h_i(\bar{x})$$

Notice that the matrix is symmetric but indefinite. The algorithm is amenable to sparse matrix computation. A 1×1 and 2×2 block pivoting strategy can be used to exploit sparsity. The next obvious question to be answered is whether the the KKT solution so computed represents a minimum. *KKT second-order sufficiency conditions for the equality constraints problem state that if* $d^T \nabla \mathcal{L}^2(\bar{x}, \bar{\mu}) d >$ [4] 0 *for all directions* d *satisfying* $\nabla h_i(\bar{x})^T d = 0$, $i = 1, \cdots, m$; *then the solution* $\bar{x}$ *is a strict local minimum.* In general, there is no straight forward way to check this condition except by forming a projected Hessian matrix (to be discussed soon). Thus, the design of the algorithm can be strengthened such that

[4] Second order necessary conditions are obtained by replacing $>$ with $\geq$.

we move to a KKT point that promises to be a minimum. Also, note that in the equality constrained NLP problem, it is not necessary to start with a feasible point. The algorithm simultaneously computes the feasible and the optimal point. An application of this approach to OPF problem will be discussed in chapter-11.

We now discuss a Sequential Quadratic Programming (SQP) methodology, an alternative to Newton method which is both popular and robust. The method avoids penalty functions which are a cause of numerical difficulties as well as slow convergence. The application of the algorithm for OPF will be discussed in chapter-11.

6.2. SQP or Projected Lagrangian Approach

Let $d_k = x_{k+1} - x_k$. Then, minimum of the following Quadratic Programming (QP) problem satisfies equation (6.24).

$$QP(x_k, \mu_k) : \ \min \nabla f(x_k)^T d_k + \frac{1}{2} d_k^T \nabla_{xx}^2 \mathcal{L}(x_k, \mu_k) d_k \tag{6.25}$$

$$\text{such that } h_i(x_k) + \nabla h_i(x_k)^T d_k = 0; \ i = 1, \cdots, m \tag{6.26}$$

The problem defined by equations (6.25-6.26) has a quadratic objective function and linear equality constraints. In the QP problem, observe that Hessian or quadratic term is simply not the Hessian of $f(x)$ but also incorporates the curvature information of the nonlinear constraints. Note that in the process of solving QP problem, we reach a beneficial KKT point. Instead of equation (6.25), if a second order Taylor series approximation for $\mathcal{L}(x_k, \mu_k)$ is used, then the following objective function will result in the QP subproblem.

$$\min \mathcal{L}(x_k) + \nabla_x \mathcal{L}(x_k)^T d_k + \frac{1}{2} d_k^T \nabla_{xx}^2 \mathcal{L}(x_k) d_k$$

Such a QP problem will solve equation (6.23).

A restriction with solving equation (6.23) is that it can not account for strict inequality constraints in computation of direction d. The problem can be solved in QP formulation by augmenting the constraint set with linearized inequality constraints.

Now, at the kth iteration of SQP following QP subproblem is solved.

$$\min \ \nabla f(x_k)^T d_k + \frac{1}{2} d_k^T \nabla_{xx}^2 \mathcal{L}(x_k, \mu_k) d_k$$

$$\text{such that } h_i(x_k) + \nabla h_i(x_k)^T d_k = 0; \ i = 1, \cdots, m_1$$

$$h_i(x_k) + \nabla h_i(x_k)^T d_k \leq 0 \ i = m_1 + 1, \ldots, m_2$$

As discussed earlier, matrix $H(x_k)$ a spd approximation to $\nabla_{xx}\mathcal{L}(x_k)$ may be used in actual computation. Note that if x_k is a feasible point then, $h_i(x_k) = 0$. In that case, equation (6.26) becomes $\nabla h_i^T(x_k)d_k = 0$, *i.e.*, feasible direction is orthogonal to the constraint gradient. In NLP, even then we need not have $h_i(x_k + d_k) = 0$, because $h_i(x_k)$ may be a nonlinear constraint. Hence, $x_k + d_k$ may not be a feasible point. This can deteriorate the convergence characteristic of the solver when the initial point is far away from the solution. To remedy this problem and to ensure global convergence, a line search is done along the direction d_k to minimize a *merit* function.

A merit function is a composite function of the objective function and the equality constraints. It penalizes violations in $h_i(x)$, thereby, emphasing both the descent and feasibility conditions. For a equality constrained problem it can be defined as $F_E(x) = f(x) + \gamma\sum_{i=1}^{m} h_i^2(x)$ where $\gamma >> 0$. It will be discussed in more detail later in the OPF chapter. The new point x_{k+1} is given by

$$x_{k+1} = x_k + \alpha_k d_k$$

The SQP solver involves two levels of iterations. Loop iterating at outer level involves formulating a QP subproblems QP(x_k, μ_k). These iterations are known as *major* iterations. Iterations required to solve the inequality constrained QP solver are called as *minor* iteration. QP Solvers will be discussed in the next chapter. We now discuss the important issue of handling indefinite Hessian.

6.3. Handling of Indefinite Hessian

A problem with the SQP formulation discussed so far is that 1) it may not be possible to compute the second derivative matrix $\nabla_{xx}^2\mathcal{L}(x_k, \mu_k)$ or 2) the second order derivative matrix $\nabla^2\mathcal{L}(x_k, \mu_k)$ might not be positive definite. Under such a situation, the direction d_k may not be a descent direction. One way to solve this problem is by using a Quasi-Newton approximation B_k to $\nabla^2\mathcal{L}(x_k)$. A Broyden-Fletcher-Goldfarb-Shanno (BFGS) update for Hessian can be defined as follows:

$$B_{k+1} = B_k + \frac{q_k q_k^T}{q_k^T p_k} - \frac{B_k p_k p_k^T B_k}{p_k^T B_k p_k}$$

$$\text{where } p_k = x_{k+1} - x_k, \; q_k = \nabla\mathcal{L}'(x_{k+1}) - \nabla\mathcal{L}'(x_k)$$

$$\text{and } \nabla\mathcal{L}'(x) \triangleq \nabla f(x) + \sum_{i=1}^{m} \mu_{k+1_i}\nabla h_i(x)$$

Note that matrix B_k need not be sparse. Alternatively, $\nabla^2\mathcal{L}+\gamma I, \gamma > 0$ can be used if $\nabla^2\mathcal{L}$ is indefinite where γ is large enough to make $\nabla^2\mathcal{L}$ positive definite. When a spd approximation to the Hessian is used, the direction obtained by solving the QP subproblem will be a descent direction.

7. Summary

The aim of this chapter was to a guide the reader from the basics of optimization to large scale optimization. Throughout the presentation, it is assumed that the function to be minimized and the constraints are at least twice differentiable. This assumption is reasonable and the successful OPF methods do use second order information to develop quadratically convergent algorithms. OPF algorithms are discussed in the chapter 11.

In the first part of the chapter, methods of unconstrained optimization are reviewed. Usually almost all nontrivial optimization problems will have constraints. The unconstrained optimization techniques become useful in two contexts *viz.* (1) converting a constrained optimization problem to an equivalent unconstrained optimization problem and (2) applying line search within the constrained minimization problem. Armijo's inexact line search has been used in power system analysis problem; it avoids too many costly function evaluations [Barret et al., 1997]. Method of Newton and an improvised Levenberg-Marquardt method which can be started from a far away starting point were also discussed. Method of Levenberg-Marquardt introduces an important concept of modifying an indefinite Hessian to a spd Hessian to guarantee descent. It improves robustness of the resulting algorithm by ascertaining a descent direction at each iteration. The concept is also used in the constrained optimization. Quasi-Newton methods which do not require computation of Hessian were also discussed. Such methods use a spd approximation of the Hessian or its inverse. A memoryless quasi-Newton update can be used in a sparse matrix computation scenario. It reduces storage requirement.

The treatment on the constrained optimization begins with the development of important KKT necessary conditions. It states that at the optimal, objective function gradient can be expressed as a linear combination of the linearly independent gradients of the constraints active at the solution. Further, the Lagrangian multipliers of the inequality constraints that are active at the solution have to be greater than or equal to zero. The Lagrangian multipliers measure the sensitivity of the objective function to perturbation in the constraints.

A constrained optimization problem can be solved by converting it to an equivalent unconstrained optimization problem by penalty function approach. Power system state estimation is an important application of this approach. Augmented Lagrangian formulations are used to limit the penalty term in such formulations. [Burchett et al., 1982] discusses one such application to OPF. Because of the numerical difficulties associated with the penalty function approach, the method of feasible directions is considered a good approach to tackle the NLP problem. Newton method and Sequential Quadratic Programming (SQP) approach belong to this class.

The world of constrained optimization is rich with many methods and algorithms. In this chapter, it is not possible to do justice to all the developments. In this presentation, we have strived to provide exposure to those methods which have been successfully used in problems like state estimation and OPF. An actual implementation may have to consider many more details. Like the SQP method for NLP, a sequential Linear Programming (SLP) algorithm can also be developed by limiting to the first order approximation of the objective function. In the context of OPF, it will be discussed in the chapter 11. At each major iteration, an SLP solver requires a LP solver. Both LP and QP solvers will be discussed in the next chapter.

8. Further Reading

[Bazaraa et al., 1993] provides an excellent treatment on both theoretical and algorithmic issues of unconstrained and constrained optimization. We have been motivated by the above text considerably and strongly suggest it as a further reading for beginners as well as those who are interested in detail. In optimization mapping from the algorithm to implementation is not straight forward [Gill et al., 1984a]. [Gill et al., 1981] is one of the best books concerning numerical implementation. [Fletcher, 1987] is also strongly recommended to get a balanced view point of practical optimization. [Gill et al., 1991] is a good starting point for those who have background of linear algebra. The authors' treatment is simple but appealing. [Gill et al., 1984b] is recommended to obtain an overview of the sparsity issues in optimization. It is the authors' view that exploiting sparsity in large scale optimization problems is still a challenge and is a open fertile area for research. For example, finding a basis of null space which is sparse is an interesting problem which if resolved satisfactorily can change the way large scale optimization problems are tackled.

Many monographs have been written on specific issues or methods in optimization. [Chvatal and Freeman, 1983] is one among many books devoted to linear programming. [Bertsekas, 1982] presents a thorough treatment on Lagrangian multiplier approach. In this chapter, we have not touched upon two important class of methods, *viz.* (1) Conjugate Gradient (CG) methods and (2) interior point methods[Morales et al., 1980]. Both these approaches hold promise for exploiting sparsity and readers are encouraged to explore them in more detail. In addition to these works, there exist numerous papers which are worth reading. An active set of Newton algorithms for large scale nonlinear programs with box constraints is dealt with in [Facchine et al., 1998]. [Gould and Toint, 2001a] provides a comprehensive list of work on quadratic programming. [Gill et al., 1997] details a SQP technique for large scale constrained optimization. Numerical methods for large-scale non-convex quadratic programming are discussed in [Gould and Toint, 2001b]. Internet is another useful source for obtaining information on the optimization methods. For example, http:www.numerical.rl.ac.uk/qp/qp.html is a QP page by Nick Gould and Philippe Toint on the wonderful world of QP.

9. Review Questions

1 Prove that a $n \times n$ matrix H is spd if and only if factorization $H = LDL^T$ exists and $D_i > 0 \; \forall \; i$.

2 If H is spd, show that H^{-1} exist, and is spd.

3 Let A be a spd matrix and $\bar{A} = A + vv^T$, a rank-1 update of A. Show that, $x^T A x \geq 0 \; \forall x \neq 0 \in \Re^n$.

4 Show that in Fibonacci search method (Figure 6.4a-6.4b),

$$b_k - \mu_k = \lambda_k - a_k$$

5 Show that direction $d_k = -\nabla f(x_k)$ is a direction of steepest descent.

6 Show that by an appropriate choice of α, it is possible to ascertain that matrix $[\alpha I + H]$ is always spd. Hence, show that Levenberg-Marquardt method always creates a descent direction.

7 Consider the solution to the following box, constrained nonlinear programming problem

$$\min \; f(x), \; x \in \Re^n; \; l_i \leq x_i \leq u_i = 1, 2, \cdots, n$$

Show that $\bar{x}$ a stationary point for the problem satisfies the following conditions:

$\bar{x}_i = l_i \Longrightarrow \frac{\partial f}{\partial x_i} \geq 0$
$l_i < \bar{x}_i < u_i \Rightarrow \frac{\partial f}{\partial x_i} = 0$
$\bar{x}_i = u_i \Rightarrow \frac{\partial f}{\partial x_i} \leq 0$

8 Show that minimizing (6.15) is equivalent to the following

$$\min f(x) + \mu \sum_{i=1}^{m} (z_i^+ + z_i^-)$$

such that $z_i^+ - z_i^- = h_i(x) = 0$; $z_i^+ \geq 0, z_i^- \geq 0, i = 1, \ldots, m$. Therefore, suggest a sequential linear programming approach to solve the problem.

9 Show that in the example 6.6, the KKT solution corresponds to the global minimum.

10 Show that Hessian in Example 6.7 is given by coefficient matrix of equation (6.19).

11 Show that in the economic dispatch problem discussed in example 6.6, the Lagrangian multiplier is given by

$$\mu = -\frac{P_L + \sum_{i=1}^{g} \frac{c_{i_1}}{c_{i_2}}}{\sum \frac{1}{c_{i_2}}}$$

12 In the economic dispatch problem discussed example 6.6, if additional inequality constraints are imposed on generation limits *viz.*,

$$P_{G_i}^{min} \leq P_{G_i} \leq P_{G_i}^{max}; \; i = 1, \cdots, g$$

Then, formulate a strategy to solve the economic dispatch problem using SUMT approach. [Hint: Start with $\mu = 0$].

13 By using the review question 7, can a more simplified approach be developed?

14 At the optimal solution of a QP problem, show that the projected gradient $Z^T(c + Qx)$ should be zero .

15 Consider an undetermined system of linear equations $Cx = d$ where C is a $m \times n$ matrix $m < n$ and matrix C has full row rank. Show that there exists a QR decomposition of C^T, *i.e.*, $C^T = Q \begin{bmatrix} R \\ 0 \end{bmatrix}$ such

that Q is a $n \times n$ orthogonal matrix and R is a $m \times m$ upper triangular matrix. Correspondingly deduce the LQ factorization of the under determined matrix C. Show that the matrix L is non-singular and columns $q_{m+1} \cdots q_n$ form the orthogonal basis of the null space of matrix C. Hence, show that $x = C^T[LL^T]^{-1}d$ is a minimal second norm solution to the problem $Cx = d$.

16 In the above problem, let the first m rows of the matrix C be linearly independent and the matrix C be correspondingly be portioned as $[C_1, C_2]$ where is C_1 is a rank m submatrix. Show that the basis of the null space is given by $\begin{bmatrix} -C_1^{-1}C_2 \\ I_{n-m} \end{bmatrix}$.

17 Consider the problem of minimizing $f(x)$ subject to constraints $h_i(x) = 0$, $i = 1, \ldots, m$. Assuming that f and h_i's are twice differentiable, show that in a SQP strategy, solution of QP subproblem in major iteration will satisfy the following KKT equations

$$\begin{bmatrix} \frac{\partial^2 f}{\partial x^2} + \sum \mu_i \frac{\partial^2 h_i}{\partial x^2} & \left[\frac{\partial h}{\partial x}\right] \\ \left[\frac{\partial h}{\partial x}\right]^T & 0 \end{bmatrix} \begin{bmatrix} x_{k+1} - x_k \\ \mu_{k+1} \end{bmatrix} = \begin{bmatrix} -\nabla f(x_k) \\ -h(x_k) \end{bmatrix}$$

18 Show that if Hessian $\left[\frac{\partial^2 f}{\partial x^2} + \sum \mu_i \frac{\partial^2 h_i}{\partial x^2}\right]$ is indefinite, then the direction $d_k = x_{k+1} - x_k$ obtained in above problem may not be a descent direction. Hence, make a case for spd approximation to Hessian.

19 Show that if a matrix has full column rank, then its null space is zero.

Chapter 7

SPARSE LP AND QP SOLVERS

Freedom and constraint are two aspects of the same necessity.
—Antonine De Saint-Exupery, in La Citadelle(1948)

The previous chapter discussed methods for unconstrained optimization. For nonlinear constrained optimization problem, two powerful and commonly used algorithms are Sequential Linear Programming (SLP) and (SQP). Such algorithms solve a Linear Programming (LP) and Quadratic programming (QP) problems at each iteration. Therefore, the efficient implementation of NLP algorithms require efficient Linear Programming (LP) and Quadratic Programming (QP) solvers. In this chapter we discuss algorithms for LP and QP. The algorithms belong to a class of methods known as *'active set'* methods which solve an equality constrained problem at each iteration where the equality constraints correspond to current guess of active constraints at the optimal. We begin with LP formulation.

1. Linear Programming

LP problem arises in many applications which include OPF, optimal relay coordination etc. The LP problem can be stated as follows:

$$\min_{x} \; c^T x; \; x \in \Re^n \tag{7.1}$$

$$\text{subject to} \;\; a_i^T x \leq b_i \; \forall \; i = 1, \ldots, m \tag{7.2}$$

Both the objective function and the constraints are linear. A standard LP problem also has constraints of the type $x_i \geq 0$. However, right now to simply notations we consider such bound constraints to present in

(7.2), *i.e.*, $-x_i \leq 0$. *The solution of LP problem always lies at a vertex.* Therefore, a LP solver is implemented in two phases. The phase I finds a feasible vertex and the phase II does the minimization. Assuming knowledge of a feasible vertex, we first describe phase II. It is followed by the explanation of phase I.

1.1. Phase II

Handling of inequality constraints is a tricky issue. We are interested in only those inequality constraints that are active at the solution, *i.e.*, are satisfied by $a_i^T x = b_i$ relationship. Since, the set of constraints active at the solution (active set) is not known beforehand, each iteration of the LP has to refine a guess of active constraints. Let the set $\mathcal{A}(x^k)$ represent the set of active constraints at k^{th} iteration, *i.e.*,

$$a_i^T x = b_i \ \forall \ i \in \mathcal{A}(x^k)$$

We presume then it contains all linearly independent constraints. As the solution of the LP problem always lies at a vertex, active set $\mathcal{A}(x^k)$ for LP problem consists of n linearly independent constraints. The Lagrangian function can be defined as follows:

$$\mathcal{L}(x^k, \mu^k) = c^T x + \mu^{k^T} (\bar{A}^{k^T} x - \bar{b}^k) \tag{7.3}$$

where $\bar{A}^{k^T} = [a_{k_1}, a_{k_2}, \ldots, a_{k_n}]^T$; $k_i \in \mathcal{A}(x^k)$ and $\bar{b}^k = [b_{k_1}, b_{k_2}, \ldots, b_{k_n}]^T$. By the KKT optimality conditions, at the optimal point, following conditions must be satisfied:

$$\mu^k = \bar{A}^{k^{-1}} c \geq 0, \tag{7.4}$$

$$\text{and } \bar{A}^{k^T} x^k = \bar{b}^k \tag{7.5}$$

$$a_i^T x^k \leq b_i^k \ \forall \ i \notin \mathcal{A}(x^k) \tag{7.6}$$

If (x^k, μ^k) satisfies equation (7.4-7.6), the LP problem is solved. Otherwise, there exists some $\mu_i^k < 0$. Recall that μ_i measures sensitivity ($\mu_i = -\frac{df}{db_i}$)of the objective function to the constraint value. Thus, if a constraint with $\mu_i < 0$ is relaxed, *i.e.*, $a_i^T x$ allowed to decrease from b_i, the objective function reduces. Consequently, we must relax such a constraint, *i.e.*, remove it from the active set. If there are more than one constraint with $\mu_i < 0$ either the constraint with most negative μ_i or with the least index is selected for deletion. Let the qth constraint be selected for the deletion from the active set. Next, we must generate a feasible descent direction to move away from the current vertex towards a vertex with lower cost. Such a direction should satisfy the following conditions.

$$a_i^T d = 0 \ \forall \ i \neq q \text{ and } i \in \mathcal{A}(x^k); \quad a_q^T d < 0$$

Let the columns of $\bar{A}^{k^{-T}}$ be given by $[s_1, s_2, \ldots, s_n]$ where the superscript $-T$ refers to inverse of a transpose matrix. Then, from the equation $\bar{A}^T\bar{A}^{-T} = I_n$, it can be seen that direction $d = -s_q$ satisfies all the above conditions. For a constraint $i \notin \mathcal{A}(x^k)$, it can be seen that

$$a_i^T(x^k + \alpha d) = a_i^T(x^k - \alpha s_q) = a_i^T x^k - \alpha a_i^T s_q$$

where, scalar α has to be greater than zero to reduce the cost function. Hence, if $a_i^T s_q \geq 0$, then such a constraint will not become active; only those constraint can get into active set which have $a_i^T s_q < 0$. For a constraint i to be active, the step required to reach it is given by $\alpha_i = \frac{a_i^T x^k - b_i}{a_i^T s_q}$. Since, we are interested in the first constraint that will become active, scalar α is selected as follows:

$$\alpha^k = \min_i \frac{a_i^T x^k - b_i}{a_i^T s_q} \; \forall \; a_i^T s_q < 0 \text{ and } i \notin \mathcal{A}(x^k).$$

If there exists no i such that $a_i^T s_q < 0$; then the problem has an unbounded solution. Next the vector x is updated: $x^{k+1} = x^k - \alpha^k s_q$. Let the index of the new active constraint be given by p. The new active set is obtained by replacing q by p in the active set. Only one column in the active set is changed at each iteration (rank one update of the matrix $\bar{A}$), therefore, $\bar{A}^{-1}$ need not be computed afresh. At the new point x^{k+1}, again the μ^{k+1} has to worked out from equation (7.4). In a nonsparse implementation, $\bar{A}^{-1}$ can be explicitly stored and updated by Sherman Morrison formula. In a sparse implementation, μ can be computed by forward backward substitution of the sparse factorization of $\bar{A}$. If all μ_is are greater than zero, the algorithm terminates. Otherwise, the same process of replacing the constraint with one having most negative Lagrangian multiplier has to be repeated. If explicit equality constraints are modeled in LP, then, the most negative μ_i should be selected from the following set $\mathcal{A} \cap \mathcal{I}$ because only an inequality constraint can be relaxed. We now discuss the Sherman Morrison Formula.

1.1.1 Sherman Morrison Formula

We discuss the solution of following problem $[A + \Delta A]X = b$ where A^{-1} of nonsparse matrix A is available. This matrix inverse modification formula for rank one changes was developed by [Sherman and Morrison, 1949] and later generalized by [Woodbury, 1950]. It is discussed in detail in [Duff et al., 1986]. Let V and W be $n \times k$ matrices and S be a $k \times k$ matrix. Then,

$$\left[A + VSW^T\right]^{-1} = A^{-1} - A^{-1}V\left[S^{-1} + W^T A^{-1} V\right]^{-1} W^T A^{-1}$$

In this formula[1], VSW^T is a rank-k matrix. Usually, k is much less than n. Working with explicit inverse is not the right way for sparse matrices. However, we may multiply the right hand vector b to obtain the following relation.

$$\left[A + VSW^T\right]^{-1} b = A^{-1}b - A^{-1}V\left[S^{-1} + W^T A^{-1} V\right]^{-1} W^T A^{-1} b$$

The steps involved in sparse implementation are as follows:

1 Solve $Ay = b$ [computes $A^{-1}b$].

2 FB substitution $A\bar{v}_i = v_i \; \forall i = 1, \ldots, k$ [computes $A^{-1}V$]

3 Sparse matrix vector procedure $W^T[\bar{v}_1, \ldots, \bar{v}_k]$ [computes $W^T A^{-1} V$]

4 Compute explicitly inverse of $Z = [S^{-1} + W^T A^{-1} V]^{-1}$

5 $y^1 = W^T y$ [$y^1 = W^T A^{-1} b$]

6 $y^2 = Zy^1$ [$y^2 = [S^{-1} + W^T A^{-1} V]^{-1} W^T A^{-1} b$]

7 $y^3 = Vy^2$ [$y^3 = V[S^{-1} + W^T A^{-1} V]^{-1} W^T A^{-1} b$]

8 Solve $Ay^4 = y^3$ [$y^4 = A^{-1}V[S^{-1} + W^T A^{-1} V]^{-1} W^T A^{-1} b$]

9 $y^5 = y - y^4$ [y^5 is the solution.]

For simple case of a rank-1 update, the formula can be represented as follows:

$$\left(A + vw^T\right)^{-1} = A^{-1} - \frac{A^{-1} v w^T A^{-1}}{1 + w^T A^{-1} v}$$

where $w^T A^{-1} v \neq -1$. Thus, the total number of computation falls down significantly.

1.1.2 Low Rank Modification

Rank one modification can take place in either of the following two ways: (1) an entry a_{ij} is replaced with a new entry or (2) an entire column is replaced. In general, if ΔA corresponds to modifying columns $i_1, i_2, \ldots i_k$ of A by $v_1, v_2 \ldots v_k$, then ΔA may be represented as

$$\Delta A = [v_1, v_2 \ldots v_k]\,[I_k] \begin{bmatrix} e_{i_1}^T \\ \vdots \\ e_{i_k}^T \end{bmatrix}$$

[1]This formula may be easily verified and is left as a review question.

where I_k is the $k \times k$ identity matrix. e_i is a column vector with 1 at the ith location and zero elsewhere. Thus, $V = [v_1, \ldots, v_k], S = I_k, W = [e_{i_1}, \ldots, e_{i_k}]$

For sparse matrices, one may update the LU factors. Updating or modifying matrix factorizations is discussed later in a section on Schurs complement. We now discuss the process of computing a feasible vertex.

1.2. Phase I

To begin with, an initial guess is made for x^1. For example, x can be initialized by setting all x_is to the upper limit. Then, the matrix $\bar{A}^{1^T} = I$. With a choice of x^1, the set of violated constraints are listed out. The process of obtaining the feasible vertex works as follows:

Set k to 1. Let vio^k represents the set of violated constraints at the k_th iteration of phase I. At the k_th iterate of the phase I, following objective function is used with phase-II procedure of LP:

$$\min_x \sum_i (a_i^T x - b_i) \; \forall i \in vio^k \tag{7.7}$$

subject to rest of the linear constraints $a_i^T x \leq b$.
The least possible value of the above objective is zero. If the optimal cost is zero, then a feasible vertex is reached and we can proceed to phase-II, wherein the only change required is replacement of the objective function by the actual $c^T x$ objective.

We now discuss the QP solver which is the heart of SQP approach.

2. Inequality Constrained QP (ICQP)

Consider the following minimization problem.

$$\begin{aligned} \text{Problem ICQP} \quad & \min c^T x + \tfrac{1}{2} x^T H x && (7.8) \\ \text{and} \quad & a_i^T x = b_i; \; i = 1, \ldots m_1; && (7.9) \\ \text{such that} \quad & a_i^T x \leq b_i; \; i = m_1 + 1, \ldots m_2 && (7.10) \end{aligned}$$

We presume that the matrix H is spd, *i.e.*, necessary precautions of handling indefinite Hessian have been taken care of. Again the solution methodology involves two phases. In the first phase, a feasible solution to the set of constraints given by (7.9-7.10) is found out. Let this feasible solution be given by x^1. Phase I of the LP can be used to compute the feasible solution. Solution of the QP problem need not lie at a vertex. On the other hand Phase I of LP finds a feasible vertex. This may entail extra work and other alternatives which compute a feasible solution (not necessarily a vertex) can be used. With the assumption that a feasible solution has been found out, we proceed to discuss the phase-II which finds the optimal solution.

2.1. Phase-II

The solution strategy described is based upon the active set strategy outlined in the LP problem. If the active set at optimal was known *apriori*, an equality constrained problem like (6.24) could have been solved.

$$\begin{bmatrix} H & \bar{A} \\ \bar{A}^T & 0 \end{bmatrix} \begin{bmatrix} x \\ \mu \end{bmatrix} = \begin{bmatrix} -c \\ \bar{b} \end{bmatrix} \tag{7.11}$$

where $\bar{A}^T x = \bar{b}$ are the binding constraints at the optimal. Unfortunately, it is not possible to predict this set beforehand and therefore, we have to iterate with each equality QP subproblem towards a more appropriate active set. The feasible point x^1 and the constraints active at this point, *i.e.*, active set $\mathcal{A}(x^1)$, provide the initialization of this phase. Iteration count k is set to 1. At the k-th step of the problem ICQP, an equality constrained QP problem is solved for the descent direction d^k with active set $\mathcal{A}(x^k)$. Then the solution point is refined from x^k to x^{k+1} by a line search and so is active set from $\mathcal{A}(x^k)$ to $\mathcal{A}(x^{k+1})$. The objective function is minimized along this direction d^k, until a constraint becomes active. This increases the number of active constraint by 1. If the step length taken is 1; then it may imply that no new constraint is added to active set. At this step, if all μ_is given by (7.11) for the active inequality constraints are greater than or equal to zero, the KKT conditions are satisfied and the problem terminates. Otherwise, there exist some $\mu_i < 0$ among the inequality constraints; hence such a constraint can be relaxed. Note that solution of QP need not lie at a vertex, *i.e.*, number of active constraints can be less than n. Also when n constraints become active at any step (as in case of LP), no additional constraint can be added without relaxing an active constraint.

The equality constrained subproblem can either be solved by applying KKT condition as discussed in equation (6.24) or by projecting the QP subproblem on to the null space of constraints in $\mathcal{A}^k$. This null space approach is now discussed.

2.2. Null Space Approach

The null space of a matrix $\bar{A}^{k^T}$ of active constraints is the set of all vectors x such that $\bar{A}^{k^T} x = 0$. If matrix $\bar{A}^{k^T}$ is square and has a full rank, then its null space is zero and the set of equations given by $\bar{A}^{k^T} x = \bar{b}^k$ has a unique solution. When the number of rows (m^k) in a $m^k \times n$ matrix are less than the number of columns (n) then, the matrix has a non-zero null space. A set of vectors $Z^k = \{z_1, \ldots, z_{m_k}\}$ spans the null space, if any vector in the null space can be written as a linear combination of the vectors of the set Z^k. If all vectors in the set Z^k are linearly independent, then such a minimum set which spans the null space is called a *basis* for the null space. For notational simplicity, we also represent by Z^k the following matrix $[z_1, \ldots, z_{m_k}]$. If a linear system $\bar{A}^{k^T} x = \bar{b}^k$ has a solution x^k then, the set $\{x^k + Z^k y, y \in \Re^{m_k}\}$ represents the solution space of the equations.

Thus, the subspace of feasible solutions is given by a set of all vectors x such that

$$x = x^k + Z^k y^k \tag{7.12}$$

By substituting equation (7.12) in equation (7.8), we formulate an equivalent unconstrained minimization problem as follows:

$$\min_{y^k} c^T x^k + \frac{1}{2} x^{k^T} H x^k + c^T Z^k y^k + x^{k^T} H Z^k y^k + \frac{1}{2} y^{k^T} Z^{k^T} H Z^k y^k \tag{7.13}$$

The solution y^k to this problem is given by following equation:

$$[Z^{k^T} H Z^k] y^k = -Z^{k^T} [c + H x^k] \tag{7.14}$$

Now a step (α^k) along direction $p^k = Z^k y^k$ is taken until the nearest constraint is reached. If α^k is less than unity, then, a new constraint is included in the working set at the next iteration. Distance to the constraint is given by the following equation:

$$\alpha^k = \min_i \left\{ 1, \frac{(b_i - a_i^T x^k)}{a_i^T p^k} \right\} \text{ and } x^{k+1} = x^k + \alpha^k Z^k y^k$$

where the index i represents constraints outside of the working set such that $a_i^T p^k > 0$. If $\alpha^k = 1$; $x^{k+1} = x^k + Z^k y^k$ is feasible, then, no additional constraint may be added to active set. Therefore, one has to evaluate if any of the constraint(s) should be removed from the working set. For this, it is necessary to estimate the Lagrangian multipliers μ^{k+1} at the point x^{k+1}. Let $\bar{A}^{k+1^T}$ denote the submatrix corresponding to the active set $\mathcal{A}(x^{k+1})$. Then, the Lagrangian multipliers can be evaluated by solving the following systems of equations:

$$[\bar{A}^{k+1}] \mu^{k+1} = -[c + H x^{k+1}] \tag{7.15}$$

Because the linear system of equations (7.15) may not be consistent, it is preferable to solve these equations in a least squares sense, *e.g.*, by QR decomposition. If for an inequality constraint $\mu_i < 0$, then it can be removed from the active set $\mathcal{A}^{k+1}$. When $\mathcal{A}^k \neq \mathcal{A}^{k+1}$, then a new basis for null space Z^{k+1} is computed and the process is repeated. However, if $\alpha = 1$ and $\mathcal{A}^k = \mathcal{A}^{k+1}$, *i.e.*, no active inequality constraint can be removed from the active set, then we conclude that optimal to ICQP problem is reached.

This approach is also known as a projection method because of the role of matrix Z^k in projecting the solution to the null space of the active set. Because, at each iteration, a set of active constraints is maintained, such a strategy is also known the *active set strategy.*

From the point of view of computational efficiency, number of changes in $\mathcal{A}^k$ at each step may be restricted. For example, if a constraint is added to $\mathcal{A}^k$, then the decision to delete a constraint may be postponed to next iteration. However, when n constraints exist in the active set, then a solution of QP is decided by constraints only as $Z^k = \{\phi\}$. Thus, when either n constraints become active or $\mathcal{A}(x^k) = \mathcal{A}(x^{k+1})$, Lagrangian multipliers must be estimated to remove inequality constraints from active set. Only one constraint may be deleted from the active set at a time. Such techniques, permit use of existing factorization through updates rather than afresh factorization.

Clearly, null space approach is an attractive proposition when the degrees of freedom (number of variables–number of active constraints) is not very large, i.e., the dimension of LSS in (7.14) is small. Even if H is large and sparse, the projected Hessian $Z^T HZ$ will not be sparse. The computational effort in evaluating the projected Hessian depends upon the choice of Z. We now discuss various representation of a null space.

2.2.1 Null Space Representation

In what follows, we drop superscript k to simplify notations. One way compute the null space basis is by sparse QR decomposition of matrix $\bar{A}$. This is equivalent to LQ factorization of $\bar{A}^T$, *i.e.*,

$$\bar{A} = Q\begin{bmatrix} R \\ 0 \end{bmatrix} \Rightarrow \bar{A}^T Q = [L, 0] \Rightarrow \bar{A}^T q_i = 0 \;\forall\; i = m+1, \cdots, n$$

where $\bar{A}^T$ is a $m \times n$ matrix, $m < n$ and the matrix Q is nonsparse having a compressed representation of it in terms of Givens rotation pair (θ, i, j). If columns of Q are partitioned on similar lines to $[L, 0]$ matrix, then the later columns provide the null space basis Z. Computed explicitly, such a basis will be nonsparse.

An alternative choice which is popularly used in the SQP approach is based upon partitioning matrix $\bar{A}^T$ as $[B, S]$ where B is full rank square submatrix of $\bar{A}^T$. The variables associated with partition B are known as *basic* variables and those associated with partition S are known as *superbasic* variables. Later on we will also introduce *nonbasic* variables. Then, the following represents the null space basis of matrix C.

$$Z = \begin{bmatrix} -B^{-1}S \\ I_{n-m} \end{bmatrix}$$

With this choice of Z, the reduced gradient $Z^T[c + Hx]$ is given by

$$\begin{bmatrix} -S^TB^{-T} & I_{n-m} \end{bmatrix} \begin{bmatrix} c + Hx \end{bmatrix} = -S^TB^{-T} \begin{bmatrix} c_1 + (Hx)_1 \\ \vdots \\ c_m + (Hx)_m \end{bmatrix} + \begin{bmatrix} c_{m+1} + (Hx)_{m+1} \\ \vdots \\ c_n + (Hx)_n \end{bmatrix}$$

To compute the reduced gradient, following linear system of equation is solved.

$$\text{If } B^T\pi = \begin{bmatrix} c_1 + (Hx)_1 \\ c_m + (Hx)_m \end{bmatrix}$$

$$\text{Then, } Z^T(c + Hx) = \begin{bmatrix} c_{m+1} + (Hx)_{m+1} \\ \vdots \\ c_n + (Hx)_n \end{bmatrix} - S^T\pi.$$

The search direction is given by

$$p = Zy = \begin{bmatrix} -B^{-1}Sy \\ y \end{bmatrix}$$

Again, the first m components of d can be obtained by solving the following sparse LSS.

$$[B] \begin{pmatrix} p_1 \\ \vdots \\ p_m \end{pmatrix} = -Sy$$

$$\text{and } p_{m+i} = y_i \text{ for } i = 1, \ldots, n - m \tag{7.16}$$

With this choice of Z, explicit computation of null space basis is not required. Only sparse LU factors of matrix B need to be stored.

2.3. Updating the Null Space Basis

Whenever, a constraint is added to active set, the dimension of null space reduces by one. Similarly if a constraint is deleted from active set, dimension of null space increases by one. Thus, the null space basis (or its representation) has to be updated whenever a constraint is added or removed from the active set. Such strategies within sparse matrix computation framework are now discussed. We first, consider the problem of addition of a bound constraint.

2.3.1 Addition of a Bound

When a bound limit becomes active (*e.g.*, bus voltage magnitude or angle in OPF), it can be enforced by simply freezing the corresponding variable. This induces the following partitioning on matrix $\bar{A}^T$

$$\bar{A}^T = \begin{bmatrix} B & S & N \\ & & I_{nb} \end{bmatrix} \tag{7.17}$$

B represents the dependent or basic variables, S the superbasic or free variables and N the frozen or nonbasic variables. The subblock I_{nb} represents the identity matrix for bounds. If only one variable is at bound, I_{nb} can be replaced by 1. With nb variables at bound, the null space representation is given by the following equation.

$$Z = \begin{bmatrix} -B^{-1}S \\ I_{n-m-nb} \\ 0_{nb} \end{bmatrix} \tag{7.18}$$

With null space basis representation (7.18), the QP search direction is given by the following equation.

$$p = Zy = \begin{bmatrix} -B^{-1}Sy \\ I_{n-m-nb}y \\ 0_{nb} \end{bmatrix} \tag{7.19}$$

From equation (7.19), it can be seen that, in the search direction p, the nonbasic variables are not allowed to change.

Initially, to minimize the complexity of discussion, we had assumed that there are no columns in N. Before a bound constraint becomes active, the bound variable will appear in either partition B or S. If the nonbasic variable belongs to partition S, then, the updating required is in only the partitioning of S and N; this does not entail any change in the LU factorization of matrix B. However, when a variable to be frozen belongs to B, then its column has to be moved to N and hence, a new variable (linearly independent column) from S has to be moved to B. Thus, one column of B is changed and the factorization of B has to be updated. Various ways of updating matrix factorization are discussed in [Gill et al., 1984b]. Sherman Morrison formula can be used to update solution vector without refactorizing B. Another alternative is the use of Schur complement.

2.3.2 Use of Schur Complement

Suppose that at each update j, a vector v_j replaces the l_jth column of B_0. Thus, at the kth update, we have matrix B_k where k-columns of

B_0 have been replaced. A key observation is that the system $B_k x = b$ is equivalent to the following linear system of equations:

$$\begin{bmatrix} B_0 & V_k \\ \bar{I}_k & \end{bmatrix} \begin{bmatrix} y \\ z \end{bmatrix} = \begin{bmatrix} b \\ 0 \end{bmatrix} \tag{7.20}$$

where

$$V_k = [v_1, v_2 \ldots v_k] \text{ and } \bar{I}_k = [e_{l_1}, e_{l_2} \ldots e_{l_k}]^T$$

Note that the rectangular matrix $\bar{I}_k$ is composed of k rows of the identity matrix corresponding to indices of columns that have been replaced. Since the equation $[\bar{I}_k]y = 0$ sets k elements of y to zero, the remaining elements of y and z together give the required solution x.

The matrix in the above equation may be factorized in several different ways. A simple form of the factorization is given by the following equation:

$$\begin{bmatrix} B_0 & V_k \\ \bar{I}_k & \end{bmatrix} = \begin{bmatrix} B_0 & \\ \bar{I}_k & C_k \end{bmatrix} \begin{bmatrix} I & Y_k \\ & I \end{bmatrix}$$

The $k \times k$ matrix C_k is the Schur complement for the partitioned matrix. The matrix Y_k can be computed by k forward backward substitution

$$B_0 Y_k = V_k \tag{7.21}$$

Then $C_k = -\bar{I}_k Y_k$ which need not be equal to $-Y_k$. Note that C_k is a $k \times k$ matrix, B_0 is a $m \times m$ matrix, V_k and Y_k are $m \times k$ matrices. In this factorization, the matrix $\begin{bmatrix} B_0 & \\ \bar{I}_k & C_k \end{bmatrix}$ is block lower triangular matrix while $\begin{bmatrix} I & Y_k \\ & I \end{bmatrix}$ is a upper triangular matrix. The scheme of solving the augmented system is as follows;

Solve $B_0 \bar{y} = b$;

Solve $C_k \bar{z} = -\bar{I}_k \bar{y}$.

Solve $\begin{bmatrix} I & Y_k \\ & I \end{bmatrix} \begin{bmatrix} y \\ z \end{bmatrix} = \begin{bmatrix} \bar{y} \\ \bar{z} \end{bmatrix}$ (A simple back substitution)

When k is not large, the effort in solving the updated system is less than a new factorization of B_k. Next, we consider the problem of adding a general linear inequality constraint to the active set which in OPF may correspond to linearized Q injection limit of a generator a line flow limit.

2.3.3 Addition of a General Constraint

When an inequality constraint reaches its limit, it has to be added in the active set. Such an addition will entail, addition of an equation to the equality constraint matrix. For such a constraint that does not enjoy the simplistic structure of a bound, the basis matrix B has to be appended with an additional row and a column. Consequently, one of the superbasic variable becomes basic and one degree of freedom is reduced. Let the new $(m+1) \times (m+1)$ basis matrix be given by B_1 and the submatrix for superbasic variables by S_1. The partition for nonbasic variables remains unchanged N. Hence, the new null space basis is given by

$$Z = \begin{bmatrix} -B_1^{-1}S_1 \\ I_{n-m-nb-1} \\ 0_{nb} \end{bmatrix} \tag{7.22}$$

It is now necessary to consider strategies to update factorization of the matrix B to account for the additional row and column.

Let the augmented B, *i.e.*, B_1 be represented by $\begin{bmatrix} B & c_1 \\ b_1^T & s \end{bmatrix}$. Let LU decomposition of the original B where U is an unit upper triangular matrix be given by $B = LU$. Then, we wish to factorize augmented B as follows.

$$\begin{bmatrix} B & c_1 \\ b_1^T & s \end{bmatrix} = \begin{bmatrix} L & \\ l_1^T & \gamma \end{bmatrix} \begin{bmatrix} U & u_1 \\ & 1 \end{bmatrix} \tag{7.23}$$

It can be verified that columns l_1, u_1 and diagonal γ can be computed by following two forward backward substitutions.

$$Lu_1 = c_1 \tag{7.24}$$

$$U^T l_1 = b_1 \tag{7.25}$$

$$\gamma = s - l_1^T u_1 \tag{7.26}$$

If the scalar γ is numerically close to zero, then the column may be nearly linearly dependent and an alternate choice of (c_1, s) may be considered.

2.3.4 Removal of a Constraint

If we simply keep on adding equality constraints to the active set, it is likely that the active set will grow and have n linear independent constraints in it. At this stage, the solution to QP problem is completely defined by the constraints alone. In fact, it corresponds to the vertex. While the solution of a LP problem always lies at a vertex, it is not necessary for a QP problem. Thus, as in case of LP, we should evaluate the Lagrangian multipliers to check if any constraint can be removed from the active set. If all $\mu_i \geq 0$, then the current solution is the optimal solution. Otherwise, the constraint with most negative μ_i of an inequality constraint can be deleted from active set. Removal of a constraint adds an additional degree of freedom.

Another possible scenario is that a step length of 1 in minor iteration of QP problem is accepted and no additional constraint is added and $m < n$. In such a case either the new estimate corresponds to the optimal or a constraint can be removed from the active set.

The Lagrangian multipliers can be computed by solving the following equations.

$$\bar{A}^{k^T} \mu = -[Hx^k + c]$$

where $\bar{A}^k$ is the constraint matrix. If the number of active constraints is less than the dimension of the problem ($m < n$) then, it is appropriate to frame the above problem as a linear least squares problem. Sparse QR decomposition can be used to solve such a problem. Solving linear least squares problem has been discussed in detail in the chapter on power system state estimation. If multiple inequality constraints have negative multipliers, then the one with the largest magnitude or least index should be removed from the active set.

If a general constraint is deleted from the active set, not much work is involved as the factorization of matrix B_0 is available. Similarly, when a bound constraint is freed, *i.e.*, removed from the active set, corresponding column is moved from the submatrix N to the submatrix S. The degree of freedom increases by one and again the basis matrix is not changed. Hence, no update in the B matrix factorization is needed.

2.3.5 Convergence Criteria for the QP problem

The QP problem converges when (1) no constraint has to be added to the active set and (2) no constraint can be deleted from the QP problem, *i.e.*, Lagrangian multiplier for all inequality constraint $\mu_i \geq 0$ and (3) reduced Hessian $Z^T HZ$ is spd. If H itself is spd then, this condition is automatically satisfied.

To conclude, active set methods are used in LP and QP solvers. Both these solvers require a two phase implementation. In the first phase, a feasible solution is computed while the second phase computes the optimal. The QP can also be solved by null space approach. The importance of these solvers is because they do the bulk of computation in SLP and SQP methods. Their application to OPF will be discussed in chapter 11.

Chapter 8

LOAD FLOW ANALYSIS

If I do not hurry my pen, this chapter will never be brought to a close.
—Paul Brunton, A Search in Secret India

1. Introduction

The starting point of any analysis of power system will be the computation of complex voltages at all the busses. Load flow analysis is a computational tool for this purpose.

An electrical circuit is governed by Kirchhoff Current Law (KCL) and Kirchhoff Voltage Law (KVL). If a RLC circuit is excited only by voltage sources, the loop equations (KVL) can be used to determine all the voltages and currents in the network. If the same circuit is excited by current sources, node equations (KCL) can be used to determine the voltages and currents in the circuit. If the excitation is sinusoidal in nature, then, the steady state solution can be obtained by solving algebraic equations.

An electrical network like a power system can be conveniently represented by a graph. If the network is analyzed by applying loop equations, then the choice of loops must correspond to the set of fundamental loops[1]. This approach is not very attractive from a computational view point, especially, when the network is large and has many non-trivial loops. The reasons are obvious; to apply KVL, we need to identify the fundamental loops in the graph. A set of fundamental loops is the smallest set of loops such that, any other non-trivial loop in the graph is a

[1]A radial system has loops only through the ground.

linear combination (modulo-2 sum) of the fundamental loops. To identify fundamental loops in a graph of n nodes with b branches, we need to identify a tree and it's links. The number of links are given by $(b-n+1)$ and so are the fundamental loops. Thus, to develop a computer aided tool based on loop equations we need (1) a procedure to identify a tree and the links, and (2) a procedure to construct the fundamental loops. All this amounts to a considerable computational work.

In contrast, from the computational view point, the application of KCL is much more attractive. If the network has n nodes other than the reference node, then the application of KCL straightaway results in n complex equations. In a network like a power system, almost all the voltages are applied with respect to ground. Hence, if there are g voltage sources, *i.e.*, g generators, then g node voltages are known. Thus, we are left with $n-g$ complex current injection equations in $n-g$ unknowns.

If the sinusoidal voltages are completely specified (magnitude and angle) and the network branches are composed of only linear elements, (*e.g.*, R,L,C) then, by solving a linear system of equations, the steady state solution can be obtained in one step. However, a power system network solver has to deal with nonlinearity because of the way in which node conditions are specified.

Typically, the electrical loads are not specified as shunt impedances, rather, they are specified by their MW and MVAR consumptions. It is difficult to specify angles of sinusoidal voltage sources (why?). Instead of specifying angle, MW injection is specified. Since, the complex power injection at i^{th} bus is given by the product $S_i = V_i I_i^*$, nonlinearity is introduced. Therefore, iterative solution in phasor domain is required.

Before proceeding to develop a KCL based circuit solver for power system network, an additional issue has to be considered, *viz.* the networks that we are dealing with are three phase networks. In this chapter, we restrict ourselves to a balanced system for which a positive sequence network analysis suffices. In the rest of the chapter, all quantities are considered in pu.

2. Load Flow Analysis Problem

Problem Definition: Given a balanced three phase network, impedance models of the network elements like transmission lines, transformers, reactors etc. and steady state models for loads, generators and any other dynamical/control elements, compute the complex steady state voltages at all the busses in the system, when the terminal conditions are specified in the following format:

1 Active power injection (P) and terminal voltage magnitude (V) at the rest of the nodes known as generator nodes or PV nodes. Phasor voltage angle and reactive power delivered by the generator at PV bus are the unknowns.

2 Active power (P) and reactive power (Q) loads at rest of nodes known as load busses or PQ busses. Voltage magnitude and angle at a PQ bus are the unknowns.

In a n node system, apparently there are $2n$ states in the system (n voltage magnitudes and n voltage angles). If there are g PV nodes then, there exist $(n - g)$ PQ nodes. In phasor analysis, one phasor (in this case a node voltage) should be chosen as a reference phasor, *i.e.*, angle of anyone of the bus can be fixed to zero arbitrarily. Usually, a PV node is chosen as a reference node. Therefore, the number of states reduce to $2n - 1$. Thus, if $2n$ quantities are specified in a network, then the system is either overdetermined or inconsistent. In a power system, total generation must be equal to total load and network losses, a quantity that can be computed *a posteriori* only. Therefore, all the generation can not be fixed *apriori*. Usually, generation at the reference bus is not fixed, so that it can account for both the losses and deficit generation. Such a bus is known as *slack* or *swing* bus.

The next section reviews steady state models for network elements like lines, transformer etc.

3. Modeling of Network Elements

A power system network is composed of transmission lines (cables), transformers, reactors, loads and generators. Modeling of these devices in Object Oriented (OO) framework is elaborated in the following sections.

3.1. Transmission Lines Modeling

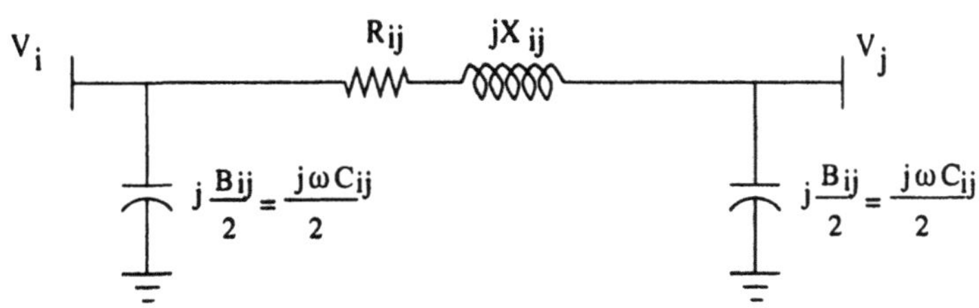

Figure 8.1. Modeling of Transmission Line

A transmission line is represented by an equivalent π circuit with a series impedance $(R + jX)$ from node i to node j and shunt capacitors at nodes i and j (Figure 8.1). In a short line of low voltage (e.g, up to 80 km), the shunt charging can be neglected. In a medium length high voltage line (e.g., up to 200 km), nominal shunt charging is divided equally at both the ends. In case of long EHV lines, a more exact model is used to accurately account for the distributed capacitance of the line. A transmission line can also be modeled by T-network. However, it creates new nodes and hence is not preferred.

In an OO system, the π model of line is mapped onto an *object* which captures steady state behavior of a line. The interface of class *branch_line* that implements a transmission line model is shown below:

```
struct branch_line {
int f_node, t_node,no_of_ckts; //node nos.
float r,x,by2, fp, fq, bp, bq, r_zero,x_zero,by2_zero,
length,kv_rat, mva_rat; //pi-model and flow attributes
char *from, *to; //bus names
branch_line():f_node(-1), t_node(-1), no_of_ckts(0),
r(INFINIT),x(INFINIT),by2(INFINIT),fp(0),fq(0),bp(0),bq(0),
r_zero(INFINIT), x_zero(INFINIT),by2_zero(INFINIT),
length(INFINIT), mva_rat(INFINIT) {}; //default state
friend ifstream & operator >> (ifstream & is,
                               branch_line & bl);
   // overload input operator
friend ostream & operator << (ostream & os,
                             branch_line & bl);
   //overloaded output operator
 };
```

Attributes like r, x, $by2$ are the primary attributes which implement the π model, while attributes for forward line flow like fp, fq are derived or secondary attributes which ease the programming. Attributes like r_{zero}, x_{zero} etc are used to model zero sequence quantities required in short circuit analysis. Short circuit analysis will be discussed in the next chapter. The end points of a line are specified by meaningful character string names using attributes $from$ and to respectively. For example `Mumbai400` will indicate a line of 400 kV class originating (or terminating) at Mumbai city. In the class *Network* (to be discussed later), these character strings are mapped onto nonnegative integers using an associative array. The integer node names are stored in *f_node* and *t_node* respectively. Thus, the end user is freed from the tedious task of assigning node numbers and tracking them later on.

It should be noted that the implementation has all attributes as *public* because any power system analysis algorithm e.g., YBUS building, needs to access entities like line resistance etc. Thus, it makes little sense to hide these details. The benefits accrued by an OO implementation for line object are as follows:

- Encapsulation of data: In this case, the entities like $r, x, by2$ have been grouped together. This could also have been achieved in a language like C also. However, the OO implementation permits association of functions to the data which is not possible in procedural implementation. Further, the default state of the object is also specified.
- Creation of an user defined data type.
- Ease in read and write operations for the programmer because of the availability of overloaded >> and << operators.

3.2. Transformer Modeling

A transformer inherently has a T model as shown in figure 8.2 (a). However, it can be approximated and by a π-model as shown in figure 8.2(b). If the magnetizing impedance $X_m \rightarrow \infty$, then the series branch is sufficient to model a transformer (see figure 8.2(c)).

3.2.1 Tap changing and Phase Shifting Transformers

Phase shifting transformers are used to control the flow across a transmission line by controlling the angle between the terminals. It can be modeled by considering a complex tap ratio of $e^{j\phi}$. For example, a three phase Y/Δ transformer can be thought of a phase shifting transformer where the phase shift is fixed to 30°. In a generalized scenario, we can have a tap which affects both the magnitude as well as angle. An admittance matrix model is now developed. The conventions are shown in figure 8.3. Note that an ideal magnitude and phase shifting transformer is connected in series with its series leakage impedance Z_{km} at nominal tap. Therefore,

$$\frac{V_k}{V_p} = a_{km}e^{j\phi_{km}} \tag{8.1}$$

As the ideal phase shifter can be considered as lossless;

$$V_k I_{km}^* = V_p(-I_{mk})^*$$

$$\text{i.e., } I_{mk} = -a_{km}e^{-j\phi_{km}} I_{km}$$

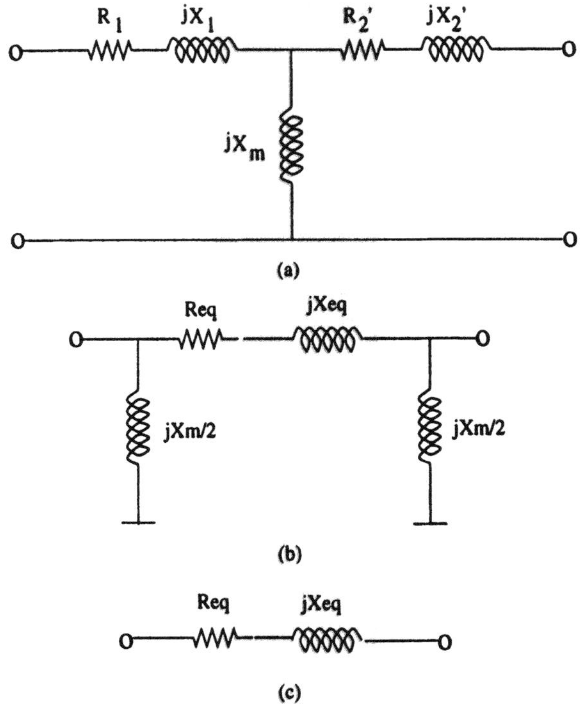

Figure 8.2. Transformer Model

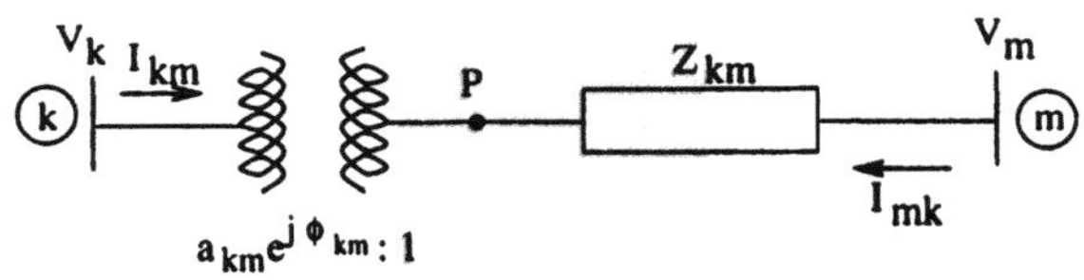

Figure 8.3. Phase Shifting Transformer

$$\text{Now, } I_{mk} = [V_m - V_p]y_{km} = [V_m - \frac{V_k}{a_{km}}e^{-j\phi_{km}}]y_{km}$$

In matrix notation, the above equations can be written as follows:

$$\begin{bmatrix} I_{km} \\ I_{mk} \end{bmatrix} = \begin{bmatrix} \frac{y_{km}}{a_{km}{}^2} & -\frac{y_{km}e^{j\phi_{km}}}{a_{km}} \\ -\frac{y_{km}}{a_{km}}e^{-j\phi_{km}} & y_{km} \end{bmatrix} \begin{bmatrix} V_k \\ V_m \end{bmatrix} \tag{8.2}$$

For a pure phase shifter $a_{km} = 1$, $\phi_{km} \neq 0$ while in case of in-phase tap changing transformers, $\phi_{km} = 0$, $a_{km} \neq 1$. When $\phi_{km} = 0$, the matrix in (8.2) is symmetric and the transformer can be modeled by a π-circuit.

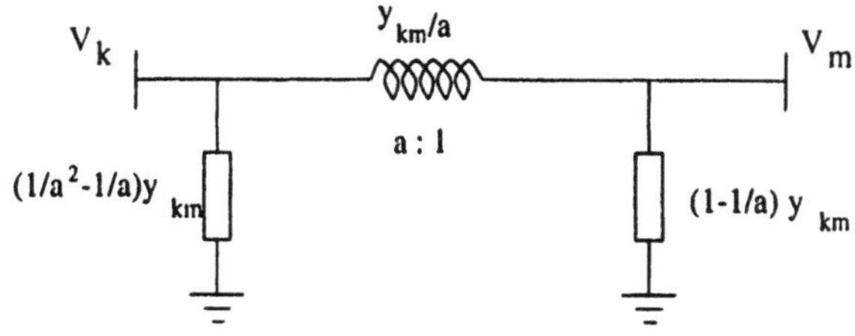

Figure 8.4. π-model of Tap Changing Transformer

3.2.2 OO modeling of Tap changing transformers

Under the assumption of $X_m = \infty$, a π model accounting for tap changer is shown in figure 8.4. If $a > 1$, *i.e.*, one wishes to boost the voltage V_k (assume V_m to be fixed by a voltage source); then, the model introduces a *fictitious* shunt capacitor at node k (assume $r << x$) because $(\frac{1}{a^2} - \frac{1}{a}) < 0$. Similarly, on the other side (m), an equivalent *fictitious* reactor is introduced because $(1 - \frac{1}{a}) > 0$. An edited version of the class that implements a transformer object is shown below. It can model[2] all the three sequence models that are used in short circuit analysis. Most of the attributes and functions are self explanatory.

```
 struct branch_X {
 int f_node, t_node,no_of_ckts;
// f_node is from node no assigned by associative array
 float r, x, r_zero,x_zero,by2_zero, mva_rat,tap,
 tap_min,tap_max, act_tap, gnd_impe1,gnd_impe2,
 kv_vol_prim, kv_vol_seco,kv_base_vol;
//gnd_impe1 is gnd impedance on primary side
 char *from, *to, *type_of_trans;
// from to character node names assigned by user
 cmplx<float> impe_PS,impe_PG,impe_SG;
 // three limbs of the pi-model are assigned values
 // depending on the sequence network being modeled.

 branch_X();  // ctor implementation similar to line
 // read operator
 friend istream & operator >>(istream &is,branch_X &x1);
 // write operator
 friend ostream & operator <<(ostream &os,branch_X &x1);};
```

[2]The terminology *modeling* is used here in the context of modeling a real life object (or even an abstract object) by a software object.

Let us now consider the modeling of three winding transformers.

3.2.3 Three Winding Transformer

Three winding transformers are used to transfer power between three different voltage levels. If H, M and L are the high, medium and low-voltage winding, then, manufactures provide leakage impedances between the winding as Z_{HM}, Z_{HL}, and Z_{ML}. While measuring impedance Z_{HM}, the low voltage winding L is kept open circuited. Impedances Z_{HL} and Z_{ML} are measured in a similar fashion. Since, the rating of the windings are different, caution has to be exercised, *i.e.*, all impedances should be converted to the system MVA base. An equivalent star network is used to model positive and negative sequences connections. It can be

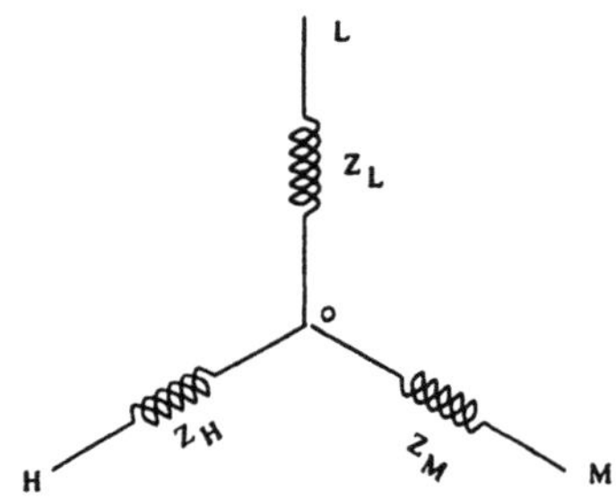

Figure 8.5. 3-Winding Transformer Positive Sequence Network

seen from the figure 8.5 that:

$$Z_H + Z_L = Z_{HL} \tag{8.3}$$

$$Z_L + Z_M = Z_{ML} \tag{8.4}$$

$$Z_M + Z_H = Z_{HM} \tag{8.5}$$

From the above equations, it follows that:

$$Z_H = \frac{1}{2}(Z_{HM} + Z_{HL} - Z_{ML}) \tag{8.6}$$

$$Z_M = \frac{1}{2}(Z_{HM} + Z_{ML} - Z_{HL}) \tag{8.7}$$

$$Z_L = \frac{1}{2}(Z_{HL} + Z_{ML} - Z_{HM}) \tag{8.8}$$

The modeling of tap on the H-winding of a 3 winding transformer is now discussed. From figure 8.6, we can write the following equations:

$$V_p = \frac{V_H}{a};$$

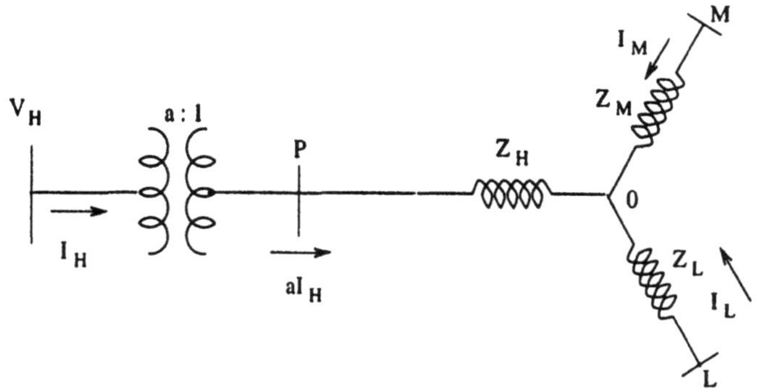

Figure 8.6. Three winding transformer with tap on H winding

$$
\begin{aligned}
aI_H &= \frac{1}{Z_H}\left[\frac{V_H}{a} - V_0\right]; \\
I_M &= \frac{V_M - V_0}{Z_M}; \\
I_L &= \frac{V_L - V_0}{Z_L}; \\
aI_H &+ I_L + I_M = 0
\end{aligned}
$$

In the matrix notation, this can be represented as follows:

$$
\begin{array}{cccc} H & M & L & 0 \end{array}
$$

$$
\begin{bmatrix}
\frac{1}{a^2 Z_H} & & & -\frac{1}{aZ_H} \\
 & \frac{1}{Z_M} & & -\frac{1}{Z_M} \\
 & & \frac{1}{Z_L} & -\frac{1}{Z_L} \\
\frac{1}{aZ_H} & \frac{1}{Z_M} & \frac{1}{Z_L} & -(\frac{1}{Z_H} + \frac{1}{Z_M} + \frac{1}{Z_L})
\end{bmatrix}
\begin{bmatrix} V_H \\ V_M \\ V_L \\ V_0 \end{bmatrix}
=
\begin{bmatrix} I_H \\ I_M \\ I_L \\ 0 \end{bmatrix}
\quad (8.9)
$$

Eliminating the node 0, leads to the following YBUS model of a 3 winding transformer with tap.

$$
\frac{1}{\Delta}
\begin{bmatrix}
Z_M + Z_L & -Z_L & -Z_M \\
-Z_L & Z_H + Z_L & -Z_H \\
-Z_M & -Z_H & Z_H + Z_M
\end{bmatrix}
\begin{bmatrix} \frac{V_H}{a} \\ V_M \\ V_L \end{bmatrix}
=
\begin{bmatrix} aI_H \\ I_M \\ I_L \end{bmatrix}
\quad (8.10)
$$

Where $\Delta = Z_H Z_M + Z_M Z_L + Z_L Z_H$. It can be verified that at a=1, the YBUS equations correspond to admittance matrix obtained by applying Y to Δ transformation.

3.3. Shunt Modeling

Shunt reactors are placed to control steady state over-voltages under light load conditions; they can be represented by the shunt R-X branch.

A shunt capacitor is used to boost a bus voltage in a stressed system. A shunt capacitor is also modeled by the same class. A C_{++} class definition for shunts is shown below.

```
struct shunt {
float r1, x1,r2,x2,r_zero,x_zero,kv_rat,mva_rat;
int bus_no; //bus no.
char *on_bus,*type; //bus name
shunt(); //ctor
friend  istream & operator >> ( istream & is, shunt &sh1);
friend  ostream & operator << ( ostream & os, shunt &sh1);
};
```

3.4. Generator or PV bus model

In the normal operating range, a voltage regulator fixes the terminal voltage of a generator bus. The steady state electrical power output is set by the mechanical input power. Hence, a generator bus is specified by it's active power injection (P) and terminal voltage magnitude (V). Any load specified at such a bus is accounted for by subtracting it from the specified P generation and the Q limits. The effect of reactive power capability curve, *i.e.*, Q limits will be considered later in section 11. The class definition for the PV node is shown below. A more detailed implementation is available in `struct.h` file.

```
struct NodePV {
char *on_bus,*type;
cmplx<float> volt, pwr_inj;
float  v_spec, v_min, v_max, q_min, q_max, p_min, p_max;
float  rated_mva;
int bus_no, PV_switch;
  //1 implies PV-bus,0 implies switched to PQ bus
NodePV(): bus_no(-1), volt(1), pwr_inj(0), v_spec(1.0),
v_min(0.9), v_max(1.05), q_min(-INFINIT), q_max(INFINIT),
PV_switch(1){};
friend  istream & operator >> (istream & is,NodePV & pv1);
friend  ostream & operator << (ostream & os,NodePV & pv1);
};
```

Note that voltage and power injection are modeled by complex data type. Various limits on voltages, P and Q generation etc have also been modeled in the class *NodePV*. A *PV_switch* is also provided for PV to PQ bus conversion. The constructor initializes the terminal voltage to 1.0 pu and angle to 0.0. Other limits are also assigned meaningful default values. This is another advantage of the object oriented approach;

whenever an object is created, it's state can also be initialized to a default value thereby, avoiding unpleasant surprises during programming (implementation).

4. Admittance Matrix Model

Since a network element is represented by its π equivalent circuit or by shunt branches, building a network model is relatively straight forward. Let y_{ij} represent the series admittance between nodes i and j. The shunts at a node-i corresponds to (1) line charging capacitances $(j\frac{B_{ij}}{2})$ and (2) shunt capacitors/reactors. Then, the equivalent current injection at node-i is equal to the sum of branch flow currents from i to j and the sum of the current in the shunt branches (Figure 8.7).

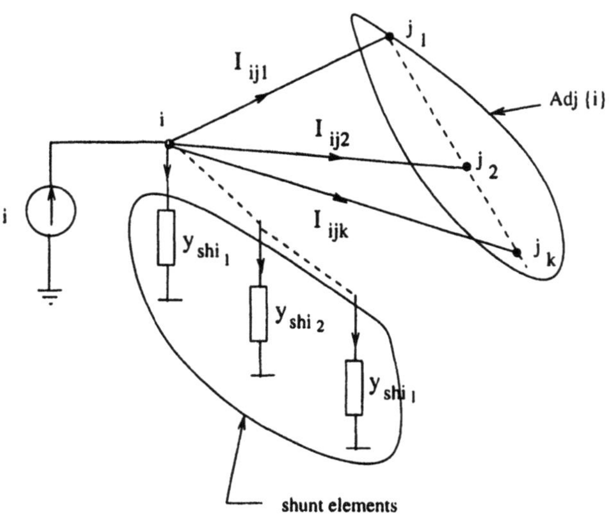

Figure 8.7. Current Injection at node i

For a node i, let $Adj(i) = \{j_1, j_2, \cdots, j_k\}$ be the set of adjacent nodes, and the number of shunts inclusive of line charging capacitors be l_i. Then, the following equations hold true:

$$I_i = y_{ij_1}(V_i - V_{j_1}) + y_{ij_2}(V_i - V_{j_2}) + \cdots y_{ij_k}(V_i - V_{j_k}) + \sum_{k=1}^{l} y_{sh_{i_k}} V_i$$
$$\text{or } I_i = \sum_{j\in Adj(i)} Y_{ij}V_j + Y_{ii}V_i$$
$$\text{where } Y_{ii} = \sum_{j\in Adj(i)} y_{ij} + \sum_{k=1}^{l} y_{sh_{i_k}}$$
$$\text{and } Y_{ij} = -y_{ij}$$

Thus, admittance matrix or YBUS model can be represented as follows:

$$\begin{bmatrix} I_1 \\ I_2 \\ \vdots \\ I_n \end{bmatrix} = \begin{bmatrix} Y_{11} & Y_{12} & \cdots & Y_{1n} \\ Y_{21} & Y_{22} & \cdots & Y_{2n} \\ \vdots & & \ddots & \vdots \\ Y_{n1} & Y_{n2} & \cdots & Y_{nn} \end{bmatrix} \begin{bmatrix} V_1 \\ V_2 \\ \vdots \\ V_n \end{bmatrix} \tag{8.11}$$

Typically, each node is directly connected to only a few nodes. Therefore, if nodes i and j are not adjacent, the element Y_{ij} is equal to 0, making matrix YBUS sparse. Due to the complex nature of admittance matrix, equation (8.11) can be written as follows:

$$[I_r + jI_m] = [G + jB]\,[V_r + jV_m] \text{ where, } [YBUS] = [G + jB]$$

In the absence of phase shifting transformers, the admittance matrix is symmetric. It can be stored using the class *SparseMatrix* of chapter 3; only, the upper triangular portion needs to be stored.

5. Interface Power Injection Equations

The power balance equation at a node i can be written as follows:

$$S_i^* = P_i - jQ_i = V_i^* \sum_{j=1}^{n} Y_{ij} V_j \tag{8.12}$$

In polar coordinates, let $Y_{ij} = |Y_{ij}|\angle\theta_{ij}$, $V_i = |V_i|\angle\delta_i$, $V_j = |V_j|\angle\delta_j$ Then, equation (8.12) leads to the following:

$$P_i = \sum_{j=1}^{n} |V_i||V_j||Y_{ij}|\cos(\delta_i - \delta_j - \theta_{ij}) \tag{8.13}$$

$$Q_i = \sum_{j=1}^{n} |V_i||V_j||Y_{ij}|\sin(\delta_i - \delta_j - \theta_{ij}) \tag{8.14}$$

Angles are specified in radians and all other quantities are specified in pu.

6. A Formal Mathematical Problem Formulation

Let the set of nodes $\{1, \cdots, g\}$ represent PV nodes and $\{g+1, \cdots, n\}$ represent the PQ nodes. Further, let the bus 1 be the slack bus. Then, the vector of unknowns is given by $[\delta_2 \cdots \delta_n, |V_{g+1}| \cdots |V_n|]$, *i.e.*, there are $2n - g - 1$ unknowns. For nodes belonging to the set $\{2, \cdots, n\}$, P injections are specified. These P injection $n-1$ equations are given by the following equation:

$$P_i^{sp} = \sum_{j=1}^{n} |V_i||V_j||Y_{ij}|\cos(\delta_i - \delta_j - \theta_{ij}) \text{ where } i \in \{2, \cdots, n\} \quad (8.15)$$

Superscript sp is used to indicate a specified value of P_i. For the $n - g$ PQ nodes belonging to set $\{g + 1, \cdots, n\}$, Q injection is also specified. Hence,

$$Q_i^{sp} = \sum_{j=1}^{n} |V_i||V_j||Y_{ij}|\sin(\delta_i - \delta_j - \theta_{ij}) \quad (8.16)$$

The load flow analysis problem amounts to solving $2n - g - 1$ nonlinear equations. It is this nonlinear nature of equations that has generated interest in designing fast and globally convergent (robust) algorithms. A nonlinear problem can have multiple solutions or no solution. Let us consider an illustrative example.

Example 8.1: A generator supplies power of 1 pu to an infinite bus through a transmission line of inductance jx. Let $|V_1| = |V_2| = 1.0$ pu and $x = 0.5$ pu. In this example, it makes sense to designate infinite bus (bus-2) as the slack bus and bus-1 as a PV bus. From figure 8.8, it is obvious that the problem has two solutions for δ. The solution, $V_1 = 1.0\angle 30^0$ represents a stable operating point while $V_1 = 1.0\angle 150^0$ represents an unstable equilibrium.

Nonlinear equations are routinely solved by iterative solvers. Since, example 8.1 has only one variable δ, it does not require an iterative solver. However, to understand the finer points of iterative solvers, we solve example 8.1 by Newton Raphson approach.

Example 8.2: Solve the example 8.1 by Newton Raphson method. The power flow equation is given by

$$P_1 = \frac{|V_1||V_2|}{x}\sin\delta = 2\sin\delta \quad (8.17)$$

Let a solution be given by δ^* and the initial point by δ^1.

$$\delta^* = \delta^1 + \Delta\delta$$

Linearizing the equation (8.17) and substituting $|V_1| = |V_2| = 1.0$ pu.

$$P_1 = 2\sin\delta^1 + 2\cos\delta^1\Delta\delta \quad (8.18)$$

In an iterative setup, we can write this equations as follows:

$$P^{sp} = P^k_{computed} + [2\cos\delta^k]\Delta\delta^k \quad (8.19)$$

$$\text{and } \delta^{k+1} = \delta^k + [2\cos\delta^k]^{-1}\Delta P^k \quad (8.20)$$

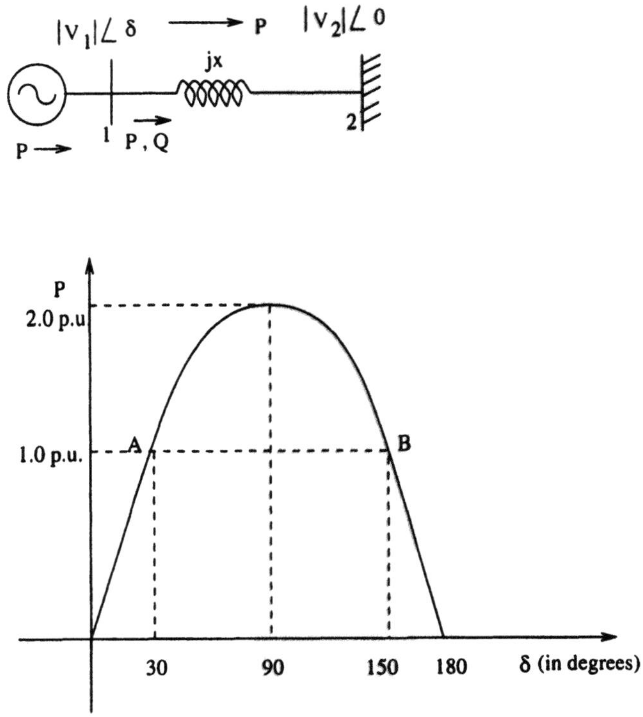

Figure 8.8. Example 8.1: Single Machine Infinite Bus

The Jacobian is given by $J_k = 2\cos\delta^k$. Table 8.1 summarize the iterations of Newton Raphson approach. The solution converges in 3

Table 8.1. Newton Raphson Approach for Single Machine Infinite Bus Problem

k	ΔP^k	δ^k	$\Delta\delta^k$	δ^{k+1}
1	1	0.0^0	28.6479^0	28.6479^0
2	0.0411	28.6479^0	1.3433^0	29.9912^0
3	2.6727×10^{-4}	29.9912^0	0.0088^0	30.0000^0

iterations to the nearest point A which is a stable operating point. Notice that a flat start ($\delta = 0.0$) has guided the solution to the stable operating point. It is interesting to see what happens if we initialize the δ to 180^0.

Example 8.3: Trace the Newton Raphson iterations with problem initialized at $\delta^1 = \pi$ radians.

Table 8.2 summarizes the progress of Newton Raphson algorithm. The

Table 8.2. Newton Raphson approach with $\delta^1 = \pi$ radians

k	ΔP^k	δ^k	$\Delta\delta^k$	δ^{k+1}
1	1	180^0	-28.6479^0	151.3521^0
2	0.0411	151.3521^0	-1.3433^0	150.0088^0
3	2.6727×10^{-4}	150.0088^0	-0.0088^0	150.0000^0

algorithm converges to the nearest solution in 3 iterations which happens to be an unstable operating point. Further, if the algorithm is initialized from $\delta = 90^0$, then Jacobian is singular (in fact zero)and therefore, $|\Delta\delta| \to \infty$, *i.e.*, the method diverges. *Thus, in a Newton Raphson method, the convergence depends upon the initial guess.* It is known to have a quadratic convergence near a solution. Practically, quadratic convergence implies that at each iteration, precision of the solution improves by one significant decimal point. Since, stable operating point corresponds to an angular difference below 90^0, load flow solvers are initialized with all δs set to zero and the unknown voltage magnitudes to either 1.0 pu or to slack bus voltage.

Example 8.4: This example illustrates that the multiple solutions need not only differ in angles. Both the voltage magnitude and angle may vary. If generation at G_1 and G_2 are identical in figure 8.9 then, by the argument of symmetry, at the solution $\delta_1 = \delta_2 = \delta$. Choosing $\delta_3 = 0^0$, *i.e.*, bus-3 as reference bus we obtain power injection equation for node-3 as follows:

$$2 \times 2|V_3|\sin\delta = 1 \Rightarrow |V_3|\sin\delta = 0.25 \tag{8.21}$$

Similarly, for the Q injection

$$2 \times 2(|V_3|\cos\delta - |V_3|^2) = 0.4 \tag{8.22}$$

Since, $|V_3|^2 = |V_3|^2\sin^2\delta + |V_3|^2\cos^2\delta$, we have from (8.22)

$$V_3\cos\delta - V_3^2\sin^2\delta - V_3^2\cos^2\delta = 0.1 \tag{8.23}$$

Substituting equation (8.21) in the above equation, we obtain the following:

$$(V_3\cos\delta)^2 - (V_3\cos\delta) + 0.1625 = 0$$

The two voltage solutions are: (1) $0.8341\angle 17.44^0$ and (2) $0.3228\angle 50.7581^0$ Clearly, the solution with lower angle $0.8341\angle 17.4461^0$ is more desirable

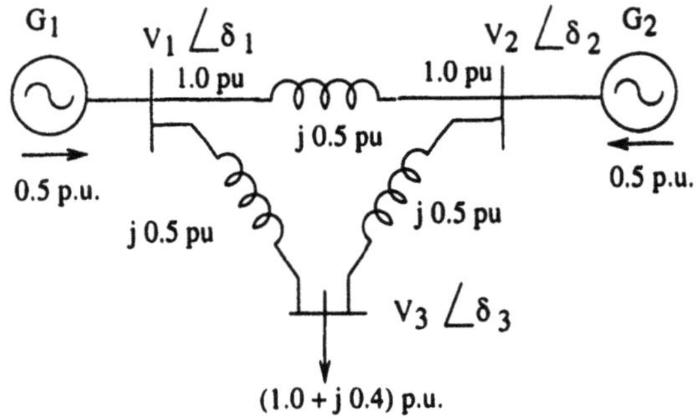

Figure 8.9. Example 8.4

and realistic than $0.3228\angle 50.7581^0$. By running the load flow program in the accompanying diskette, it can be seen that the program converges to a steady state solution 0.8341 pu. Thus, in this case, Q injection is given by $Q_1 = Q_2 = 0.408$ pu. In case (2), $Q_1 = Q_2 = 0.200$ pu. In case (1), an additional 0.416 pu reactive power injection helps in boosting the voltage magnitude $|V_3|$, while in case (2), Q generation exactly matches Q demand, but the lack of additional reactive power support causes dip in the voltage $|V_3|$. In general, all load flow programs are initialized to a flat start, *i.e.*, $|V_i| = 1.0$ pu for PQ nodes and all angles are set to zero degrees. *This example also shows that a generator is not only a source for active power but it also provides reactive power support.* With this background, we now discuss, Newton Raphson (NR) approach for the multidimensional problem.

7. Newton Raphson Load Flow Algorithm

At each iteration k of the algorithm, a guess to the solution vector $\begin{bmatrix} \delta^k \\ |V|^k \end{bmatrix}$ is refined until, the maximum absolute mismatch between the specified and the computed power injection (active and reactive) falls below a specified tolerance. At a node i, when the active power injection is linearized around a $(\delta^k, |V|^k)$ point, we obtain the following:

$$P_i^{sp} = P_i(\delta^k, V^k) + \left[\frac{\partial P_i}{\partial \delta}\right]^T [\Delta\delta^k] + \left[\frac{\partial P_i}{\partial |V|}\right]^T [\Delta|V^k|] \tag{8.24}$$

The active power injections are specified for nodes $i \in \{2, \cdots, n\}$, bus 1 being the slack bus. Since, generator voltage magnitudes are fixed, $\Delta|V_i| = 0$; $i = 1, \cdots, g$ and hence, they can also be eliminated. This leads to $n-1$ linear equations in $2n-1-g$ variables. The additional $n-g$ equations are obtained by linearizing the equations for Q injections which are specified at the PQ busses ($i = g+1, \cdots, n$). Thus,

$$Q_i^{sp} = Q_i(\delta^k, V^k) + \left[\frac{\partial Q_i}{\partial \delta}\right]^T \Delta\delta^k + \left[\frac{\partial Q_i}{\partial |V|}\right]^T \left[\Delta|V|^k\right] \tag{8.25}$$

The mismatches of power are defined as

$$\Delta P_i = P_i^{sp} - P_i(\delta^k, V^k)$$

$$\Delta Q_i = Q_i^{sp} - Q_i(\delta^k, V^k)$$

In what follows, all the quantities $(\delta^k, |V|^k)$ are evaluated at iteration k. Henceforth, the superscript k has been dropped for convenience. Equations (8.24) and (8.25) can be represented in matrix notation as follows:

$$\begin{bmatrix} \frac{\partial P_2}{\partial \delta_2} & \cdots & \frac{\partial P_2}{\partial \delta_n} & \vdots & \frac{\partial P_2}{\partial |V_{g+1}|} & \cdots & \frac{\partial P_2}{\partial |V_n|} \\ \vdots & & \vdots & \vdots & \vdots & & \vdots \\ \vdots & \ddots & \vdots & \vdots & \vdots & \ddots & \vdots \\ \frac{\partial P_n}{\partial \delta_2} & \cdots & \frac{\partial P_n}{\partial \delta_n} & \vdots & \frac{\partial P_n}{\partial |V_{g+1}|} & \cdots & \frac{\partial P_n}{\partial |V_n|} \\ \cdots & \cdots & \cdots & \cdots & \cdots & \cdots & \cdots \\ \frac{\partial Q_{g+1}}{\partial \delta_2} & \cdots & \frac{\partial Q_{g+1}}{\partial \delta_n} & \vdots & \frac{\partial Q_{g+1}}{\partial |V_{g+1}|} & \cdots & \frac{\partial Q_{g+1}}{\partial |V_n|} \\ \vdots & & \vdots & \vdots & \vdots & & \vdots \\ \vdots & \ddots & \vdots & \vdots & \vdots & \ddots & \vdots \\ \frac{\partial Q_n}{\partial \delta_2} & \cdots & \frac{\partial Q_n}{\partial \delta_n} & \vdots & \frac{\partial Q_n}{\partial |V_{g+1}|} & \cdots & \frac{\partial Q_n}{\partial |V_n|} \end{bmatrix} \begin{bmatrix} \Delta\delta_2 \\ \vdots \\ \vdots \\ \Delta\delta_n \\ \cdots \\ \Delta|V_{g+1}| \\ \vdots \\ \vdots \\ \Delta|V_n| \end{bmatrix} = \begin{bmatrix} \Delta P_2 \\ \vdots \\ \vdots \\ \Delta P_n \\ \cdots \\ \Delta Q_{g+1} \\ \vdots \\ \vdots \\ \Delta Q_n \end{bmatrix} \tag{8.26}$$

The coefficient matrix in equation (8.26) is known as the Jacobian and is denoted by J^k. If the sequence of matrices $\{J^1, J^2, \cdots, J^k, \cdots\}$ is non-singular, and the limit

$$\lim_{k\to\infty} \left\{ \begin{bmatrix} \Delta P \\ \Delta Q \end{bmatrix}^1, \begin{bmatrix} \Delta P \\ \Delta Q \end{bmatrix}^2, \cdots, \begin{bmatrix} \Delta P \\ \Delta Q \end{bmatrix}^k, \cdots \right\} = 0 \tag{8.27}$$

then, it follows that the sequence $\left\{ \begin{bmatrix} \Delta\delta^k \\ \Delta|V^k| \end{bmatrix} \right\}$ also converges to zero. Thus, convergence of the algorithm can be checked by either mismatch in P and Q or corrections in δ and $|V|$. Usually, the mismatch in the right hand side is monitored for checking the convergence. Partial derivatives required in equation equation (8.26) can be computed easily from equations (8.15,8.16) and are summarized below.

$$\frac{\partial P_i}{\partial \delta_j}_{i \neq j} = |V_i||V_j|Y_{ij}|\sin(\delta_i - \delta_j - \theta_{ij}) \quad (8.28)$$

$$\frac{\partial P_i}{\partial \delta_i} = -\Sigma_{j\neq i}|V_i||V_j|Y_{ij}|\sin(\delta_i - \delta_j - \theta_{ij}) \quad (8.29)$$

$$\frac{\partial P_i}{\partial |V_j|} = |V_i||Y_{ij}|\cos(\delta_i - \delta_j - \theta_{ij}) \quad (8.30)$$

$$\frac{\partial P_i}{\partial |V_i|} = \Sigma_{j\neq i}|V_j||Y_{ij}|\cos(\delta_i - \delta_j - \theta_{ij}) + 2|V_i||Y_{ii}|\cos(-\theta_{ii}) \quad (8.31)$$

$$\frac{\partial Q_i}{\partial \delta_j} = -|V_i||V_j||Y_{ij}|\cos(\delta_i - \delta_j - \theta_{ij}) \quad (8.32)$$

$$\frac{\partial Q_i}{\partial \delta_i} = \Sigma_{j\neq i}|V_i||V_j||Y_{ij}|\cos(\delta_i - \delta_j - \theta_{ij}) \quad (8.33)$$

$$\frac{\partial Q_i}{\partial |V_j|} = |V_i||Y_{ij}|\sin(\delta_i - \delta_j - \theta_{ij}) \quad (8.34)$$

$$\frac{\partial Q_i}{\partial |V_i|} = \Sigma_{j\neq i}|V_j||Y_{ij}|\sin(\delta_i - \delta_j - \theta_{ij}) + 2|V_i||Y_{ii}|\sin(-\theta_{ii}) \quad (8.35)$$

It can be observed that an element of the Jacobian is zero if, the corresponding $|Y_{ij}| = 0$, *i.e.*, the structure of matrix J depends upon the structure of YBUS. Hence, Jacobian is also a sparse matrix. Since, $\frac{\partial P_i}{\partial \delta_j} \neq \frac{\partial P_j}{\partial \delta_i}$, the Jacobian is not symmetric. Each iteration of the Newton Raphson method also requires the solution of a large sparse linear system of equations. As the Jacobian changes at each iteration, the method requires a LU factorization at each iteration. To reduce the computational effort, following approximations are made:

1 Calculate Jacobian after every ρ iterations ($\rho > 1$). This is not a very successful approximation.

2 Use a decoupled formulation. Let the Jacobian be represented as

$$J = \begin{bmatrix} J_1 & J_2 \\ J_3 & J_4 \end{bmatrix};$$

where submatrices J_1, J_2, J_3, J_4 are given by,
$J_1 = \left[\frac{\partial P}{\partial \delta}\right]^T, J_2 = \left[\frac{\partial P}{\partial |V|}\right]^T, J_3 = \left[\frac{\partial Q}{\partial \delta}\right]^T, J_4 = \left[\frac{\partial Q}{\partial |V|}\right]^T$

If x/r ratio of the line is high, *i.e.*, $\theta_{ij} \approx 90^o$ and the Jacobian is evaluated with all voltage magnitudes set to 1.0 pu and angles set to zero, then the Jacobian is almost decoupled. The standard active-reactive decoupling technique is based on the fact that, in normal situations corresponding to a stable operating state of a power system, the changes in active power injections depend primarily on the phase angle variations and not on the variation in the voltage magnitude. Hence, submatrix J_2 can be neglected. Similarly, the variations in reactive power injections are more sensitive to the voltage magnitude variations rather than the angles and hence J_3 can also be neglected. This leads to the following decoupled equations:

$$[J_1]\,[\Delta\delta] = [\Delta P] \tag{8.36}$$

$$[J_4]\,[\Delta|V|] = [\Delta Q] \tag{8.37}$$

As stated earlier, initially decoupling does give a very good approximation of Jacobian (J_2 and J_3 are nearly zero). Therefore, in a decoupled formulation, there are two independent updates per iteration. The first update is for angles (8.36) followed by the update of Q mismatch. With this updated Q mismatch, an update of voltage magnitudes is obtained (8.37). Then, both P and Q mismatches are updated to check for overall convergence. This completes one iteration. In contrast, in a coupled formulation, both the P and Q mismatches are computed at the start of an iteration and then, simultaneously an update of angles and voltage magnitudes is obtained. Thus, there is only one update per iteration. It appears that with flat start, the primary benefit is obtained in the initial stages by more frequent mismatch vector updates.

In the equation (8.36,8.37), for a node-i, variable $|V_i|$ features in all the row entries $\frac{\partial P_i}{\partial \delta_j}$ and $\frac{\partial Q_i}{\partial |V_j|}$. Hence, it makes sense to scale each row of Jacobian and corresponding right hand side by $|V_i|$. Decoupling has an advantage that the size of the coefficient matrix to be factorized reduces. Further simplifications are possible if while computing the Jacobian, voltage magnitudes are chosen to be 1.0 pu (review question 4-5) and the angles are fixed to zero radians. With these additional approximations, equations (8.36) and (8.37) can be written as follows:

$$[J']\,[\Delta\delta] = \left[\frac{\Delta P}{|V|}\right]$$

$$[J'']\,[\Delta|V|] = \left[\frac{\Delta Q}{|V|}\right]$$

The matrices J' and J'' can be obtained from negative of the suspectance matrix (B) of YBUS network (equations 8.28,8.29,8.34, 8.35). It follows from equation (8.29) that diagonal elements of $\frac{\partial P_i}{\partial \delta_i}$ do not contain contribution from shunt elements. In fact, for computation of matrix J', we omit representation of those network elements that predominantly affect MVAR flows, *i.e.*, shunt reactances off nominal transformer taps, and line charging. Further, in the formulation of matrices J' and J'', following approximations are made [Stott and Alsac, 1974]:

1 Omit from matrix J'', the angle shifting effects of phase shifting transformers.

2 Neglect series resistances in the calculation of the elements of J'. The authors' have faced convergence difficulties when in the matrix J'' line resistances were neglected. Therefore, in the formulation of matrix J'' resistance should not be neglected.

With these additional approximations, matrices J' and J'' are referred as B' and B'' respectively. Thus, Fast Decoupled Load Flow (FDLF) solves the following equations iteratively until maximum absolute mismatch is below a tolerance value.

$$[B'] [\Delta\delta] = \left[\frac{\Delta P}{|V|}\right] \tag{8.38}$$

$$[B''] [\Delta|V|] = \left[\frac{\Delta Q}{|V|}\right] \tag{8.39}$$

Both matrices B' and B'' are large and sparse. Matrix B'' is also symmetric. If phase shifting transformers are absent or accounted in other ways, matrix B' is also symmetric. With these assumptions, the resulting Load Flow method is known as Fast Decoupled Load Flow (FDLF) method.

8. FDLF Algorithm

An algorithm for FDLF is as follows:

1 Read the system data, *i.e.*, network data, load data and generator data. Choose a slack bus, and the convergence criteria (absolute maximum permitted mismatch ϵ and number of iterations).

2 Check if the network is connected; if not, *i.e.*, it has more than one component (area or island), then the option of load flow for each of components should be looked into.

3 Initialize unknown voltages and angles (flat start).

4 Compute symmetric matrices B' and B'' matrices and factorize (LDL^T decomposition) them. The matrices can be overwritten by their respective L factors. The tools developed in chapter 4 can be directly reused.

5 Set $k = 1$ and compute ΔP^k, ΔQ^k.

6 Compute the max $[|\Delta P^k|, |\Delta Q^k|]$ mismatch.

7 While the absolute max mismatch $> \epsilon$, and $k <$ maximum iteration. do

Solve $[B'][\Delta\delta^k] = [\frac{\Delta P^k}{|V|^k}]$

$\delta^{k+1} = \delta^k + \Delta\delta^k$

Compute ΔQ^k

Solve $[B''][\Delta|V|^k] = [\frac{\Delta Q^k}{|V|^k}]$

$|V|^{k+1} = |V|^k + \Delta|V|^k$

k=k+1

and compute ΔP^k, ΔQ^k

8 If converged, print voltage magnitude and angle of all the busses. Compute the branch flows. Mark abnormal operating conditions if any like under-voltages, overloading of lines etc.

9 Else if not converged, troubleshoot the data, check the load generation balance, choice of slack bus, x/r ratio of lines, reactive power problems etc.

The advantages of FDLF method are as follows:

- Matrices B' and B'' are sparse symmetric, and with sufficiently large diagonal entries. Hence, the LSS developed in chapter 4 can be used directly.

- The analyze and factorize phases have to be performed only once. At subsequent iterations, only mismatch vectors ΔP^k and ΔQ^k change and therefore, only forward-backward solution is required.

FDLF method is known to work well on transmission systems where x/r ratio is high (preferably greater than 3).

9. OO Implementation of FDLF

Implementation of the power system analysis programs like Load Flow Analysis (LFA), Short Circuit Analysis (SCA) and Optimal Power Flow (OPF) requires modeling of the electrical network. The application functions can then be implemented as specific methods on a network object. An edited interface of the class *Network* is shown below.

```
template<class T> class Network {
private :
        // DATA TYPE FOR EACH TYPE OF EQUIPMEMNNT
 branch_line *line;
 branch_X *Xmer;
 shunt *shunts;
 generator *gen;
            //DATA  TYPE FOR THE TYPE OF NODES
 node_PQ  *PQ_bus;
 node_PV  *PV_bus;
            // NUMBER OF VARIOUS EQUIPMENT
 int lines, trans, ngens,nshunts,nodes,size_ybus,loads,
 pv_bs,slack_bus_no;
 float tolerance,slack_bus_volt,total_gen,total_load;
 vector<float> tap_vect;
 char * infile,* resultfile,* ybusfile; * name_of_system
                    // FUNCTIONS TO HELP FORM YBUS
 void read_data();
 void write_ybus();
 void add_lines(spmat<T> * A); // add line data to ybus
 void add_lines_b1(spmat<T> * A); // to create B'
 void add_xmers(spmat<T> * A);
 void add_shunts(spmat<T> *A);
 void add_gens(spmat<T> *A);
 vector<float> initialise();  // flat start
            // COMPUTE DELTA P AND DELTA Q FLOWS
 vector<float> delta_Q_cal(const vector<float> &inj_power,
 vector<cmplx<float> > &V,
 const sparse_matrix<cmplx<float> >
 & Ybus, int &iterations,int &pv_to_pq, int &pq_to_pv);

 vector<float> delta_P_cal(const vector<float>
 & inj_power, vector<cmplx<float> > &V,
 const sparse_matrix<cmplx<float> >
 & Ybus, int &iterations,int &pv_to_pq, int &pq_to_pv);
           // FORMATION OF B' AND B'' MATRICES
 int B1(spmat <float> * B11,const spmat<cmplx<float> >
                                                *ybus_p);
 int B2(spmat <float> * B11,const spmat<cmplx<float> >
 *ybus_p, const sparse_matrix<cmplx<float> > & Ybus);

public:
inline  Network() {read_data();};
```

```
//default ctor, reads and intialize from file data
   ~Network();
 Network(branch_line *A, branch_X *B, shunt *C,
 node_PV *D, node_PQ *E, int & nl, int & nX,
 int & nS, int & nPV, int & nPQ) ; //ctor 1

 spmat<T> *   create_ybus_p();
 // create positive sequence ybus
 spmat<T> *   create_ybus_B1();
 // create B' matrix

 asso_arr<T>  name_to_no;
 int  size() { return size_ybus;};
 int  m() { return nodes;};
 void FDLF();
 };
```

A network is an aggregation of various apparatus like lines, transformers etc. Notice that user defined data types for network elements e.g., *branch_line* for line, *NodePV* for PV nodes etc simplify the structure of the program. The main capability of class *Network* is that it implements the admittance matrix model (for all sequence networks) on which load flow and short circuit analysis applications are based.

The class *Network* is a general purpose class and it caters to many functionalities. Unlike the apparatus classes, the attributes of the class *Network* are *private*, *i.e.*, the outside world has no access to it. Even the methods that are implemented have been grouped into *public* and *private* category. Creation & initialization of a network object and it's destruction requires that constructor and destructor be *public*. Similarly, methods that cater to a higher level functionality like formation of YBUS, FDLF and short circuit analysis are public. To implement these functionalities, a programmer implementing say a FDLF method can use the *private* functions e.g., *delta_P_cal*. However, the programmer cannot invoke private methods from the main program. They can at most be nested in the public methods of the class *Network*.

A noteworthy aspect is the implementation of associative array which maps the power system busses defined by meaningful character strings to nonnegative unique integers.

9.1. Associative Array

Associative array is used to map character strings to non-negative integers. The nodes (or busses) of the power systems are best described by character strings. However, procedural languages like FORTRAN could only handle integers while subscripting arrays. Therefore, application engineers had to provide mapping of bus names to integers making interpretation of network data cumbersome. However in C_{++} , it is possible to implement an associative array that maps the character string to non-negative integers. Overloading operator [ ] permits then to automate the job of assigning and indexing by character strings. Thus, one gets the simplistic feel of using character string index. Another application of associative array is in Network Topology Processing and will be discussed in chapter-10.

The authors have presented one design for the class *Network*. Alternative designs which may be more modular and less bulky are possible. Readers are encouraged to try out various designs. For example, the class *Network* can have a limited functionality like building of YBUS. Then, class for load flow analysis and short circuit analysis can be derived from the network. Advantage of such a scheme would be that weight of the class will reduce and it will be become more cohesive. Methods for formulation of B' and B'' etc can be relegated to derived class. Obviously, it is difficult to develop a good Object Oriented Design (OOD) in first attempt. However, an iterative/evolutionary approach is possible. In the next two sections, we discuss two important issues *viz.* steady state models for loads and generators in load flow analysis.

10. Load Modeling

Electrical loads are sensitive to various parameters like voltage, frequency etc. For example, the active and reactive power consumption of an impedance load is proportional to square of the voltage magnitude. The power consumed by a current source is proportional to terminal voltage. Electrical machine loads like induction motor vary with line frequency. Thus, a constant P and Q load model is not generic enough; at a transmission system level, a composite load should be modeled. Active and reactive power loads can be modeled as functions of bus voltage magnitudes and system frequency deviation. The functions considered are as follows:

$$P_{L_i} = P_{Lo_i}(1 + F_{1_i}\Delta F)(A_{1_i} + A_{2_i}|V_i| + A_{3_i}|V_i|^2) \tag{8.40}$$

$$Q_{L_i} = Q_{Lo_i}(1 + F_{2_i}\Delta F)(R_{1_i} + R_{2_i}|V_i| + R_{3_i}|V_i|^2) \tag{8.41}$$

The coefficient F_1 and F_2 represent the frequency sensitive component of the loads. If the frequency is assumed fixed at the nominal value, $\Delta F = 0$. Constants (A_3, R_3), (A_2, R_2) and (A_1, R_1) refer to the constant impedance (Z), current (I) and power (P) components of the loads with $\Sigma A_i = \Sigma R_i = 1$. Hence, this model is also referred to as a ZIP model.

Additionally, to model non standard loads exponential terms $A_{4_i}V^{ep}$ and $R_{4_i}V^{eq}$ can also be included in the above model. The range of coefficients *ep* and *eq* is usually between $1.4 - 1.6$. The compensation at a node can also be modeled by the minimum and maximum limits of Q and the set values. The class for modeling PQ node is shown below.

```
struct NodePQ  {
 char *on_bus,*type;
 cmplx<float> load;
 float  q_comp, q_comp_min, q_comp_max, actual_comp;
 float cp_r, cc_r, ci_r; // load model for real power
 float cp_i, cc_i, ci_i; // load model for reactive power
 float F1, F2; // modeling of frequency dependence
 int bus_no;
 NodePQ() :bus_no(-1), load(0),q_comp(0),q_comp_min(0),
 q_comp_max(0), cp_r(1.0), cc_r(0.0), ci_r(0.0), cp_i(1.0),
 cc_i(0.0), ci_i(0.0), F1(0), F2(0){};
// default load model const. power type
 friend istream & operator >> (istream & is,NodePQ & pq1);
 friend ostream & operator << (ostream & os,NodePQ & pq1);
};
```

In a load flow program, if loads are modeled by equation (8.40,8.41), then the specified power itself becomes a function of voltage magnitude and thus varies from iteration to iteration. The load modeling does not disturb the Jacobian in Newton Raphson approach or the matrices B' and B'' in FDLF. It affects only the right hand side of the mismatch equation. It is expected that load modeling will improve the convergence characteristic of a load flow program e.g., in a planning study where realistic loads may not be available. This can be explained as follows. Let us presume that a system under investigation corresponds to stressed condition, wherein, low voltages are expected. If the loads are modeled as voltage dependent, clearly, as iterations proceed P_i^{sp} will reduce due to reduction in constant impedance and constant current load contributions. If the voltages are low, one expects the computed power to reduce as well. Thus by load modeling, maximum $|\Delta P|$ will reduce and hence, convergence will improve.

11. Generator Reactive Power Capability Curves

The load flow algorithm discussed so far is incomplete because, it does not account for the reactive power capability curves. The reactive power capability curve of a generator is shown in figure 8.10. The constraint represented by the arc of the circle cutting the P-axis, corresponds to

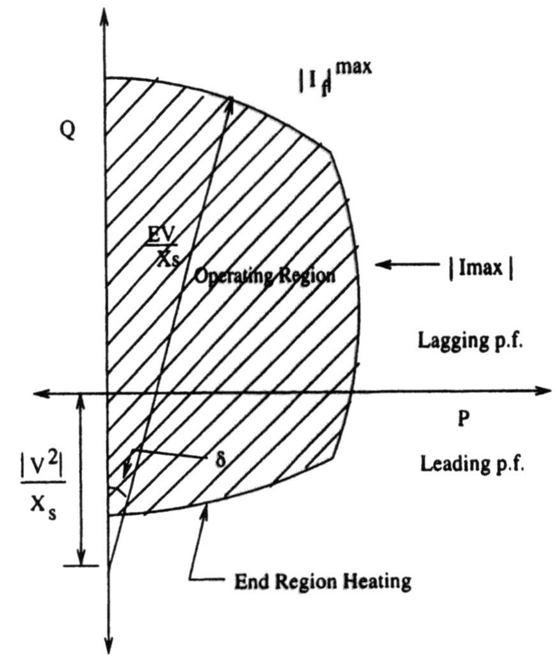

Figure 8.10. Reactive Power Capability of Generator

the armature current capability limit which can be stated as follows:

$$I_r^2 + I_m^2 \leq |I_{max}|^2$$

where I_r and I_m are the real and imaginary parts of the armature current. The arc of the circle cutting +Q axis arises due to the field current rating, *i.e.*, $|I_f| \leq |I_f|^{max}$. The power delivered by a cylindrical pole synchronous machine can be described by the following equations.

$$P = \frac{|E||V|}{X_s} \sin\delta \tag{8.42}$$

$$Q = \frac{|E||V|}{X_s} \cos\delta - \frac{|V|^2}{X_s} \tag{8.43}$$

X_s is the synchronous reactance; and E and V are internal and terminal voltages respectively. Equations (8.42 and 8.43) represent a circle, with center at the Q-axis with an offset of $-\frac{|V|^2}{X_s}$ and radius $\frac{|E||V|}{X_s}$.

$$P^2 + \left(Q + \frac{|V|^2}{X_s}\right)^2 = \left(\frac{|E||V|}{X_s}\right)^2$$

Since, the magnitude of internal or open circuit voltage $|E|$ is proportional to field current I_f, the limit is dictated by the maximum field current. The intersection of the two arcs, defined by maximum field current limit and armature current limit, determine the rated operating point where both the field and armature circuits are optimally utilized. This operating point lies in the first quadrant where the generator is delivering both active and reactive power (lagging power factor operation). Such is the normal operating mode of a generator.

However, under light load conditions, generator may be required to absorb reactive power because of 1) the presence of significant amount of line charging MVARs and 2) lack of reactive power demand at load centers. The absorption capability of a generator is limited due to end region heating effect (curve intersecting the $-Q$ axis). Also, it can be seen from equations (8.43) that when the generator is made to absorb reactive power, $|E|$ is less than $|V|$. Equation (8.42) states that for constant electrical power output, $\sin\delta$ is inversely proportional to $|E|$. Therefore as $|E|$ decreases, δ increases towards 90° limit, thereby imposing limit on reduction of $|E|$. For the reasons of steady state stability, maximum permissible value of δ is much less than 90°. At $\delta = 90^o$, generator absorbs $\frac{|V|^2}{X_s}$ pu reactive power.

From the point of view of load flow algorithm, there are two important issues to be considered:

1 For any given generator, there exists an upper limit $Q^{max}(P)$ and a lower limit $Q^{min}(P)$ on reactive power that a generator can deliver. Clearly, a load flow solution that violates any of these constraints results in an infeasible solution.

2 When a reactive power capability limit is violated, it implies that the load flow problem should be recasted so that a feasible solution results. It also implies that the specification of the terminal voltage at a PV node is not appropriate.

Obviously, at a bus P_i, Q_i and $|V_i|$ can not be specified *apriori* as the resulting system of equations will be either overdetermined or inconsistent. However, to obtain feasible solution, Q_i limit has to be enforced. Therefore, in a load flow algorithm program when a Q limit violation is detected, the PV bus is converted to a PQ bus, *i.e.*, Q is specified to the nearest limit and the fixed terminal voltage magnitude constraint is released. If at a later stage, computed Q falls within the limit, PQ to PV conversion can be performed. Alternatively, a specified voltage can be modified based upon the local $\frac{\partial P}{\partial V_i}$ sensitivity and/or other heuristics. It can be concluded that in load flow analysis, modeling reactive power capability of a generator plays a significant role.

In general, an algorithm that solves the following set of equalities and inequalities will solve the load flow problem.

$$P_i^{sp} = \sum_{j=1}^{n} |V_i||V_j||Y_{ij}| \cos(\delta_i - \delta_j - \theta_{ij}) \; i = 2, \cdots, n \tag{8.44}$$

$$Q_i^{sp} = \sum_{j=1}^{n} |V_i||V_j||Y_{ij}| \sin(\delta_i - \delta_j - \theta_{ij}) \; i = g+1, \cdots, n \tag{8.45}$$

$$Q_i^{min} \leq Q_i \leq Q_i^{max} \; i = 1, \cdots, g \tag{8.46}$$

Specifying a minimization objective function while solving equations (8.44-8.46) leads us to the optimal power flow problem.

12. Summary

Following are some of the important observations on load flow analysis:

1 The nonlinearity of the load flow analysis problem arises mainly because of the way the terminal conditions are specified. Voltage sources are not specified by the phasor magnitude and angle; instead magnitude of voltage and active power injection is specified. As the problem is nonlinear, it has multiple solutions. Newton Raphson approach provides a good starting point for engineering a robust solver.

2 The solution to which a load flow problem converges depends upon the initialization. Usually a flat starts allows us to reach a desirable solution which is characterized by near nominal voltage magnitudes and low angles.

3 Computationally, decoupled load flow is an attractive proposition. Fast Decoupled Load Flow avoids repeated LDL^T factorization. The approximations made in FDLF should be viewed as good heuristics to avoid repeated factorizations.

4 The linear system solver developed in chapter 4 can be used for engineering a FDLF solver. Power apparatus like lines, transformers, reactors etc are naturally modeled as objects in an OO environment. On the other hand, some objects arise because of the algorithmic reasons like PV and PQ nodes. A class *Network* aggregates the branch objects like lines, transformers etc. Readers should use the `struct.h` file to understand the C_{++} design and implementation. In the object oriented paradigm , objects like matrices and vectors, provide computational tools which can be handled with easy interfaces; aggregation

of data objects like line, transformers create a network object (see file `Network.h`) on which methods like FDLF, short circuit analysis can be written. This modularity simplifies programming.

13. Further Reading

Load flow analysis is one of the first problems to be tackled using digital computers. Many books [Stagg and El-Abiad., 1968; Brown, 1975; Pai, 1979; Kusic, 1986; Wallach, 1986; Singh, 1986] have also discussed the application in detail. [Monticelli, 1999] also discusses both the modeling of power apparatus as well as exploitation of sparsity. The view taken in [Barret et al., 1997] is generic enough to incorporate OPF constraints and makes a very interesting reading. A CRIC decoupling technique is discussed in this book. [Sauer and Pai, 1995] presents the relationship between a detailed power system dynamic model and a standard load flow model. They show two examples where singularity in load flow Jacobian implies singularity of the system dynamic state Jacobian. The treatment in [Kundur, 1994] on transmission line, transformer and generator modeling is very educative. [Kundur, 1994; Arrillaga and Bruce, 1998] also provide an excellent treatment on modeling HVDC transmission system in the load flow program. [Pandit et al., 2000b] considers the problem of incorporating the AGC controls in the fast decoupled load flow in an OO framework. [Fuerte-Esquivel and Acha, 1997; Noroozian and Andersson, 1993; Gotham and Heydt, 1998] deal with incorporating FACTS in the load flow analysis. [Sachdev and Medicherla, 1977] has dealt with the issue of incorporating second order information in load flow. However, the class of second order load flow methods have not been able to replace the first order approximation of Taylor series based methods due to the computational complexity. In the evaluation of steady state voltage stability limits, one encounters divergence due to Jacobian ill-conditioning. [Ajjarapu and Christy, 1992] presents a method of finding continuum of power flow solutions starting at some base loads and leading to the steady state voltage stability limit (critical point) of the system. As a consequence, divergence due to ill-conditioning is not encountered in the system.

14. Review Questions

1 For a lossless network, can slack bus generation be fixed *apriori*? Similarly, comment on the Q-generation of the slack bus.

2 Show that a connected graph of n nodes, b branches has a tree of $n-1$ branches, and $b-n+1$ links (fundamental loops). Describe a

procedure to trace *path* from a node i to j and hence construct an algorithm for computing fundamental loops in the network.

3 Show that in the figure 8.11,

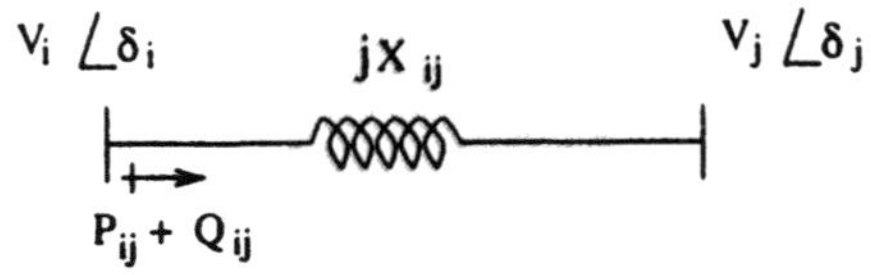

Figure 8.11. A Simple Transmission Line Model

$$P_{ij} = \frac{|V_i||V_j|}{X_{ij}} \sin(\delta_i - \delta_j)$$

$$Q_{ij} = \frac{|V_i|}{X_{ij}}(|V_i| - |V_j| \cos(\delta_i - \delta_j))$$

Under the assumption that $|\delta_i - \delta_j|$ is small, show that ΔP_{ij} is proportional to $\Delta\delta_{ij}$ and less sensitive to $\Delta|V_i|$ and $\Delta|V_j|$. Similarly, show that ΔQ_{ij} is proportional to $(\Delta|V_i| - \Delta|V_j|)$ and less sensitive to $\Delta\delta_i - \Delta\delta_j$.

4 For an EHV system, assume that $\frac{x}{r}$ is high and that transformer taps are all set to 1. Hence, the approximate power flow equation are given as follows:

$$P_i = \sum_{i \neq j} P_{ij} = \sum_{i \neq j} \frac{|V_i||V_j|}{x_{ij}} \sin(\delta_i - \delta_j);$$

$$Q_i = \sum_{i \neq j} \left(\frac{|V_i|}{x_{ij}}(|V_i| - |V_j| cos(\delta_i - \delta_j)) - \frac{B_{ij}}{2}|V_i|^2 \right) + \frac{|V_i|^2}{x_{sh_i}}$$

Further, assume that (1) $V_i = V_j = 1.0$ pu, *i.e.*, actual voltages in pu do not deviate significantly from 1.0 pu and (2) $\delta_i - \delta_j$ is small. Show that the Jacobian J_1 in the Newton Raphson load flow is a constant with elements specified as follows:

diagonal for node $i = \sum_{j \in Adj(i)} \frac{1}{x_{ij}}$;

off diagonal entry for bus i to bus j connection $= -\frac{1}{x_{ij}}$

Hence, justify the assumptions made in forming matrix B' in fast decoupled load flow.

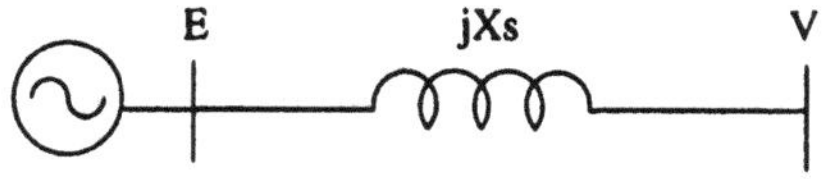

Figure 8.12. Synchronous Machine Model

5 "The matrix B' should be singular because the summation of columns of B' equals to zero". Can you identify the inaccuracy in the statement. What about the nature of B''?

6 Solve example 8.1 by NR approach with (a) $\delta^1 = 89^o$ and $\delta^1 = 91^o$. What conclusions do you draw?

7 Compute Q_G for Example 8.2 and 8.3. Does the unstable operating point result in abnormal reactive power generation or absorption?

8 Show that in the Newton Raphson method,

$$\lim_{k\to\infty} \begin{bmatrix} \Delta P^k \\ \Delta Q^k \end{bmatrix} \to \mathbf{0}$$

is equivalent to checking $\lim_{k\to\infty} \begin{bmatrix} \Delta\delta^k \\ \Delta V^k \end{bmatrix} \to \mathbf{0}$.

9 Why is the steady state model of a generator, as shown in figure (8.12) is not used in load flow analysis. Why is the internal voltage not accounted in load flow analysis?

10 Argue that load flow Jacobian at a stable equilibrium point can not be singular. [Hint: Argue that in the vicinity of a stable solution there can not be any other solution. What would happen if the Jacobian is singular?]

11 Steady state voltage at 50% distance for a long transmission line has to be computed. Suggest modifications in the equivalent π model for load flow study.

12 Formulate the load flow problem in rectangular coordinates. How many times can the power injection equations be differentiated?

13 In the context of distribution system load flow analysis, comment on the validity of approximations in FDLF.

14 **DC Load Flow Problem:** [Barret et al., 1997] points out that precision brought about by the use of complete active-reactive decoupling is not justified for transmission system planning studies given

the uncertainties which affect the data knowledge. Show that under the assumptions that (1) all voltage magnitudes are nearly 1 pu (2) difference of angles between adjacent busses is small and (3) that high voltage transmission lines can be modeled by simple reactance, the load flow calculations can be done by solving a $n-1 \times n-1$ single linear system of equations:

$$[P] = [X][\delta]$$

Let n^{th} node be the slack bus then, show that $X(i,j) = \frac{-1}{x_{ij}}$ if node i and j are adjacent and (2) $X(i,i) = \Sigma_{i \in Adj(i)} 1/x_{ij}$. Will the matrix [X] be singular?

15 The active power loss in a system is given by $\sum_{i \neq j} |I_{ij}|^2 r_{ij}$. Similarly, the reactive power loss can be described by $\sum_{i \neq j} |I_{ij}|^2 x_{ij}$. Does there exist any correlation between the two? Comment on whether line charging effect should be included in the reactive power loss computation.

16 Is it advisable to derive a class for three winding transformer from the class of transformer. Justify your choice.

17 A three winding transformer has $Z_{HL} = 0.45$ pu, $Z_{HM} = 0.15$ pu and $Z_{ML} = 0.30$ pu on the same MVA base. Calculate Z_H, Z_M and Z_L for the Y model of the transformer.

Chapter 9

SHORT CIRCUIT ANALYSIS

Everyone on earth runs after power. None pays heed to the source

—Kabir

1. Introduction

Short circuit analysis is primarily required to compute the following:

1 Three phase fault level at one or more nodes (busses) in the system. In turn, the 3ϕ fault level is used to decide the short circuit current interruption capacity of circuit breakers. The issue will be discussed in more detail in section-5.

2 Evaluate fault currents for faults on transmission lines and the corresponding contribution from the adjacent busses. Such current pairs are used in the coordination of overcurrent relays. This application will be discussed in detail in section 7.

In other words, the main use of short circuit calculations is in designing the protection system. Also, fault currents to earth induce currents in the neighboring conductive circuits e.g., metallic piping and telecommunication circuits, which can affect their correct operation and safety.

1.1. Dynamics and Short Circuit Calculations

A fault involves a sudden change in system configuration. It results in the excitation of transients in the network, generator and other components of the system. If these transients die down in time, then the system settles to a steady state. A detailed study of faults would necessitate dynamic modeling (using differential equations) of system components. However, conventional short circuit studies involve several simplifying assumptions:

1 Network transients are neglected; the network is assumed to be in sinusoidal steady state.

2 Stator transients of the generator are neglected.

3 Slowly changing states associated with the electromechanical system (speed and rotor angle of a generator) are assumed to frozen at their prefault values.

4 While the dynamics of generator field winding and damper winding flux is not represented by differential equations, they are accounted for by representing the generator as a fixed voltage source behind a reactance whose value is chosen depending on the time scale of interest.

The consequence of the first two assumptions is that the d.c. and/or high frequency oscillatory transients are neglected in the calculation of response. Further, following assumptions on data are frequently made in the short circuit calculations (refer question 2):

1 transformers are assumed to be operating at nominal taps;

2 series resistance and shunt capacitance of transmission lines are ignored;

3 if maximum fault currents are to be computed, then all generating units are considered to be in service.

Remark 1: Performance of a protection system including the behavior of current transformers, relaying logic and circuit breaker response may need detailed modeling of network transients. A program like Electro-Magnetic Transients Program (EMTP) is suitable for this purpose.
Remark 2: The behavior of distance relays under power swings requires modeling of generator flux and electromechanical dynamics. Transient stability program can be used for this purpose.

Short circuit analysis is usually done using sequence networks. The next section reviews the same.

2. Sequence Networks

The method of symmetrical components was developed in late 1913 by Charles L. Fortescue of Westinghouse, while he was investigating mathematically the operation of induction motors under unbalanced conditions

[Fortescue, 1918]. As shown by Fortescue, an unbalanced 3ϕ network can be analyzed by considering three sequence networks *viz*, positive (+,1), negative (-,2), and zero (0) sequence networks (figure 9.1). In figure 9.1

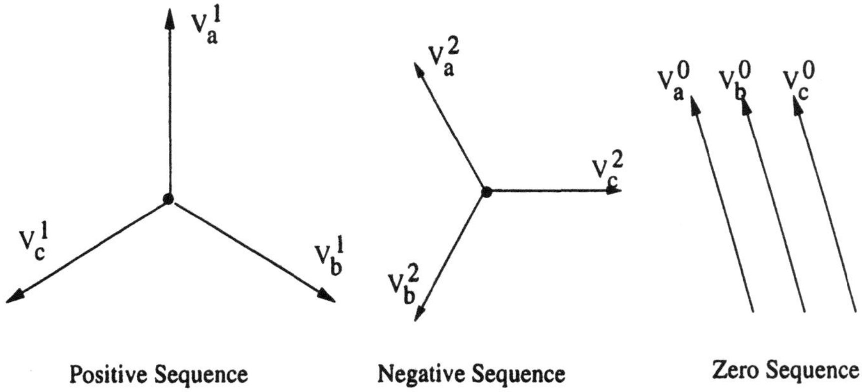

Figure 9.1. The 3-sequences in a 3ϕ systems

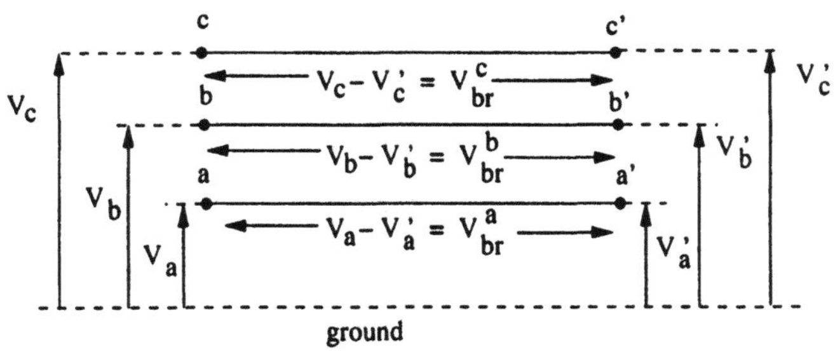

Figure 9.2. 3ϕ Transmission System Voltage & Current Conventions

a, b, c refer to the phase quantities and $0, 1, 2$ correspond to the zero, positive and negative sequence quantities. Figure 9.2 shows a 3ϕ transmission line, with Z_s as the self impedance and Z_{m_1}, Z_{m_2} as the mutual impedances.

$$[V_{br}^{abc}] = [Z_{br}^{abc}][I_{br}^{abc}]; \qquad \text{where} \tag{9.1}$$

$$[Z_{br}^{abc}] = \begin{bmatrix} Z_s & Z_{m_1} & Z_{m_2} \\ Z_{m_2} & Z_s & Z_{m_1} \\ Z_{m_1} & Z_{m_2} & Z_s \end{bmatrix}; \ [I_{br}^{abc}] = \begin{bmatrix} I_{br}^a \\ I_{br}^b \\ I_{br}^c \end{bmatrix}; \ [V_{br}^{abc}] = \begin{bmatrix} V_{br}^a \\ V_{br}^b \\ V_{br}^c \end{bmatrix} \tag{9.2}$$

For a transposed transmission line $Z_{m1} = Z_{m2}$. The matrix has a cyclic symmetry and so has it's inverse $[Y_{br}^{abc}] = [Z_{br}^{abc}]^{-1}$. It can be shown that linear transformation given by equation (9.4) diagonalizes $[Z_{br}^{abc}]$. In terms of sequence components, we can write,

$$[V_{br}^{abc}] = [T][V_{br}^{012}]; \quad [I_{br}^{abc}] = [T][I_{br}^{012}] \text{ where,} \tag{9.3}$$

$$T = \begin{bmatrix} 1 & 1 & 1 \\ 1 & a^2 & a \\ 1 & a & a^2 \end{bmatrix}; \ a = e^{j\frac{2\pi}{3}}; \ V_{br}^{012} = \begin{bmatrix} V_{br}^0 \\ V_{br}^1 \\ V_{br}^2 \end{bmatrix}; \ I_{br}^{012} = \begin{bmatrix} I_{br}^0 \\ I_{br}^1 \\ I_{br}^2 \end{bmatrix} \tag{9.4}$$

Equation (9.2) can be written in terms of sequence coordinates as follows:

$$[V_{br}^{012}] = [T^{-1} Z_{br}^{abc} T][I_{br}^{012}]$$

$$[Z^{012}] = [T^{-1} Z_{br}^{abc} T]$$

$$[Z^{012}] = \begin{bmatrix} Z_s + Z_{m_1} + Z_{m_2} & & \\ & Z_s + a^2 Z_{m_1} + a Z_{m_2} & \\ & & Z_s + a Z_{m_1} + a^2 Z_{m_2} \end{bmatrix} \tag{9.5}$$

Thus, in the sequence components frame, the network is decoupled, *i.e.*,
$V_{br}^0 = Z_0 I_{br}^0$ where $Z_o = Z_s + Z_{m_1} + Z_{m_2}$
$V_{br}^1 = Z_1 I_{br}^1$ where $Z_1 = Z_s + a^2 Z_{m_1} + a Z_{m_2}$
$V_{br}^2 = Z_2 I_{br}^2$ where $Z_2 = Z_s + a Z_{m_1} + a^2 Z_{m_2}$
Further, it can be shown that the corresponding sequence admittance matrix is also decoupled, *i.e.*,

$$[I_{br}^{012}] = [Y_{br}^{012}][V_{br}^{012}];$$

$$[Y_{br}^{012}] = [T]^{-1}[Z_{br}^{abc}]^{-1}[T] = [T^{-1} Z_{br}^{abc} T]^{-1} = [Z_{br}^{012}]^{-1}$$

Impedances Z_0, Z_1 and Z_2 are referred to as the zero, positive and negative sequence impedances.

A 3ϕ admittance matrix model for power system in phase coordinates can be expressed as follows:

$$\begin{bmatrix} I_1^{abc} \\ \vdots \\ \vdots \\ \vdots \\ I_n^{abc} \end{bmatrix} = \begin{bmatrix} Y_{11}^{abc} & Y_{12}^{abc} & \cdots & \cdots & Y_{1n}^{abc} \\ Y_{21}^{abc} & Y_{22}^{abc} & \cdots & \cdots & Y_{2n}^{abc} \\ \vdots & \vdots & \ddots & \vdots & \vdots \\ \vdots & \vdots & \vdots & \ddots & \vdots \\ Y_{n1}^{abc} & Y_{n2}^{abc} & \cdots & \cdots & Y_{nn}^{abc} \end{bmatrix} \begin{bmatrix} V_1^{abc} \\ V_2^{abc} \\ \vdots \\ \vdots \\ V_n^{abc} \end{bmatrix} \tag{9.6}$$

where each entry in the Y-matrix is itself a 3×3 matrix with a cyclic structure, $[V_i^{abc}] = [V_i^a, V_i^b, V_i^c]^T$; $[I_i^{abc}] = [I_i^a, I_i^b, I_i^c]^T$. V_i refers to the voltage of a node i and I_i refers to the current injection at a node i.

The rules for the formation of 3ϕ YBUS are identical to that of YBUS formation as discussed in previous chapter with the exception that 1×1 element is replaced by corresponding 3×3 matrix. A 3ϕ shunt reactance can be described by a 3×3 diagonal matrix. To obtain the sequence components of the network, a block diagonal transformation defined by equation (9.7) is used:

$$\begin{bmatrix} V_1^{abc} \\ V_2^{abc} \\ \vdots \\ \vdots \\ V_n^{abc} \end{bmatrix} = \begin{bmatrix} [T] & & & & \\ & [T] & & & \\ & & \ddots & & \\ & & & [T] & \\ & & & & [T] \end{bmatrix} \begin{bmatrix} V_1^{012} \\ V_2^{012} \\ \vdots \\ \vdots \\ V_n^{012} \end{bmatrix} \qquad (9.7)$$

Thus, in the sequence coordinates, the admittance model is given by the following equations:

$$\begin{bmatrix} I_1^{012} \\ I_2^{012} \\ \vdots \\ \vdots \\ I_n^{012} \end{bmatrix} = \begin{bmatrix} T^{-1}Y_{11}^{abc}T & T^{-1}Y_{12}^{abc}T & \cdots & \cdots & T^{-1}Y_{1n}^{abc}T \\ T^{-1}Y_{21}^{abc}T & T^{-1}Y_{22}^{abc}T & \cdots & \cdots & T^{-1}Y_{2n}^{abc}T \\ \vdots & \vdots & \ddots & & \vdots \\ \vdots & \vdots & & \ddots & \vdots \\ T^{-1}Y_{n1}^{abc}T & T^{-1}Y_{n2}^{abc}T & \cdots & \cdots & T^{-1}Y_{nn}^{abc}T \end{bmatrix} \begin{bmatrix} V_1^{012} \\ V_2^{012} \\ \vdots \\ \vdots \\ V_n^{012} \end{bmatrix} \qquad (9.8)$$

It can be easily seen that

$$[Y_{ij}^{012}] = T^{-1}[Y_{ij}^{abc}]T = \begin{bmatrix} Y_{ij}^0 & & \\ & Y_{ij}^1 & \\ & & Y_{ij}^2 \end{bmatrix}$$

There is no coupling between the zero, positive and negative sequence components of a balanced network because 3×3 matrices $[Y_{ij}^{012}]$ and $[Y_{ii}^{012}]$ are diagonal matrices. By permuting the rows and columns in such a way that all the zero sequence, positive sequence and negative sequence quantities are grouped together, a 3ϕ admittance matrix can be described by three decoupled (block diagonal) sequence matrices as follows:

$$\begin{bmatrix} I_{bus}^0 \\ I_{bus}^1 \\ I_{bus}^2 \end{bmatrix} = \begin{bmatrix} YBUS^0 & & \\ & YBUS^1 & \\ & & YBUS^2 \end{bmatrix} \begin{bmatrix} V_{bus}^0 \\ V_{bus}^1 \\ V_{bus}^2 \end{bmatrix} \qquad (9.9)$$

In equation (9.9), each of the sequence admittance matrix represents the corresponding sequence network.

Remark 3: In load flow analysis, it is assumed that the network is balanced. A synchronous machine produces balanced (positive sequence) voltage. If loads are balanced and the transmission lines are transposed, a balanced network assumption is valid. Therefore, positive sequence network analysis suffices.

Example 9.1: In the above formulation we have not considered the mutual inductance between two 3ϕ circuits e.g., two 3ϕ circuits on the same tower. Under the assumption that all the lines are transposed, it is reasonable to treat all the mutual impedances between circuits 1 and 2 as equal. For example, the mutual inductance of phase a of circuit-2 with phases a, b and c of circuit-1 be equal to say α. The model of such a transmission line accounting for the mutual coupling between two circuits on the same tower in the phase coordinates is given by the following equation:

$$\begin{bmatrix} V_{br}^{a^1} \\ V_{br}^{b^1} \\ V_{br}^{c^1} \end{bmatrix} = \begin{bmatrix} Z_s & Z_m & Z_m \\ Z_m & Z_s & Z_m \\ Z_m & Z_m & Z_s \end{bmatrix} \begin{bmatrix} I_{br}^{a^1} \\ I_{br}^{b^1} \\ I_{br}^{c^1} \end{bmatrix} + j\alpha \begin{bmatrix} 1 & 1 & 1 \\ 1 & 1 & 1 \\ 1 & 1 & 1 \end{bmatrix} \begin{bmatrix} I_{br}^{a^2} \\ I_{br}^{b^2} \\ I_{br}^{c^2} \end{bmatrix} \tag{9.10}$$

where superscripts 1 and 2 refer to circuit-1 and circuit-2 respectively. Applying the sequence transformation leads to the following equation:

$$\begin{bmatrix} V_{br}^{0^1} \\ V_{br}^{1^1} \\ V_{br}^{2^1} \end{bmatrix} = \begin{bmatrix} Z_s + 2Z_m & & \\ & Z_s - Z_m & \\ & & Z_s - Z_m \end{bmatrix} \begin{bmatrix} I_{br}^{0^1} \\ I_{br}^{1^1} \\ I_{br}^{2^1} \end{bmatrix} + \begin{bmatrix} 3j\alpha & 0 & 0 \\ 0 & 0 & 0 \\ 0 & 0 & 0 \end{bmatrix} \begin{bmatrix} I_{br}^{0^2} \\ I_{br}^{1^2} \\ I_{br}^{2^2} \end{bmatrix} \tag{9.11}$$

where the 1st superscript refers to the sequence number and the second superscript to the circuit number. Thus, it can be concluded that mutual coupling in positive sequence (or negative sequence) network of a transmission line is zero. However, it cannot be neglected in zero sequence network for parallel transmission lines. The equation (9.11) can be written as follows:

$$\begin{aligned} V_{br}^{0^1} &= (Z_s + 2Z_m)I_{br}^{0^1} + 3j\alpha I_{br}^{0^2} \\ V_{br}^{1^1} &= (Z_s - Z_m)I_{br}^{1^1} \\ V_{br}^{2^1} &= (Z_s - Z_m)I_{br}^{2^1} \end{aligned} \tag{9.12}$$

A method to account for mutual coupling in zero sequence YBUS matrix will be discussed section 3.1.1.

A justification of the above observations using Ampere's law ($\oint \vec{H} \cdot \vec{dl} = i$) can be based on the following arguments. Figure 9.3 shows two 3ϕ transmission lines running parallel to each other in a close vicinity. To compute the mutual inductance, circuit-1 is excited while circuit-2 is not. Now, the equivalent sequence mutual inductance between two circuits

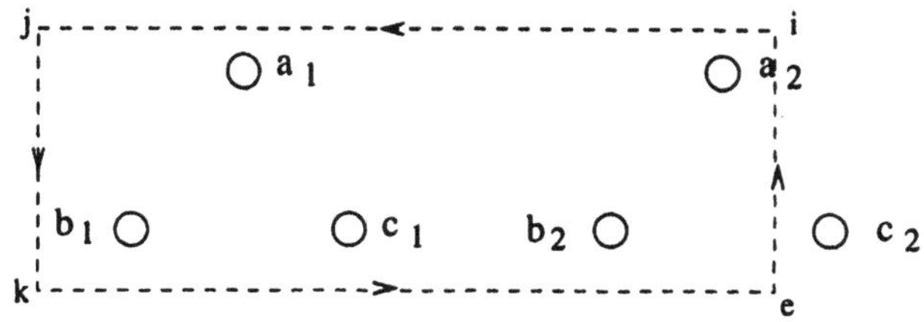

Figure 9.3. Double Circuit Transmission Line and a Closed Contour Enclosing Circuit 1

will be given by the ratio of flux linking circuit-2 due to corresponding sequence excitation on circuit-1. If the circuit-1 carries a positive sequence or negative sequence current, then $I_{a_1} + I_{b_1} + I_{c_1} = 0$. Since, the mmf linking circuit-2 is zero, mutual flux is negligible. Hence, positive and negative sequence mutual impedances are negligible.

Now, if the circuit-1 is excited by zero sequence currents then, the net mmf linking the path is given by $3I_0^1$ and hence, non-trivial mutual coupling will be established.

Clearly, as seen in equation (9.9), each of the sequence components can be solved separately. A power system as described by equation (9.9) can be compactly shown as follows:

$$[I^{012}] = [YBUS^{012}][V^{012}] \tag{9.13}$$

Remark 4: With the appropriate choice of a reference node[1], it can be ascertained that $[YBUS^{012}]$ is invertible.

Usually, ground is chosen as the reference node. If there does not exist any shunt connection to ground, then YBUS will be singular, because sum of all the columns will be zero. Shunts and generator reactances (*i.e.* x'_d, x''_d etc.) contribute only to the diagonal element and not to off-diagonal elements and hence, with their inclusion, the YBUS matrix becomes non-singular. The equation (9.13) can alternatively be expressed as follows:

$$[V^{012}] = [ZBUS^{012}][I^{012}] \tag{9.14}$$

where, $[ZBUS^{012}] = [YBUS^{012}]^{-1}$. While YBUS is sparse, ZBUS is not.

Computationally, sequence networks merit attention because of the decoupling that they provide in the YBUS matrix. For example, in load

[1]Throughout this chapter we consider ground to be the reference node, (unless otherwise stated) *i.e.*, reference node is assumed to be at zero potential.

flow analysis, we restrict to only positive sequence network. Also, sequence components forms the vocabulary of protection engineers. They use sequence voltages and currents for relay setting e.g., earth fault relay.

In the next section, we discuss the modeling of various network elements in sequence domain.

3. Sequence Models for Network Elements

A network consists of static elements like lines, transformers etc. and rotating elements like synchronous machine, induction motor etc.

3.1. Sequence Impedance of Static Elements

The positive sequence modeling of transmission lines and transformers is identical to that of load flow analysis. Both transmission lines and transformers are static elements. Positive and negative sequence impedances are identical for static elements. The matrix like in equation (9.2) represents, the impedance-matrix for a 3ϕ transmission line in the abc coordinates. It can be seen from equation (9.5) that sequence networks are decoupled. Further, if the lines are transposed then, $Z_{m_1} = Z_{m_2} = Z_m$.

3.1.1 Zero Sequence Modeling of Transmission Lines

In a positive or negative sequence network, earth or return wire does not carry current because $I_a + I_b + I_c = 0$. Consequently, earth or return wire impedance does not play a role in these networks. In contrast, the zero sequence impedance of a transmission line depends upon the configuration as well as the return path for zero sequence currents e.g., through earth, ground wire etc. Chapter-11 and 12 of [Blackburn, 1993] considers this issue in detail. In case of parallel or nearby transmission lines on the same tower, mutual coupling resulting from positive and negative sequence currents is negligible. However, for the zero sequence network it can be as high as 50-70% of the self impedance. This can have a significant effect on fault current and therefore, on protective relaying. If in the zero sequence network, m mutually coupled lines exist; then, they can be represented as follows:

$$\begin{bmatrix} \Delta V_{l_1} \\ \Delta V_{l_2} \\ \vdots \\ \Delta V_{l_m} \end{bmatrix} = \begin{bmatrix} z_{l_1 l_1} & z_{l_1 l_2} & \cdots & z_{l_1 l_m} \\ \vdots & z_{l_2 l_2} & \cdots & \vdots \\ \vdots & & \ddots & \vdots \\ z_{l_m l_1} & z_{l_m l_2} & \cdots & z_{l_m l_m} \end{bmatrix} \begin{bmatrix} I_{l_1} \\ I_{l_2} \\ \vdots \\ I_{l_m} \end{bmatrix} \tag{9.15}$$

where ΔV_{l_i} is the voltage drop in i-th line, and I_{l_i} is the current in the direction of voltage drop. Superscript zero has been dropped for convenience. The Z-matrix of primitive impedances in equation (9.15) is referred to as Z_m. Impedance $z_{l_i l_i} = r_{l_i l_i} + jwL_{l_i l_i}$ is the self impedance of the line. Note that changing the direction of current and voltage drop simultaneously does not change the Z matrix. Reactance $z_{l_i l_j} = jwL_{l_i l_j}$ is the mutual coupling between the line i and j. If the inverse exist for the impedance matrix in (9.15), (*e.g.*, when $r_{l_i l_i} \neq 0$) then, the branch currents can be computed as follows:

$$[I_l] = [Z_m]^{-1}[\Delta V_l] \tag{9.16}$$

Now the voltage drop across a line can be expressed in terms of vector of node voltages as follows:

$$[\Delta V_l] = [P][V] \tag{9.17}$$

where P is a $m \times n$ matrix such that for a line number k connecting nodes i to j and $P(k,i) = 1, P(k,j) = -1$. Therefore, from equation (9.16) and (9.17), we can express line currents as a function of node voltages as follows:

$$[I_l] = [Z_m]^{-1}[P][V] = \widehat{Y}_m[V] \tag{9.18}$$

Therefore, in presence of mutually coupled lines, a procedure to compute zero sequence YBUS is as follows:

1 Form YBUS^0 matrix without accounting for the lines that have mutual coupling.

2 From the matrix Z_m compute matrix Z_m^{-1} for those lines having mutual coupling.

3 Compute matrix P and $\widehat{Y}_m$.

4 For each line k in the set of mutually coupled lines, connecting node i_k to j_k where the ordering of nodes is consistent with convention of voltage drop across the line.

   (a) Add k^{th} row of $\widehat{Y}_m$ matrix to i_k^{th} row of YBUS^0.

   (b) Subtract k^{th} row of $\widehat{Y}_m$ matrix from j_k^{th} row of YBUS^0.

   (c) Account line charging in the usual manner.

Example 9.2: Compute the bus impedance matrix for the network shown in figure 9.4[2]. Since there are no shunt connections to ground in

[2]This example has been dealt with using ZBUS approach in [Pai, 1979].

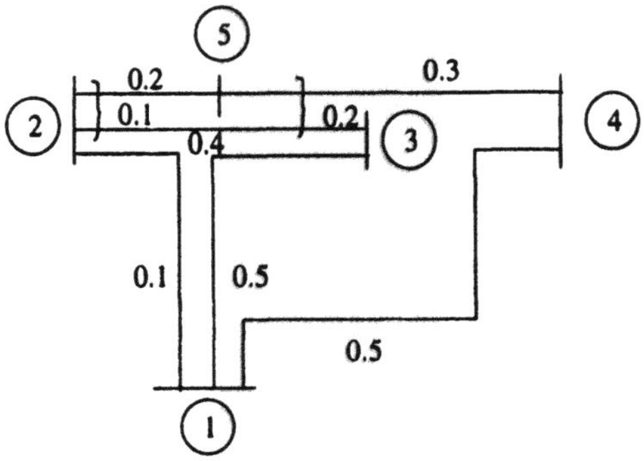

Figure 9.4. **A 5-Bus Example**

this case, bus 1 can be taken as the reference bus.

Solution:

Step-1: Lines 2-1, 3-1, 4-1 are considered in formation of $YBUS^0$ as they have no mutual coupling. This leads to the $YBUS^0$ as shown.

$$YBUS = \begin{array}{c} \\ 2 \\ 3 \\ 4 \\ 5 \end{array} \begin{array}{c} \begin{array}{cccc} 2 & 3 & 4 & 5 \end{array} \\ \begin{bmatrix} 10 & & & \\ & 2 & & \\ & & 2 & \\ & & & 0 \end{bmatrix} \end{array} \tag{9.19}$$

Step-2 The matrix Z_m is as follows

$$\begin{bmatrix} \Delta V_{25} \\ \Delta V_{54} \\ \Delta V_{23} \end{bmatrix} = \begin{bmatrix} 0.2 & 0 & 0.1 \\ 0 & 0.3 & 0.2 \\ 0.1 & 0.2 & 0.4 \end{bmatrix} \begin{bmatrix} I_{25} \\ I_{54} \\ I_{23} \end{bmatrix} \tag{9.20}$$

Step-3: Computation of matrix $\widehat{Y}_m$ is as follows:

$$\begin{bmatrix} I_{25} \\ I_{54} \\ I_{23} \end{bmatrix} = \begin{bmatrix} 6.1538 & 1.5385 & -2.3077 \\ 1.5385 & 5.3846 & -3.0769 \\ -2.3077 & -3.0769 & 4.6154 \end{bmatrix} \begin{bmatrix} 1 & & & -1 \\ & & -1 & 1 \\ 1 & -1 & & \end{bmatrix} \begin{bmatrix} V_2 \\ V_3 \\ V_4 \\ V_5 \end{bmatrix} \tag{9.21}$$

$$\begin{bmatrix} I_{25} \\ I_{54} \\ I_{23} \end{bmatrix} = \begin{bmatrix} 3.8462 & 2.3077 & -1.5385 & -4.6154 \\ -1.5385 & 3.0769 & -5.3846 & 3.8462 \\ 2.3077 & -4.6154 & 3.0769 & -0.7692 \end{bmatrix} \begin{bmatrix} V_2 \\ V_3 \\ V_4 \\ V_5 \end{bmatrix} \tag{9.22}$$

Step-4: Following the MATLAB® we denote row-i of a matrix A by A(i;:). Then the operators of step-4 can be shown as following. Note that 1st row of YBUS represent node no. 2 etc.

1 YBUS(1,:)=YBUS(1,:)+$\widehat{Y}_m$(1,:)

2 YBUS(4,:)=YBUS(4,:)−$\widehat{Y}_m$(1,:)

3 YBUS(1,:)=YBUS(1,:)+$\widehat{Y}_m$(3,:)

4 YBUS(2,:)=YBUS(2,:)−$\widehat{Y}_m$(3,:)

5 YBUS(3,:)=YBUS(3,:)−$\widehat{Y}_m$(2,:)

6 YBUS(4,:)=YBUS(4,:)+$\widehat{Y}_m$(2,:)

The final ZBUS=YBUS^{-1} is as follows:

$$Z_{Bus} = \begin{bmatrix} 0.0862 & 0.0372 & 0.0319 & 0.0660 \\ 0.0372 & 0.2074 & 0.1064 & 0.0532 \\ 0.0319 & 0.1064 & 0.2340 & 0.1170 \\ 0.0660 & 0.0532 & 0.1170 & 0.2085 \end{bmatrix} \tag{9.23}$$

3.1.2 Zero Sequence Modeling of Transformers

The zero sequence impedance of a transformer depends upon (1) type of the transformer construction, e.g., whether a three phase transformer is of core type, five limbed construction or a bank of three single phase transformers and (2) the connection of the windings, e.g., $Y-\Delta$, $\Delta-\Delta$, grounding connections etc. For a core type construction, return path for in-phase flux in the limbs is not available, except via insulating medium and the transformer tank. Hence, large reluctance reduces leakage flux and consequently zero sequence leakage impedance is about 85-90% of positive sequence impedance. The zero sequence model of a transformer is a π-circuit as shown in figure 9.5. The values to be assigned to

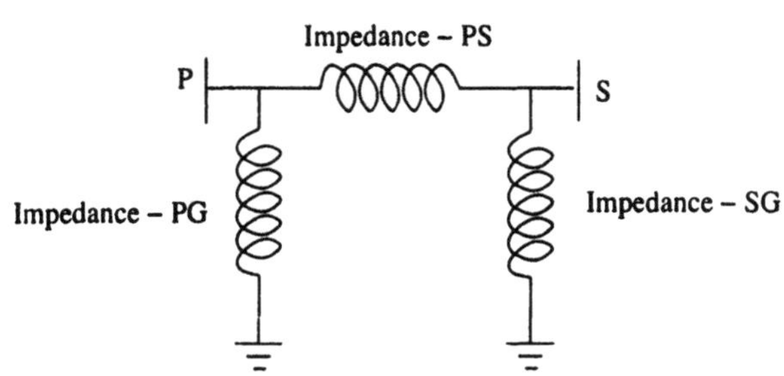

Figure 9.5. Zero Sequence Transformer Model

the limbs depend upon the type of winding and is summarized in table 9.1. In the table, Y denotes a star winding, D a delta winding, G

Table 9.1. Zero Sequence Impedance for 3ϕ Transformer

Type of Transformer	impedance-PS	impedance-PG	impedance-SG
YGD	∞	$r_0 + jx_0$	∞
YGID	∞	$r_0 + jx_0 + 3z_n$	∞
DD	∞	∞	∞
YGYG	$r_0 + jx_0$	∞	∞
YGIYGI	$r_0 + jx_0 + 3(z_{n_1} + z_{n_2})$	∞	∞
YGY	∞	∞	∞
YD	∞	∞	∞
DY	∞	∞	∞
YY	∞	∞	∞
DYG	∞	∞	$r_0 + jx_0$
DYGI	∞	∞	$r_0 + jx_0 + 3z_n$
YYG	∞	∞	∞

ground and I impedance. Thus, a transformer marked YGI,YGI has a 3ϕ star winding on both primary and secondary. Both the windings are grounded through appropriate grounding impedances. Similarly, Y,D transformer has primary star (not grounded) and secondary delta winding. The model of a transformer depends upon the type of connections. No generic formulae can be predefined. Hence, in a short circuit program, a library of π-models for different types of 3ϕ transformers is provided. The user during data entry provides the type of transformer, e.g., YY; the overloaded operator >> requests the various parameters and sets the impedance-PS, impedance-PG and impedance-SG. The short circuit analysis program uses appropriate model as per the connection. If a new transformer connection is to be modeled, a new model can be incorporated in `struct.h` file.

3.1.3 Modeling of Three Winding Transformer

Positive and negative sequence models of a three winding transformer has been discussed in chapter 8. In this section, we consider the zero sequence modeling. The zero sequence circuits for various transformer connections are shown in [Blackburn, 1987]. The equivalent circuit in figure 9.6b shows the generic model of a three winding transformer used for computer modeling. The new equivalent impedances are summarized in table 9.2.

$$Z'_{HG} = Z_H + 3Z_{NH} + \frac{Z_L Z_M}{Z_L + Z_M} \qquad (9.24)$$

Table 9.2. Three winding transformer-Zero Sequence Impedances

Type (H,M,L)	Z'_{HM}	Z'_{ML}	Z'_{LH}	Z'_{HG}	Z'_{MG}	Z'_{LG}
YGI,D,D	∞	∞	∞	eqn. 9.24	∞	∞
YGI,YGI,D	eqn. 9.27	∞	∞	eqn. 9.25	eqn. 9.26	∞
YGI,Y,D	∞	∞	∞	$Z_H + Z_L + 3Z_N$	∞	∞

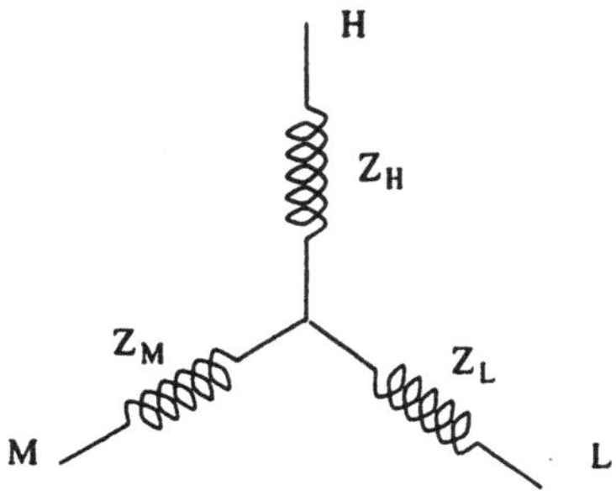

Figure 9.6a. Positive and Negative Sequence Model

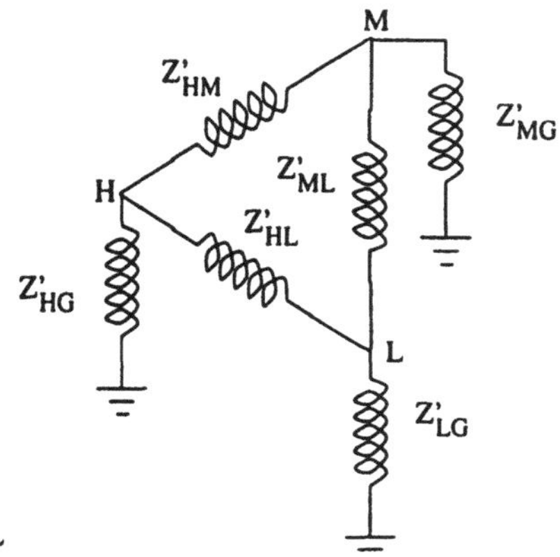

Figure 9.6b. Generic Model of a three winding transformer

$$Z'_{HG} = \frac{\Delta}{(Z_M + 3Z_{NM})} \tag{9.25}$$

$$Z'_{MG} = \frac{\Delta}{3Z_{NH} + Z_H} \tag{9.26}$$

$$Z'_{HM} = \frac{\Delta}{Z_L} \tag{9.27}$$

where $\Delta = (3Z_{NH} + Z_H)Z_L + Z_L(Z_M + 3Z_{NM}) + (Z_M + 3Z_{NM})(Z_H + 3Z_{NH})$ and Z_{NH}, Z_{NM} denotes grounded reactance from Y type H winding and M winding respectively.

3.1.4 Load Modeling

A common practice is to model a load as an equivalent admittance in the positive sequence network. The load model is as shown in figure 9.6a. As loads are supplied through transformers with neutral insulated on the network side, they are not usually, considered in the zero sequence admittance matrix. In the short circuit program provided along with this

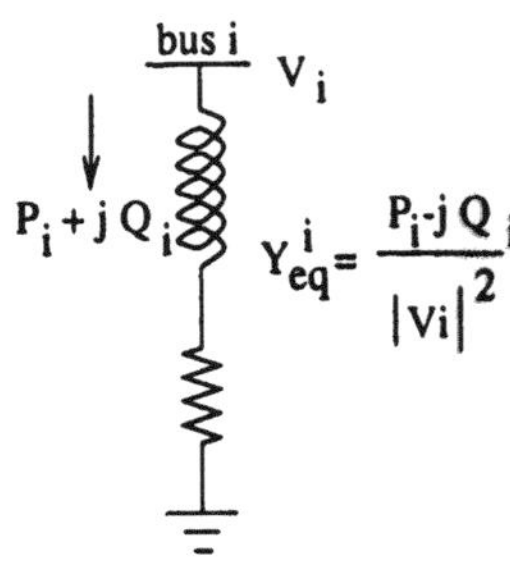

Figure 9.7a. Load Model

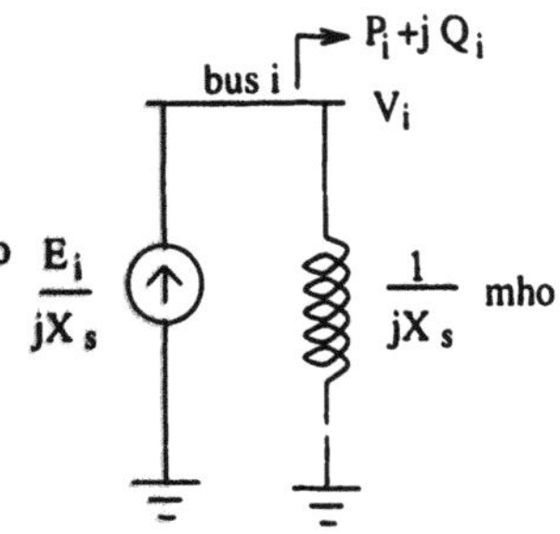

Figure 9.7b. Generator Model

book, user is given the option of either including all or none of the loads as shunt admittance in zero sequence network.

3.2. Sequence Impedances of Rotating Machines

Unlike load flow analysis, modeling of generators and loads in the positive sequence network in short circuit analysis is slightly different. As discussed in the introduction, generator is represented by emf behind a suitable reactance (jX_s); [Barret et al., 1997] provides the following guidelines for the selection of generator impedance:

1 If the time of study corresponds to $t < 0.10$ secs, *i.e.*, less than 5 cycles for a 50 Hz system, then the reactance to be taken into account is the *subtransient* reactance.

2 If the time of the study correspond to $0.10 < t < 1$ seconds, then the *transient* reactance is used. [Kusic, 1986] suggests that transient reactance should be used from 0.1 to 6 seconds interval.

3 The steady state reactance is used beyond the transient period.

The current delivered by the generator is given by:

$$I_i = \frac{P_i - jQ_i}{V_i^*}$$

For a unloaded system $I_i = 0$, and for a loaded system I_i can be computed from load flow study. Hence, the internal emf is computed as follows:

$$E_i = V_i + jX_s I_i$$

With the knowledge of the internal emf E_i, the Norton equivalent can be computed as shown in figure 9.7b. The shunt admittances of figure

9.7a and 9.7b can be incorporated in the YBUS. The net current injected at PQ node is zero leading to the network structure as shown in figure 9.8. The prefault network equations are defined as follows:

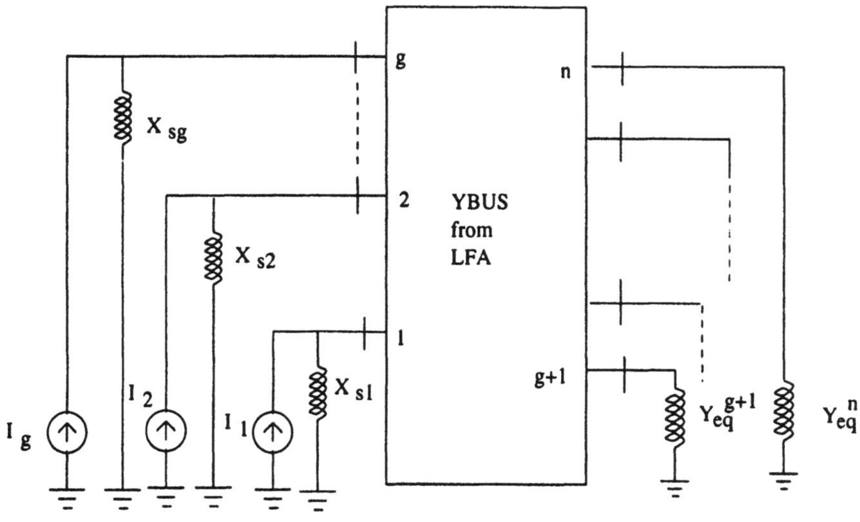

Figure 9.8. Positive Sequence YBUS Model for Short Circuit Analysis

$$[YBUS^1][V^1] = I^1 \text{ or} [V^1] = [ZBUS^1][I^1] \tag{9.28}$$

where, $I^1 = [\frac{E_1}{jX_{s1}}, \frac{E_2}{jX_{s2}}, \cdots, \frac{E_g}{jX_{sg}}, 0, \cdots, 0]^T$. It is conveniently assumed that the first g busses are generator busses.

For rotating machines, positive and negative sequence impedances cannot be equal. For example, in case of a synchronous machine, negative sequence currents produce a rotating mmf in opposite direction to the rotor mmf. Hence, the negative sequence reactance corresponds to the flux which rotates at twice the synchronous speed with respect to the rotor. Negative sequence impedance is 70-95% of the subtransient reactance. However, it can be approximated by subtransient reactance. For a salient pole machine, it is taken as a mean between the direct axis X_d'' and quadrature axis X_q'' values [Blackburn, 1993], or

$$X_2 = \frac{X_d'' + X_q''}{2}$$

Zero sequence impedance of a synchronous machine is only a small percentage (0.1-0.7 of x_d'') of the positive sequence impedance. X_o varies so critically with armature winding pitch that an average value can hardly be given[Blackburn, 1993]. Since, synchronous machines generate only positive sequence voltages, the internal voltages used in negative and zero sequence network is zero. Hence, the negative sequence impedance is grounded and so is zero sequence impedance if the generator has star grounded connection. If the star point is grounded through an impedance Z_g, then $3Z_g$ will have to be added to the zero sequence impedance of generator before incorporating it as a shunt in the YBUS^0.

For *asynchronous* machines e.g., induction motor, the transient state of the current is damped quickly (1-2 cycles) and subsequently machine behaves as a passive element with impedance of value $Z = kV_{LL}^2/S_{MVA}$ where the rated line to line voltage is the numerator and 3ϕ rated MVA is the denominator. Zero sequence impedances can be treated on similar lines as synchronous machines.

Sequence impedance matrix ZBUS has a close relationship with the Thevenin equivalent at a given bus. This relationship is now established.

4. Thevenin Equivalent Circuit At a Bus

We begin with the familiar positive sequence network. To compute the Thevenin equivalent at a bus, we require to compute (1) open circuit Thevenin voltage source and (2) the Thevenin impedance. In fact, open circuit voltage source is directly available from the prefault load flow study. The positive sequence Thevenin equivalent circuit is shown in figure 9.9. It models the network prior to the fault; the switch k is open. Closing the switch k, introduces the fault with impedance Z_f in series with Z_{th}. To compute Z_{th}^1, all the voltage sources are short circuited

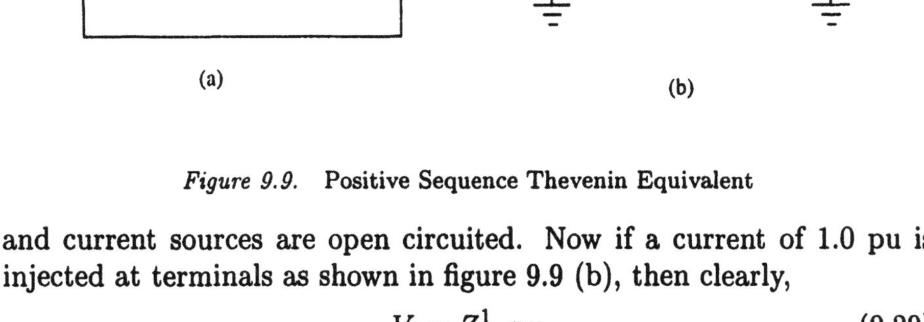

Figure 9.9. Positive Sequence Thevenin Equivalent

and current sources are open circuited. Now if a current of 1.0 pu is injected at terminals as shown in figure 9.9 (b), then clearly,

$$V_i = Z_{th}^1 \text{ pu} \tag{9.29}$$

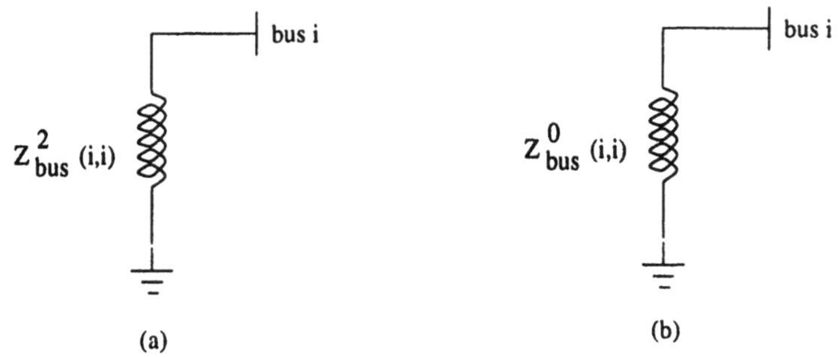

Figure 9.10. Negative and Zero Sequence Thevenin Circuit

Hence, Z_{th}^1 can be measured by injecting a current of 1.0 pu at bus I_i with all other current injections as zero, *i.e.*, in equation (9.28) $[I^1] = [e_i]$. In the vector notation, this corresponds to a current injection of e_i where, vector e_i represents the i-th column of identity matrix.

Therefore, from equation (9.28) and (9.29), $Z_{th}^1 = ZBUS^1(i,i)$.

For negative and zero sequence networks, Thevenin voltage source is zero. By the similar arguments, it can be seen that $Z_{th}^2 = ZBUS^2(i,i)$, and $Z_{th}^0 = ZBUS^0(i,i)$. Hence, the corresponding Thevenin circuits are given by the figure 9.10 (a) and (b). The next section considers the various computational methods for computing the Thevenin impedances.

4.1. ZBUS Vs YBUS approach

There are two common ways to organize computations in short circuit analysis. They are:

- ZBUS approach;
- YBUS approach.

ZBUS building algorithm has been covered in [Stagg and El-Abiad., 1968; Brown, 1975; Pai, 1979; Grainger and William D. Stevenson, 1994] etc. We restrict the discussion to the YBUS approach. The YBUS approach is preferred in large scale computation because it is directly amenable to sparsity exploitation. It requires computation of sparse sequence admittance matrices $[YBUS^{012}]$ and a sparse linear system solver that can handle complex matrices. To compute Thevenin impedance $Z[i,i]$, a current of 1.0 pu of appropriate sequence is injected at the node i. Then, the solution of the equations (9.30) gives the i-th column Z_i of the $ZBUS$.

$$[YBUS^0]\,[Z_i^0] = [e_i] \tag{9.30}$$

$$[YBUS^1]\,[Z_i^1] = [e_i] \tag{9.31}$$

$$[YBUS^2]\,[Z_i^2] = [e_i] \tag{9.32}$$

where e_i is the i-th column of the identity matrix. The positive and negative sequence admittance matrices have a structure identical to the admittance matrix of load flow although the generator is modeled in more detail as a Norton's equivalent current source. The transient or subtransient reactance is accounted in the diagonal element of the YBUS. In case of zero sequence network, transformers (*e.g.*, Δ/YG connection) may introduce open circuit between the nodes. This can change the structure of zero sequence admittance matrix. Alternatively, an open circuit can be approximated by a reasonably large impedance value. Then, structurally, ZBUS^0 is identical to ZBUS^1 and ZBUS^2. The algorithmic steps in developing sequence admittance matrices are identical to that of admittance matrix building in load flow analysis.

To illustrate the various steps involved in short circuit analysis, an illustrative example is solved.

Example 9.3: The following table 9.3 summarizes the voltage, power

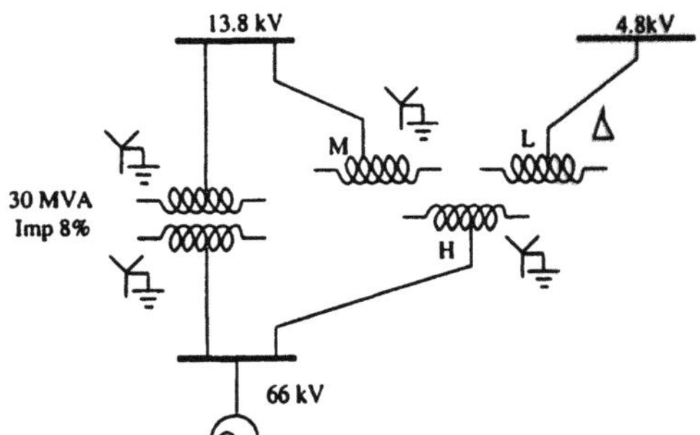

Figure 9.11a. Example 9.3

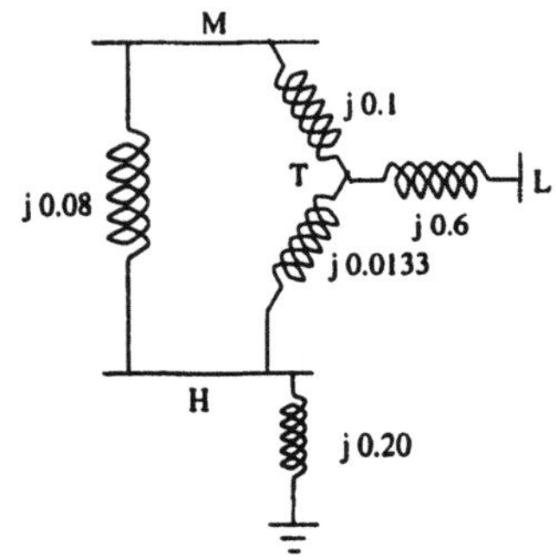

Figure 9.11b. Positive and Negative Sequence Network

and impedance data for the three windings transformer. The figure

Table 9.3. Three winding transformer data of example 9.3

Impedance	Rated Voltage kV	Rated MVA	% Impedance
	Line to Line	3ϕ	Base Name Plate Rating
Z_H	66	18	j0.8
Z_M	13.8	18	j6
Z_L	4.8	5	j10

9.11a represents connection of the transformer in a subsystem. The positive sequence equivalent circuit of 3 winding transformer is shown in figure 9.11b. The equivalent generator at 66 kV bus has a subtransient reactance of 0.2 pu on the 30 MVA base.

1 If a 3ϕ solid short circuit occurs on 4.8 kV bus, find the short circuit current as seen at the tertiary winding.

2 If a SLG fault takes place at 13.8 kV bus, compute the fault current as seen on the 13.8 kV side. Assume that the positive, negative and zero sequence impedances are equal.

Solution: The first step is to convert all % impedances on to the same base MVA, say, 30 MVA. Therefore,

$$\begin{aligned} Z_H &= j\tfrac{0.8}{100} \times \tfrac{30}{18} = j0.0133 \text{ pu} \\ Z_M &= j\tfrac{6}{100} \times \tfrac{30}{18} = j0.1 \text{ pu} \\ Z_L &= j\tfrac{10}{100} \times \tfrac{30}{5} = j0.6 \text{ pu} \end{aligned}$$

The sequence network for positive sequence network is shown in figure 9.11b. The circuit contains a fictitious node T; it can be eliminated either by network reduction or by simply applying a Y/Δ transformation. For a 3 winding transformer, the second approach can be easily incorporated in the implementation of a class for 3 winding transformer. Applying Y/Δ conversion, the equivalent network is obtained as shown in figure 9.12a. From figure 9.12a, the positive sequence matrix is computed and given below:

$$YBUS^{1,2} = \begin{matrix} \\ H \\ M \\ L \end{matrix} \; j \begin{bmatrix} H & M & L \\ -27.60 & 21.158 & 1.4423 \\ 21.158 & -21.350 & 0.1923 \\ 1.4423 & 0.1923 & -1.6347 \end{bmatrix} \tag{9.33}$$

Readers are encouraged to verify that the same YBUS results by first considering the 4 node network of figure 9.11b and then applying network reduction to eliminate node T. The ZBUS for the positive and negative sequence matrices is given by

$$ZBUS^{1,2} = [YBUS^{-1}]^{1,2} = j \begin{matrix} \\ H \\ M \\ L \end{matrix} \begin{bmatrix} H & M & L \\ 0.2 & 0.2 & 0.2 \\ 0.2 & 0.2469 & 0.2055 \\ 0.2 & 0.2055 & 0.8124 \end{bmatrix} \tag{9.34}$$

Then, the 3 ϕ short circuit fault current at L is given by

$$I_f = \frac{1.0}{j0.8124} = -j1.2309 \text{ pu}$$

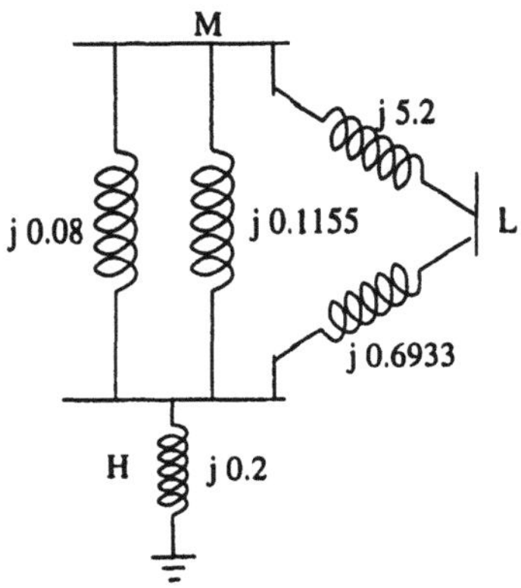

Figure 9.12a. Positive Sequence Network with Node-0 Eliminated

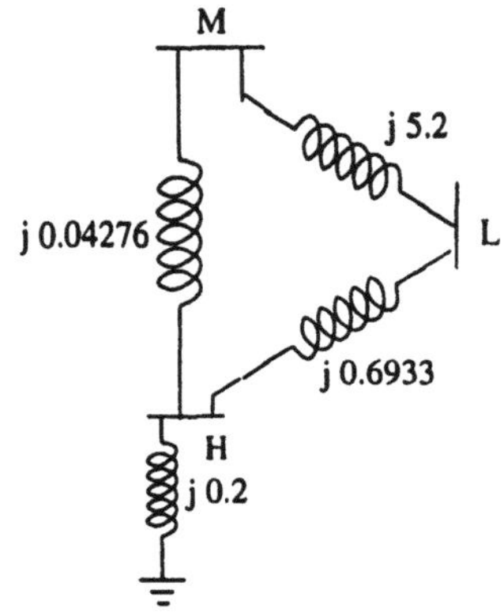

Figure 9.12b. Equivalent Network

SLG fault at bus M: The zero sequence admittance network for the overall network is shown in figure 9.13a. By applying Y/Δ transformation, the equivalent network is shown in figure9.13c. The corresponding matrices are as follows:

$$YBUS = j \begin{array}{c} \\ H \\ M \\ L \end{array} \left[\begin{array}{ccc} H & M & L \\ -27.60 & 21.158 & 0 \\ 21.158 & -21.350 & 0 \\ 0 & 0 & 0 \end{array} \right] \tag{9.35}$$

Notice that the zero sequence network has a different structure from the positive sequence network. To compute the zero sequence Thevenin equivalent impedance at M, we inject 1.0 pu current at M and solve for the voltage at M, *i.e.*,

$$\begin{bmatrix} -27.60 & 21.158 & 0 \\ 21.158 & -21.350 & 0 \\ 0 & 0 & 0 \end{bmatrix} \begin{bmatrix} V_H \\ V_M \\ V_L \end{bmatrix} = \begin{bmatrix} 0 \\ 1 \\ 0 \end{bmatrix} \tag{9.36}$$

$$\Rightarrow \begin{bmatrix} V_H \\ V_M \end{bmatrix} = \begin{bmatrix} -27.60 & 21.1158 \\ 21.1158 & -21.350 \end{bmatrix}^{-1} \begin{bmatrix} 0 \\ 1 \end{bmatrix} = j \begin{bmatrix} 0.1473 \\ 0.1925 \end{bmatrix} \tag{9.37}$$

$$Z^{th}_{(M,M)} = j0.1925 \text{ pu}$$

Note that, the zero sequence YBUS matrix is not invertible. To circumvent this numerical difficulty, it may be worth while to take a sufficient large value for impedance Z'_{ML}, Z'_{HL}. Choosing 10^6 pu as an approxi-

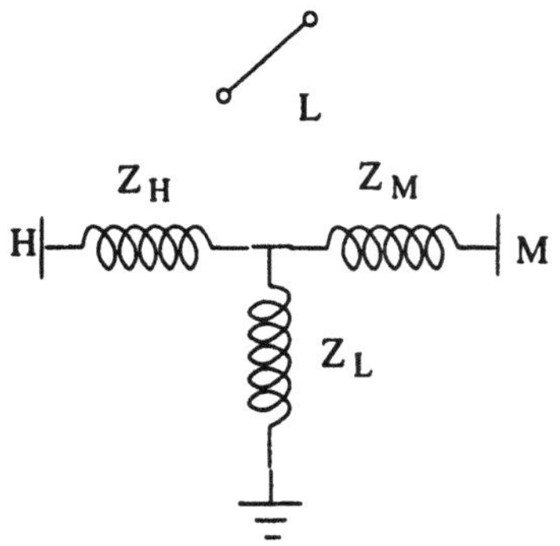

Figure 9.13a. 0-Sequence Model

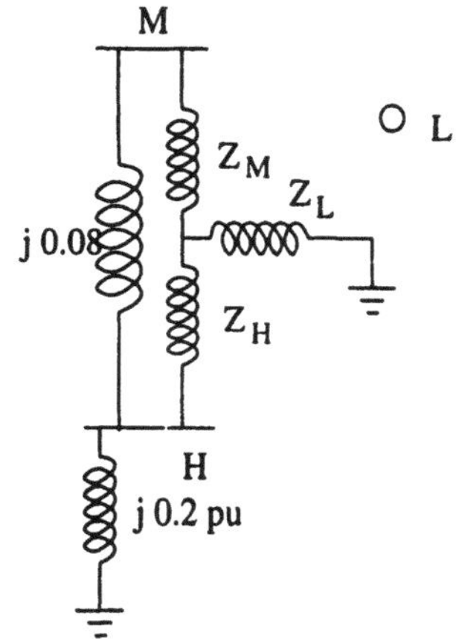

Figure 9.13b. 0-Sequence Network

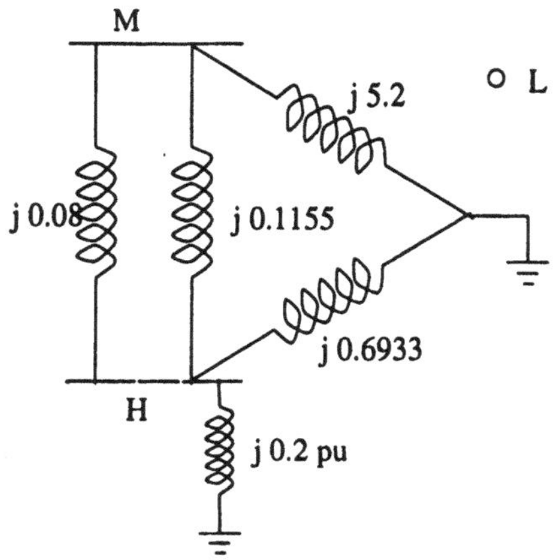

Figure 9.13c. Equivalent 0-Sequence Network

mation of infinite impedance, we get the YBUS matrix as follows:

$$Y_{BUS} = j\begin{bmatrix} -27.60 & 21.1158 & 10^{-6} \\ 21.1158 & -21.350 & 10^{-6} \\ 10^{-6} & 10^{-6} & -2\times 10^{-6} \end{bmatrix} \tag{9.38}$$

$$Z_{BUS} = j\begin{bmatrix} 0.1508 & 0.1494 & 0.1501 \\ 0.1494 & 0.1949 & 0.1722 \\ 0.1501 & 0.1722 & 5\times 10^{5} \end{bmatrix} \tag{9.39}$$

Thus, instead of measuring an impedance of $j0.1925$, a value 0.1949 has been obtained. Hence I_{SLG} for a fault at bus M is given by: $\frac{3.0}{2\times 0.2469+0.1949} = 4.3567$ pu (true value 4.3712).

Results of a short circuit program are commonly expressed in terms of the short circuit MVA because of the closeness to circuit breaker rating. The next section reviews the same.

5. Short Circuit MVA [Blackburn, 1993]

Three phase short circuit MVA is defined as follows:

$$MVA_{SC} = \frac{\sqrt{3} I_{3\phi} kV}{1000} \tag{9.40}$$

where kV refers to system line to line voltage in kV and $I_{3\phi}$ is the three phase fault current in amperes. Short circuit MVA data provides the equivalent impedance of the power system up to a bus. Usually, two values corresponding to (1) three phase fault and (2) single line as ground fault are computed. Given the short circuit MVA, this impedance can be calculated by the following equations:

$$Z_{\Omega} = \frac{kV \times 1000}{\sqrt{3} \times I_{3\phi}} = \frac{kV^2}{MVA_{SC}} \Omega$$

If 3ϕ system MVA base is given by MVA_{base} and the line to line base voltage in kilovolts is given by kV, then

$$Z_{base} = \frac{kV^2}{MVA_{base}} \Omega$$

The corresponding per unit impedance value is given by the following equation:

$$Z_{pu} = \frac{MVA_{base}}{MVA_{SC}} \tag{9.41}$$

Similarly, for a single phase to ground fault, short circuit MVA is defined as follows:

$$MVA_{\phi GSC} = \frac{\sqrt{3} kV I_{\phi G}}{1000} \tag{9.42}$$

where $I_{\phi G}$ is the single line to ground fault current in amperes and kV is the system line to line voltage in kilovolts. The current $I_{\phi G}$ is given by the following equation:

$$I_{\phi G} = \frac{\sqrt{3} kV}{Z_1 + Z_2 + Z_0} \times 1000 \ A \tag{9.43}$$

where Z_1, Z_2 and Z_0 are positive, negative and zero sequences impedances respectively. It follows from (9.42) and (9.43) that

$$Z_1 + Z_2 + Z_0 = \frac{3MVA_{base}}{MVA_{\phi GSC}} \text{ pu} \tag{9.44}$$

Knowing Z_1 and Z_2 and three phase short circuit MVA, zero sequence impedance Z_0 can be calculated using equation (9.44) and can be adjusted to restrict short circuit MVA to a desired level.

In short circuit analysis, from the computational viewpoint, the crucial issue is the solution of equation (9.30). The solution of complex linear system of equations is now discussed.

6. LDL^T Decomposition of Sequence Admittance Matrices

In chapter 4 on sparse linear system solver, the discussion was restricted to solution of linear system of equations with only real matrices. Special attention is required if the matrices to be factorized have complex data type. A complex matrix can be represented as follows:

$$A = B + iC \tag{9.45}$$

Complex matrix A can be classified into the following three categories:

1 C_1 : Matrix A is a complex symmetric $n \times n$ matrix with matrices B and C real symmetric and positive definite;

2 C_2 : Matrix A is a Hermitian matrix, *i.e.*, $A = A^*$, where A^* is *conjugate transpose* of A.

3 C_3 : Matrix A does not belong to category C_1 and C_2.

If a matrix is complex Hermitian (category C_2), it is well known that Cholesky decomposition can be performed. Because of diagonal dominance of intermediate partially factorized matrices, diagonal pivots are stable. Unfortunately, complex symmetric matrices do not admit to such simple methods. Consider the admittance model formulation required in short circuit analysis.

$$[I] = [YBUS]\,[V] \tag{9.46}$$

Let $[YBUS] = [G + jB]$, $[I] = [I_r + jI_i]$ and $[V] = [V_r + jV_i]$. Though not elegant, one way to solve equation (9.46), is to convert the problem into a real linear system of equations.

$$\begin{bmatrix} I_r \\ I_i \end{bmatrix} = \begin{bmatrix} G & -B \\ B & G \end{bmatrix} \begin{bmatrix} V_r \\ V_i \end{bmatrix} = \begin{bmatrix} G & B \\ B & -G \end{bmatrix} \begin{bmatrix} V_r \\ -V_i \end{bmatrix} \tag{9.47}$$

It is known that such a scheme can be inefficient in both storage and time [Press et al., 1992]. One good option to factorize symmetric complex matrix is to use a version of Bunch's diagonal pivoting method for equation (9.47) [Bunch and Kaufman, 1977]. The essential step at each stage of the elimination is to choose 1×1 or 2×2 blocks for pivoting. This choice is dictated by the magnitude of the pivot which affects numerical stability of implementation.

In an important result, Steven M Serbin [Serbin, 1980] showed that if a complex matrix has Symmetric Positive Definite (spd) like characteristics, it can be exploited to expedite the solution. This characteristics corresponds to category C_1 matrices. Serbin showed that for category C_1 matrix, LDL^T decomposition can be obtained in a numerical stable manner using diagonal pivots alone. Thus, it was concluded by [Pandit et al., 2000b] that:

- conventional wisdom of solving sparse spd/ Hermitian matrices by analyze, factorize breakup can be applied for solving such complex symmetric matrices also.
- the LDL^T algorithm for sparse spd linear system solver can be used by replacing data type *float* to data type *complex*, without affecting numerical stability.

Therefore, the linear system solver of chapter 4 is used in both the short circuit analysis and FDLF programs. In fact in C_{++} , by defining data type as a variable using template, the very same implementation is used for FDLF and fault analysis program.

$$SparseMatrix < complex < float > > YBUS$$

The above construct defines a complex symmetric matrix for short circuit analysis. $SparseMatrix < float > \; B1$ defines a real matrix for FDLF. Then $YBUS.factorize()$ factorizes the YBUS matrix and $B1.factorize()$ factorizes the matrix $B1$. Similarly, $solver(YBUS, e_i)$ solves the equation (9.46) $solver(B_1, \Delta P)$ solves the mismatch equation in FDLF. The solution of equation (9.46) gives us the Thevenin sequence impedance and the prefault load flow Thevenin voltage source. The calculation of fault current is now reduced to the proper interconnection of sequence networks. These are summarized in appendix A. In rest of the chapter, we discuss the application of short circuit analysis.

One of the important applications of short circuit analysis is in coordination of the overcurrent relays. The next section discusses this application.

7. Overcurrent Relay Coordination

An overcurrent relay compares the fault current with a reference or pickup current. If the ratio of the fault current to pickup current is greater than unity (a more realistic range is 1.5-2.0), then relay issues a trip decision. Usually, inverse characteristic is used, *i.e.*, as ratio of the fault current to pickup current increases, the time of operation of relay decreases (figure 9.14). In radial systems, this feature provides a natural *discrimination* between faults on the line which is to be primarily protected by the relay and faults on the downstream lines. If the pickup current is chosen such that (1) it is less than the minimum fault current on adjacent line (down stream) and (2) more than pick up current of the adjacent (down stream) relay, then the relay will not only respond to fault on the adjacent (downstream) line but also act after the down stream relay has had it's opportunity. Thus, it can *backup* the operation of corresponding *primary* relay. Now, each primary/backup relay pair has to be coordinated so that, a backup relay responds only if primary relaying system has failed. The minimum time interval required between operation of primary and backup relay is known as Coordination Time Interval (CTI). Pickup current alone can not provide the necessary freedom to maintain the CTI. For providing this time gradation, additionally, a time multiplier setting (TMS) is also provided. By changing the TMS, the curve can be moved (scaled) up or down. For example, an IEC standard inverse curve is given by following equation.

$$t = TMS \times \left(\frac{0.14}{(I/I_s)^{0.02} - 1} \right) \tag{9.48}$$

where I_s is the pickup current. Thus, for the same pickup current, increasing the TMS setting by a factor of two will double the relay operation time.

Each relay, has to perform both the *primary* and the *backup* protection roles. This can be explained by figure 9.15, which shows a typical radial network with directional relays [Damborg and Venkata, 1984]. Use of directional relay implies that a fault can be fed from both the ends. Here, A, B, C, D are the buses or *nodes* of the system. As relay number 9 is located on line AB, near bus A, line AB is the *primary* line for relay 9, with bus B as it's *remote bus.* Bus C and bus D are it's *second remote buses.* Lines emanating from the remote bus are called *remote lines.* For faults at $F1$ and $F2$, relays 11 and 5 act as *primary* relays; with relays 9 and 6 acting as *backup* relays. For a fault at $F3$, relay 9 and 10 act as *primary* relays while relays 5 and 6 act as *backup* relays. For fault at $F4$, relay 11 backs up relay 7. Thus, relay 11 has to coordinate with relay 9 and 6 for faults at $F1$ and $F2$, while it also has to coordinate with relay 7 for fault at $F4$.

In overcurrent relay coordination application, the following information is required during relay coordination:

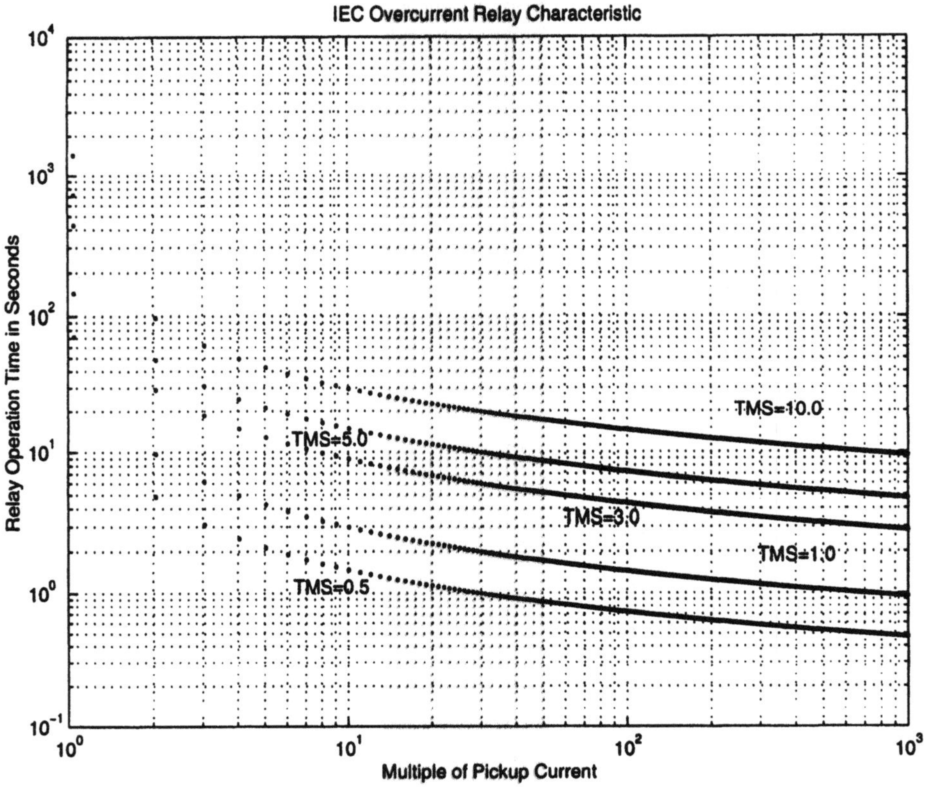

Figure 9.14. Overcurrent Relay Characteristics

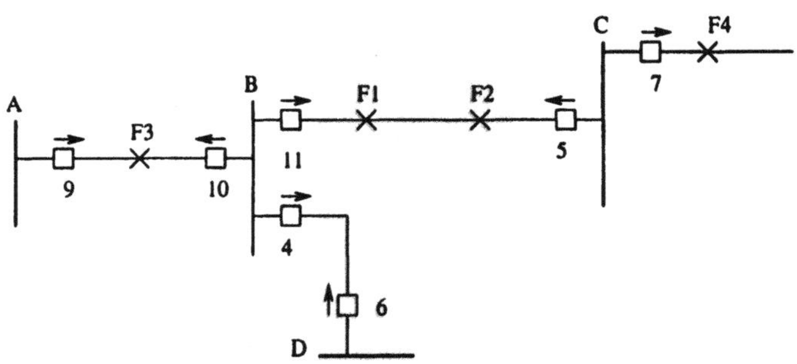

Figure 9.15. Typical Radial System with Directional Relays

1 Fault current at the faulted line for faults at both the remote bus and intermediate points at the line. This information is used to decide the pickup currents of the instantaneous and inverse time overcurrent units.

2 Currents in the adjacent lines for deciding the pickup current of backup relay as well as evaluating primary, backup coordination.

Let the fault be created at bus i and the sequence bus fault currents be given by I_f^0, I_f^1 and I_f^2 respectively. Note that a fault current I_f can be modeled as a current injection of $-I_f$ in the YBUS/ZBUS network model. Then by superposition, the post fault voltages at bus i are given by the following equations:

$$V_i^{f,1} = V_i^{pref,1} - Z_{ii}^1 I_f^1 \tag{9.49}$$

$$V_i^{f,2} = 0 - Z_{ii}^2 I_f^2 \tag{9.50}$$

$$V_i^{f,0} = 0 - Z_{ii}^0 I_f^0 \tag{9.51}$$

Superscript f refers to post fault condition and $pref$ to the prefault condition. To compute the post fault node voltage at an adjacent bus j, we require transfer impedance $Z_{ji}^{0,1,2}$. These are already available from the solution of the following linear system of equations:

$$[YBUS^{012}][Z_i^{012}] = [e_i^{012}]$$

Similarly, by superposition fault voltage at bus j is given by,

$$V_j^{f,1} = V_j^{pref,1} - Z_{ji}^1 I_f^1 \tag{9.52}$$

$$V_j^{f,2} = -Z_{ji}^2 I_f^2 \tag{9.53}$$

$$V_j^{f,0} = -Z_{ji}^0 I_f^0 \tag{9.54}$$

Thus, the post fault sequence currents in branch j to i on which primary relay is installed will be given by the following equations.

$$I_{ji}^1 = \frac{V_j^{pref,1} - V_i^{pref,1} + (Z_{ii}^1 - Z_{ji}^1) I_f^1}{r_{ij}^1 + jx_{ij}^1} \tag{9.55}$$

$$I_{ji}^2 = \frac{(Z_{ii}^2 - Z_{ji}^2) I_f^2}{r_{ij}^2 + jx_{ij}^2} \tag{9.56}$$

$$I_{ji}^0 = \frac{(Z_{ii}^0 - Z_{ji}^0) I_f^0}{r_{ij}^0 + jx_{ij}^0} \tag{9.57}$$

As mentioned in section-1, it is a common practice to neglect the shunt capacitance in fault calculation. However, from a computational view point, if shunt capacitances are available then, the effect of line charging can be easily accounted for in the diagonal elements of the sequence admittance matrices. Positive sequence shunt capacitances are almost always available. If the zero sequence shunt capacitances are also available, then it is worth while to model them in the short circuit analysis. In such a case, the branch current expressions (9.55-9.57) can be modified suitably. For example, for the positive sequence current, the term $\frac{jB^1_{ji}}{2}V_j^{f,1}$ should be added to (9.55) where, $V_j^{f,1}$ is obtained from (9.52). The line currents can be computed by the following linear transformation.

$$[I_{ij}^{abc}] = [T][I_{ij}^{012}]$$

The line currents I_{ji}, $j \in Adj(i)$ give the fault contribution of the 1st tier of lines feeding the fault. They contain line fault currents for the primary relays. The currents for the backup lines of a primary line can be calculated by considering the next tier of nodes adjacent to j, *viz*, set of nodes $k \in Adj(j)$. The corresponding equations are as follows:

$$I^1_{kj} = \frac{V_k^{pref,1} - V_j^{pref,1}}{r^1_{kj} + jx^1_{kj}} + \frac{(Z^1_{ji} - Z^1_{ki})I^1_f}{r^1_{kj} + jx^1_{kj}} \tag{9.58}$$

$$I^2_{kj} = \frac{(Z^2_{ji} - Z^2_{ki})I^2_f}{r^2_{kj} + jx^2_{kj}} \tag{9.59}$$

$$I^0_{kj} = \frac{(Z^0_{ji} - Z^0_{ki})I^0_f}{r^0_{kj} + jx^0_{kj}} \tag{9.60}$$

The line currents can be calculated by simple linear transformation

$$[I_{kj}]^{abc} = [T][I_{kj}]^{012}$$

The pairs $\{I_{k,j}, I_{j,i}\}$ where $k \in Adj\{j\}$, provide backup-primary current pairs for a primary relay at $j \in Adj\{i\}$. Note that all the impedances used above have been computed by a single linear system solution. Fault on a line can be treated by creation of an intermediate bus for computational purpose. At all times, a power system has some load connected to it. However, it may be reasonable to assume that fault currents are much higher than load currents *e.g.*, in overcurrent protection. If such an assumption is acceptable, then fault analysis can be done for an unloaded system. For an unloaded system, prefault condition is defined by setting (1) all prefault voltage magnitudes to 1.0 pu, (2) active and reactive power injections to zero and hence (3) the prefault branch currents to zero. On the other hand, a short circuit circuit study can also be conducted under loaded conditions. In such a case, a load flow has to be carried out on the prefault system to obtain the prefault voltages. For any given load condition, the fault current will also depend upon the number of generating units in service.

8. OO Implementation

The steps involved in short circuit analysis are as follows:

1 Read the system data, type of fault, fault impedance, study to be done on loaded/unloaded system etc.

2 Compute positive, negative and zero sequence YBUS of the system. If only 3ϕ fault is considered, YBUS^1 is sufficient. If LL fault is considered, YBUS^2 is also required. To deal with earth faults YBUS^0 is also required. In YBUS^0, open circuit can be approximated by large impedance, *e.g.*, 10^6 pu. Compute LDL^T decomposition, YBUS^1, YBUS^2 and YBUS^0. The LSS of chapter 4 can be reused.

3 For the list of faulted nodes, compute the sequence driving point impedances by solving equation (9.30). For a fault at bus i, the right hand side corresponds to the i-th column of the identity matrix. Computation of short circuit impedances for more than one bus requires a *for* loop. Depending upon the type of fault, three linear system solvers corresponding to the three sequence networks need to be solved.

4 Compute the short circuit sequence currents, the line currents, short circuit MVA etc.

While discussing the OO implementation of the Short Circuit Analysis program, we consider that the load flow implementation as discussed in chapter 8 is already available. Thus, the class for network can be reused because, it aggregates lines, transformers etc. Objects for lines, transformers etc have already been designed to model the positive, negative and zero sequence models. A suitably edited version of the class is reproduced below.

```
template<class T> class Network {
private :
              // CLASS FOR EACH TYPE OF EQUIPMEMNNT
   branch_line *line;
   branch_X *Xmer;
   shunt *shunts;
   generator *gen;
   vector<float> rated_bus_vol;
                     // CLASS FOR THE TYPE OF BUSES
   node_PQ  *PQ_bus;
   generator *ngens;
                      // FUNCTIONS TO HELP FORM SC YBUS
```

```
   void add_lines(spmat<T> * A); // add line data to ybus
   void add_lines_zero(spmat<T> * A); // add line data to
                                         zero sequence bus
   void add_xmers(spmat<T> * A);
   void add_xmers_zero(spmat<T> * A);
   void add_shunts(spmat<T> *A);
   void add_shunts_zero(spmat<T> *A);
   void add_gens(spmat<T> *A);
   void add_gens_zero(spmat<T> *A);
   void add_eq_load_admit(spmat<T> *A);
           // FUNCTIONS  FOR SHORT CIRCUIT ANALYSIS
void SLG_cal(const char * bus_name,
const vector<cmplx<float> > &voltage, vector<cmplx<float> >
& impe,ofstream & fout, vector<cmplx<float> > &z1,
vector<cmplx<float> > & z2, vector<cmplx<float> > & z0);

void LL_cal(const char * bus_name,
const vector<cmplx<float> > &voltage, vector<cmplx<float> >
& impe, vector<cmplx<float> > &z1,
vector<cmplx<float> > & z2);

void LLG_cal(const char * bus_name,
const vector<cmplx<float> > &voltage, vector<cmplx<float> >
& impe, vector<cmplx<float> > &z1,
vector<cmplx<float> > & z2, vector<cmplx<float> > & z0);

void three_phase_cal(const char * bus_name, const
vector<cmplx<float> >& voltage,vector<cmplx<float> >& impe,
vector<cmplx<float> > & z1);

vector<cmplx<float> > voltage_cal(vector<cmplx<float> >& cnt,
vector<cmplx<float> > &impe);

vector<cmplx<float> > current_cal(vector<cmplx<float> > & cnt);
void  print_flow(ofstream & fout,vector<cmplx<float> > &V);

public:
inline  Network() {read_data();};
//default ctor, reads and intialize from file data
   ~Network();
 Network(branch_line *A,  branch_X *B, shunt *C,node_PV *D,
 node_PQ *E, int & nl, int & nX, int & nS, int
```

```
 & nPV, int & nPQ) ;

 spmat<T> *   create_ybus_p();
// create positive sequence ybus
 spmat<T> *   create_ybus_n();
// create negative sequence ybus
 spmat<T> *   create_ybus_z();
// create zero sequence ybus
 asso_arr<T>  name_to_no;
 void short_circuit_study();
 };
```

The study of class *Network* reveals the following:

- Functions like add_lines() add the positive sequence line model to the YBUS and corresponding functions add_lines_zero() are used to add zero sequence model to the zero sequence YBUS. Similar comments hold for transformers and shunts. In case of generators, add_gens() adds the positive sequence impedances to the positive sequence YBUS. Similar statement holds good for adding load admittances by add_eq_load_admit(). These functions are called inside public functions like create_ybus_zero() which creates a zero sequence YBUS.
- Functions like SLG_cal() implement the fault current calculation for SLG fault. Similar comments apply for rest of the faults.
- Functions like voltage_cal() and current_cal() generate branch flows for applications like relay coordination.
- The core computation in the short circuit analysis is the solution of sparse and symmetric linear system of equations. For this purpose, the linear system solver of chapter 4 is reused by specifying complex data type in the template specification.
- The complete short circuit analysis is implemented by the function short_circuit_analysis().

9. Summary

Conceptually, a 3ϕ short circuit analysis is simple creation and application of positive sequence Thevenin equivalent circuit. When unbalanced faults are encountered, Thevenin equivalent of negative and zero sequence network is also required. The computation of sequence impedances can be done by either the ZBUS building algorithm or by inversion of the YBUS matrix. If only driving point and transfer impedances for a few nodes is required, sparse linear system solver implementation is preferred. In the YBUS matrix approach, generators are modeled as equivalent current sources with source admittance corresponding to either subtransient, transient or steady state. The choice depends upon the time interval of the study following the initiation of the fault. Modeling of the positive and negative sequence networks is quite straightforward. However, the zero sequence network requires special care. [Kundur, 1994] states that zero sequence impedance of a line may be 2-3 times positive sequence impedance. For a transformer, the zero sequence impedance also depends upon the core construction. Handling of zero sequence impedance for transformers requires special care. It is best supported by writing library functions. Readers are requested to check the `struct.h` file for the same.

Fault level can be different for unloaded and loaded system. If the fault level has to be computed considering loads, prefault bus voltages at the faulted nodes is required. These should be computed using a load flow program. The equivalent load admittance can then be computed and treated as shunt admittance load in the positive and negative sequence admittance matrices by the program. As far as zero sequence network is concerned, our approach is to give user the choice of incorporating loads as shunt impedances.

Short circuit analysis is primarily required in (1) choosing CB rating and (2) relay coordination. Hence, short circuit analysis results are often given by 3ϕ short circuit MVA and 1ϕ short circuit MVA rating. The overcurrent relay coordination application requires computation of branch currents for a large number of faults. By applying superposition theorem, branch contribution of fault currents can be superimposed on prefault currents.

10. Further Reading

A lot has been written on short circuit analysis. Books like [Stagg and El-Abiad., 1968; Brown, 1975; Pai, 1979; Kusic, 1986; Wallach, 1986; Singh, 1986] have also discussed the application in detail. The zero sequence network modeling is an important issue in short circuit studies. [Blackburn, 1993] is devoted to only the study of sequence networks. The treatments by [Blackburn, 1987; Barret et al., 1997] are recommended for further details. [Kusic, 1986] has also considered incorporation of mutually coupled lines. [Damborg and Venkata, 1984] discusses in detail the problem of overcurrent and distance relay coordination. [Kundur, 1994] has discussed sequence networks in the context of protection. [Anderson, 1973] considers machine models and fault analysis in detail. Chapter 7 of IEEE Std 399-1990 [Smith, 1990] considers short circuit study procedure, preparation of input data, interpretation of results with test data and examples.

11. Review Questions

1 Show that for fault analysis a prefault load flow study gives the Thevenin equivalent voltage source at the bus of interest.

2 Justify the assumptions made in short circuit analysis. Why is effect of shunt charging and transformer taps neglected in short circuit analysis? Should shunt charging be neglected in cables and/or EHV transmission lines?

3 Show that pu short circuit impedance equals inverse of the pu short circuit (MVA) interrupting capacity. A short circuit study on a 66 kV system shows the following 1) MVA_{SC}=600 MVA and 2) $MVA_{\phi GSC}$=650 MVA. If the MVA base is 100 MVA, compute X_1, X_2 and X_0 in pu.

4 Under the assumption at $Z_1 = Z_2 = Z_0$, show that, a)I_{LL}=0.866 $I_{3\phi}$ and b)$I_{SLG} = I_{3\phi}$.

5 For a SLG fault, can I_{SLG} be greater than $I_{3\phi}$ at some locations?

6 Derive from first principles, the zero sequence equivalent for three winding 3ϕ transformers of table 9.2 and 9.3.

7 In a cylindrical rotor synchronous machine, the net mmf produced in air gap by zero sequence armature currents is zero. Hence, the equivalent inductance defined by net flux linkages to current is zero. Thus, zero sequence reactance should be zero. Comment on these statements.

8 In a synchronous machine, if the rotor field circuit is open circuited and effect of damper winding neglected then, what would be the negative sequence impedance as measured at the armature terminals?

9 To analyze the effect of negative sequence armature currents on the flux linkages, we can consider the rotor field circuit as short circuited. The armature mmf rotates at synchronous speed in the opposite direction to the rotor, appearing as a double frequency ($2w_s$) mmf wave with respect to the rotor. Consequently, an emf is induced in the rotor. The resultant currents in the field winding are inductive in nature and hence produce a demagnetizing flux (see figure 9.16). Hence, the

net flux linkage reduces and so does the equivalent reactance (X_2) defined by net flux linkages per phase to the corresponding phase current. Comment!

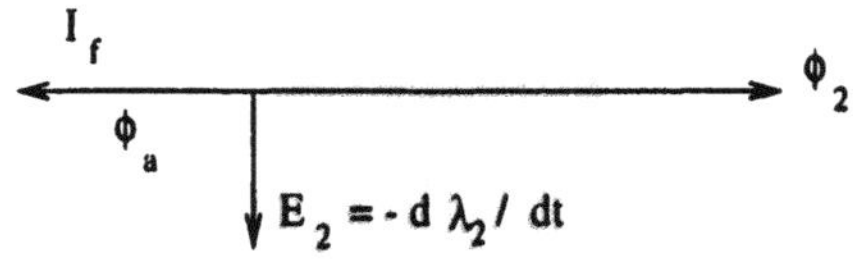

Figure 9.16. Induced current in rotor winding due to negative sequence excitations

10 With reference to problem (9), the rotor has currents of frequency $2w_s$. Show that because of the rotation of rotor, two rotor mmf waves of equal magnitude but speed $3w_s$ and $-w_s$ will be seen in the air gap. This implies that harmonic voltages will be induced in armature.

11 Let $i_a(t) = \sum_{n=1}^{\infty} |I_n| \cos nw_0 t$
$i_b(t) = \sum_{n=1}^{\infty} |I_n| \cos n(w_0 t - \frac{2\pi}{3})$
$i_c(t) = \sum_{n=1}^{\infty} |I_n| \cos n(w_0 t + \frac{2\pi}{3})$
Show that a) 3rd harmonics currents behave like zero sequence components and b) 2nd harmonics behave like the negative sequence components. How will the fifth and seventh harmonic components behave?

12 A fault has to be created at 50% length of a transmission line. The short circuit analysis program is capable of only computing fault currents at the busses. What modification in the data file will you require to get the fault currents?

13 Solve example 9.3 using the short circuit analysis program on the diskette. Also, study the 24-node system available on the diskette.

14 Comment on the following statement. "In comparison to phase relays a separate ground relay helps in clearing high impedance ground faults".

Chapter 10

POWER SYSTEM STATE ESTIMATION

When using a mathematical model, careful attention must be given to the uncertainties in the model.

—Richard P. Feynman: What Do Your Care What Other People Think

1. Introduction

Secure and economic operation is an important requirement of modern power systems. This is achieved through a hierarchical control system incorporating both local and centralized control. The control of the power system can be classified into following:

- **Protection:** It is a real-time control system which detects an undesirable event, *e.g.*, a fault, underfrequency, undervoltage etc. Fault protection is achieved by relays that either raise an alarm or trip a circuit breaker, thereby isolating the faulty equipment. Time scale of operation is in the order of few cycles.

- **Load Frequency & Voltage Control:** The next level of control is meant to maintain load generation balance by regulating mechanical input power of generators. It is achieved by governor control and Automatic Generation Control (AGC). Due to the rotational inertia of electrical machines, the time scale of operation is large, *i.e*, of the order of seconds. Both the protection system and AGC act without human intervention. The primary role of AGC system is to maintain load generation balance, regulate system frequency and tie line flows. Though AGC serves a critical role in steady state system operation, it does not account for network constraints like limits on loading of lines,

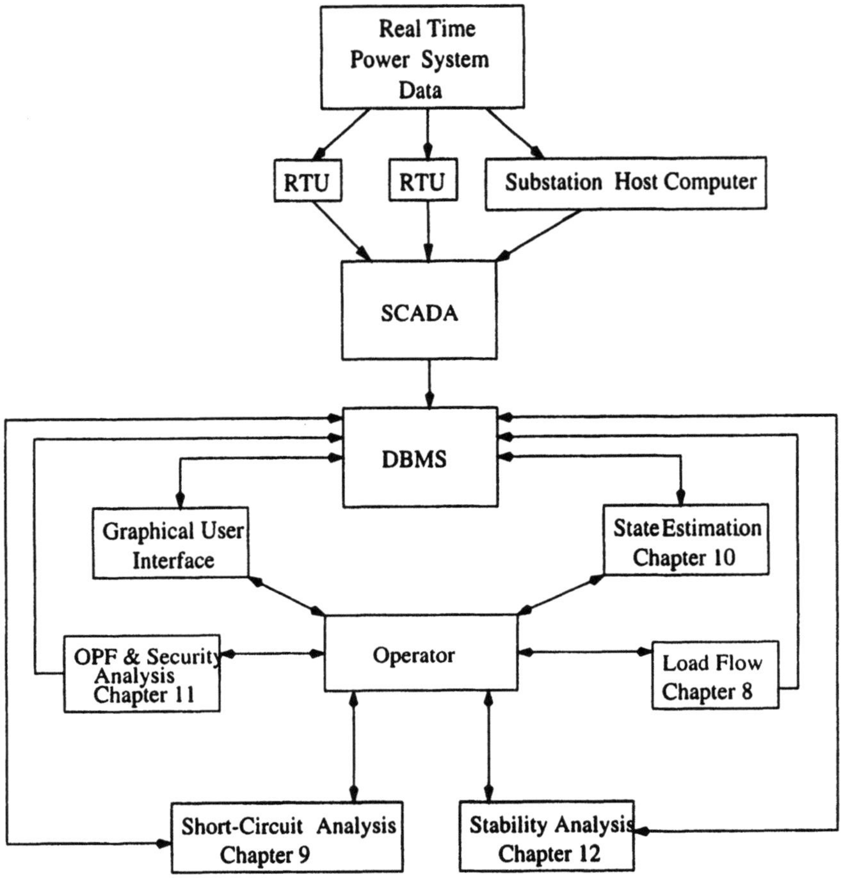

Figure 10.1. Online Control of a Power System

voltage constraints etc. Automatic Voltage Regulators (AVRs) are installed on the generators to maintain a constant terminal voltage under changing load conditions.

- **Energy Management System:** The online steady state control is achieved by Energy Management System (EMS). The snapshot of the power system is captured in regular intervals (say in minutes) at an Energy Control Center (Figure 10.1). Remote Terminal Units (RTUs) (Figure 10.2) placed at the substation level gather digital data (circuit breaker position) and analog data (voltage, current, active power, reactive power etc.) and transmit it through Supervisory Control and Data Acquisition System (SCADA) to the energy control center. At an energy control center, an online model of the power system is constructed. This online model is used to evaluate security level and implement control actions for secure and economic operation. At this level, decision making requires, human participation via the system operator.

The concept of *steady state security* describes the ability of the system to perform satisfactorily, *i.e.*, without loss of a load, and violations of apparatus limits in case of disturbance or contingencies.

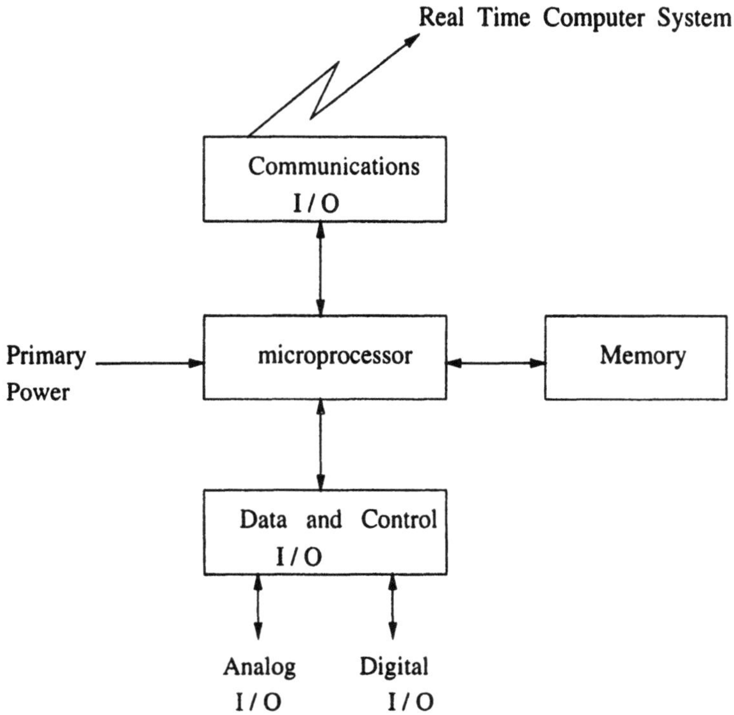

Figure 10.2. Remote Terminal Unit

1.1. Steady State Security

From security considerations, a power system is categorized into four states [Liacco, 1978; Stott et al., 1987], *viz*, 1) normal state 2) alert state 3) emergency state and 4) restorative state. Figure 10.3 shows the various state of a power system. In the *normal* or *secure* state, all bus voltage magnitudes, line flows, generator Q limits etc. are within the prescribed limits. All loads are met; further, in the event of a contingency there will not be any violation or loss of load. Contingency(ies) may push the system into an *alert* state, where, an additional disturbance will cause either (partial) loss of service or violation of operating limits. An alert state may be correctable, *i.e.*, it may be possible to move back to secure state by preventive control action. In the *emergency* state, all loads are supplied but operating limits are violated. A correctable emergency corresponds to a situation when through control action, violations can be corrected without loss of load. In a non-correctable emergency, violations can not be corrected without loss of load. When, loss of service takes place, a system is led into the *restorative* state. In the restorative state, a system operator tries to bring the system back to normal state.

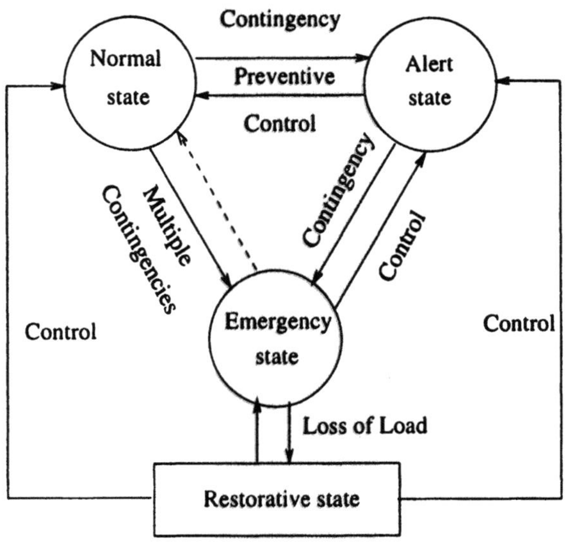

Figure 10.3. Static Security Levels

Figure 10.3 illustrates various possible transitions in the power system.

Evaluation of power system security level requires the following (Figure 10.4):

- **Monitoring of the system:** In this mode the system information, *e.g.*, line flows, Circuit Breaker (CB) position, bus voltages etc are gathered and displayed in energy control center.

- **Estimation:** All real life measurements contain noise. Hence, state of the power system, *i.e.*, all of the complex bus voltages can not be simply computed by tools like a load flow program. Instead an estimation problem has to be solved to filter out noise. Filtering of noise requires *redundancy* in the measurements.

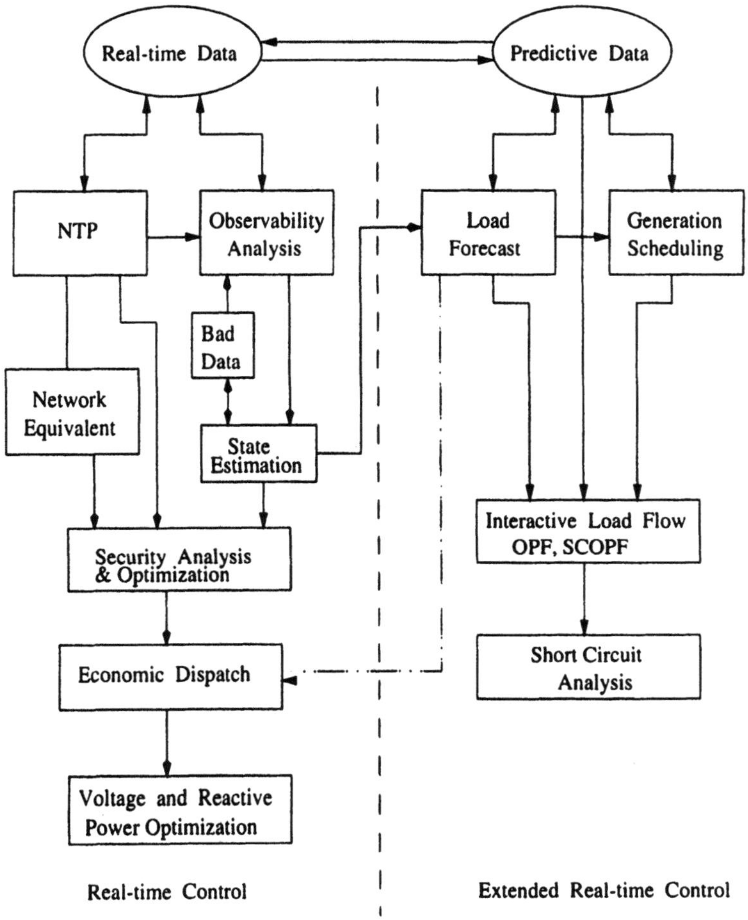

Figure 10.4. An Energy Control Center Scenario

- **Analysis** and **decision:** making tools like contingency analysis, operator request load flow, optimal power flow, load forecasting, unit commitment etc. are used to evaluate the security of the power system and device control strategies to maintain system in optimal and secure mode. Two possible control strategies exist. In *preventive* control, the system operator plans for the future (*e.g.*, 10-15 minutes ahead), while in *corrective* control, current abberations are rectified.

1.2. Overview of the Estimation Process

Estimation of the power system state requires the following:

1 **Network Topology Processing:** With the knowledge of the real time circuit breaker status, substation topology and power apparatus connectivity, the topology of the network has to be developed.

2 **Observability Analysis:** Analysis of placement of transducers to ascertain whether the estimation problem can be carried out.

3 **State Estimation:** It is a computational tool to filter out noise from the measurements and estimate the power system state, *i.e.*, phasor voltages at all the nodes. In general, higher the redundancy in data, better is the quality of estimation.

4 **Bad Data Processing:** The residuals obtained from step-3 are analyzed to detect and eliminate bad or faulty measurements. Step 2-4 are then repeated to improve the quality of the estimation.

Example 10.1: Consider a simple system as shown in figure 10.5. To develop the topology, we need to know the status of switches k_1, k_2 and k_3. Let all the switches be closed. Then the electrical circuit has two loops. Now the state of the system can be completely described by the knowledge of loop currents I_1 and I_2. Hence, state estimation problem estimates I_1 and I_2. If only one ammeter X_1 or X_2 or X_3 is present, then the system is unobservable because redundancy is less than one. If all the three ammeters are in one branch, redundancy is greater than one but still the system is unobservable. For the configuration shown, system is observable with a redundancy factor of 1.5. When currents I_1 and I_2 are measured by ammeters X_1 and X_2, respectively, the values read are 1.1 and 0.95. With measurements for both I_1 and I_2, the system is observable but elimination of noise is not possible. Now let there be an independent measurement of current $I_3 = I_1 + I_2$ by an ammeter X_3. Let the measured value be 2.02. Note that the sum of currents measured by X_1 and X_2 is not equal to X_3. This is an indication of *noise.* Table 10.1 illustrates various possible solutions to the problem. Thus, there is no unique solution to the problem. The problem of state estimation is to find that estimate which closely fits all the measurements. In addition to noise, if there are grossly erroneous measurements then they can be eliminated prior to estimation by simple consistency checks. We now discuss the development of network topology processor.

2. Network Topology Processing

The role of a network topology processor is to develop the electrical network topology similar to the single line diagram in a load flow analysis. It furnishes information from which the admittance matrix

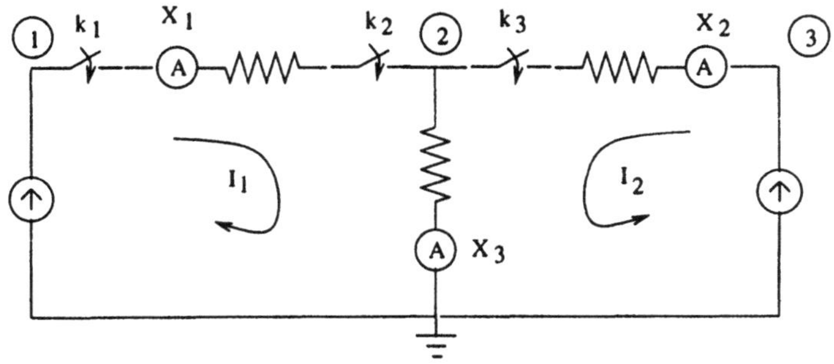

Figure 10.5. Example 10.1

Table 10.1. Noise and Estimation Problem

Measurements	Estimated State
X_1 and X_2	X_1=1.1, X_2=0.95
X_1 and X_3	X_1=1.1, X_2=0.92
X_2 and X_3	X_1 = 1.07 and X_2=0.95
Least Square Estimate	X_1=1.09
Solve $\begin{bmatrix} 1 & 0 \\ 0 & 1 \\ 1 & 1 \end{bmatrix} \begin{bmatrix} x_1 \\ x_2 \end{bmatrix} = \begin{bmatrix} 1.1 \\ 0.95 \\ 2.02 \end{bmatrix}$	X_2=0.94

model $[I] = [Y][V]$ of the network can be developed. Network topology is determined from the real time circuit breaker status, substation topology and offline connectivity data of transmission lines, transformers, reactors. This analysis can be done by viewing the power system as a hierarchical network (figure 10.6a). At the topmost layer, we observe a network of substation interconnected through *circuits*, with each substation having it's own topology. A substation can have various configurations like, ring bus arrangement (figure 10.6b), double breaker and $1\frac{1}{2}$ (section II) breaker topology. The selection of a substation topology is dictated by cost, reliability requirement, etc. For example, a ring bus arrangement has a single breaker per feeder, while a $1\frac{1}{2}$ circuit breaker (CB) arrangement requires 3 breakers for every 2-feeders. In general, more breakers per feeder permit more paths to reach a feeder from a bus and hence provides higher level of flexibility at the corresponding cost. The substation topology also affects the complexity of protection system.

There are two types of node in a substation. The ends of a CB define *physical* nodes while an *electrical* node corresponds to a set of physical

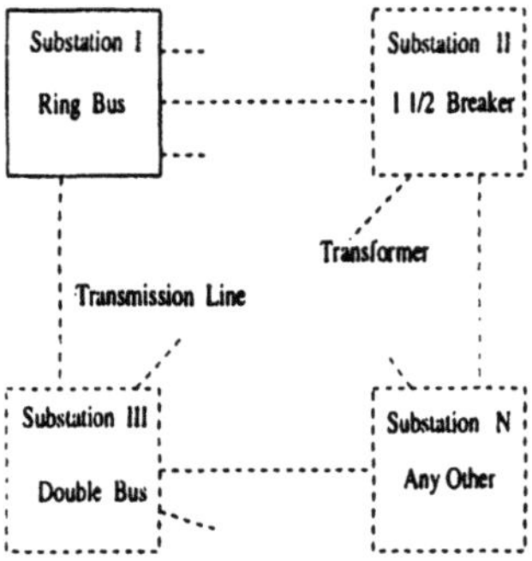

Figure 10.6a. Hierarchical Power System

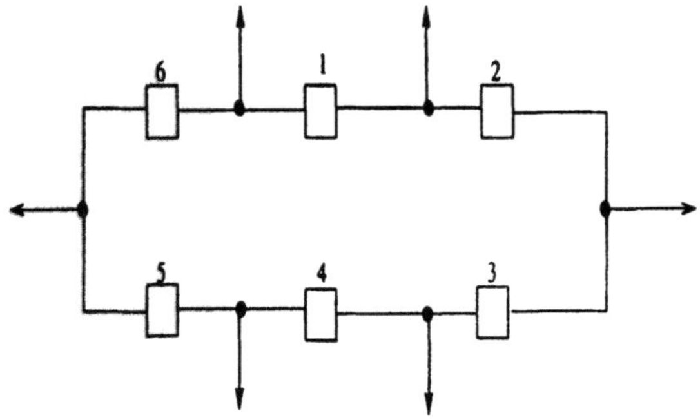

Figure 10.6b. A Ring Main Substation Topology

nodes interconnected by closed CBs. Ideally, a closed CB provides a path with zero electrical impedance. Generally, closing all of the CBs in a substation will result in a single electrical node. However, depending upon the status of CBs, a substation may split into one or more electrical nodes. For example, in the ring bus arrangement (Figure 10.6b), opening breakers 1 and 4 splits the substation into two electrical nodes. NTP involves graph theoretic computations.

2.1. Algorithm

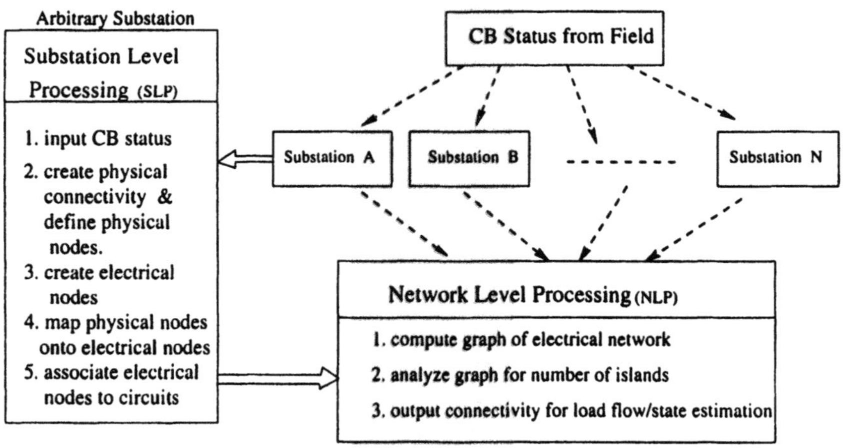

Figure 10.7. Typical Process Flow for Network Topology Processing

NTP involves two levels of processing [Pandit et al., 2001a]:

- Substation level processing;

- Network level processing.

2.1.1 Substation Level Processing

Substation level processing involves creation of the electrical graph of a substation by analyzing the CB status. Then, the graph is analyzed for components; each component represents an electrical node. The electrical graph of the substation is computed as follows:

1 Mark the physical nodes of the substation.

2 Between any two physical nodes connect an edge only if a closed CB connects them. An edge establishes a zero impedance path between a pair of physical nodes.

3 Analyze the substation graph for components. A component is a maximal connected subgraph. In a component, all physical nodes are connected by closed CBs. Therefore, it represents a single electrical node. Thus, the components provide mapping of physical nodes to electrical nodes in a substation.

Example 10.2: Consider a substation having $1\frac{1}{2}$ CB arrangement as shown in figure 10.8a. The darkened boxes indicate closed CBs. The configuration has 8 physical nodes. With the current CB status, it can be seen that node-1 is isolated from 8, thereby creating two electrical nodes. The same conclusion can be drawn by building the graph of the system shown in figure 10.8b. The graph is composed of two components *viz.* bus-1 and rest of the busses $\{2, 3, \ldots, 8\}$. The following example

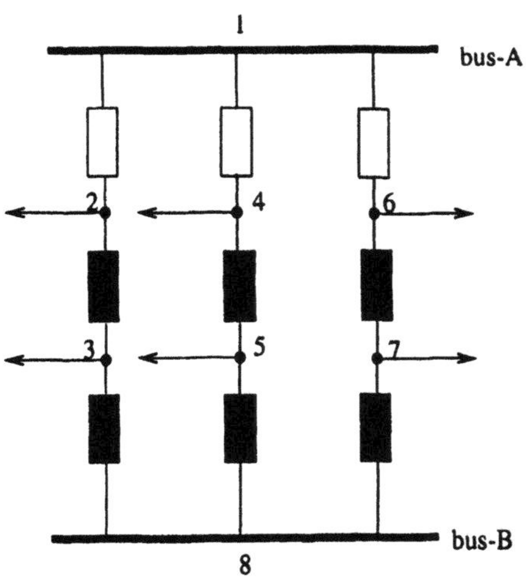

Figure 10.8a. $1\frac{1}{2}$breaker topology

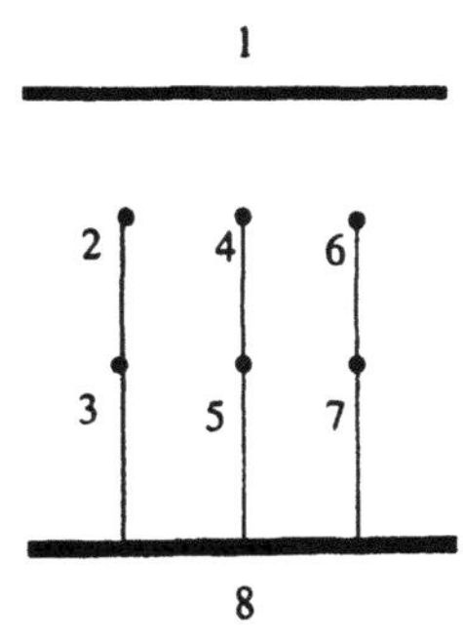

Figure 10.8b. Substation Graph

describes a method to compute the components of a graph.

Example 10.3: Compute the components in figure 10.8b by graph theoretic computation.

We begin with an arbitrary node say node 2. Let us mark node 2 as processed and define a set *comp* to store component to which it belongs. Recall that class *set* was developed in chapter 3. Initialize $comp = \{2\}$. Since, adjacent nodes of 2 belong to the same component as 2, we add $Adj\{2\}$ to *comp*, *i.e.*,

$$comp+ = Adj\{2\} = \{2, 3\}$$

Next, we check up if any node in set *comp* is not processed (in this case node 3). Since, node 3 is not processed, by the very same logic, nodes adjacent to 3 must belong to the same component as 3, therefore, we add $Adj\{3\}$ to *comp*. The result is

$$comp+ = Adj\{3\} = \{2, 3, 8\}$$

Now node 3 is marked as processed. In the next step, $Adj\{8\}$ is added to *comp*. Therefore,

$$comp+ = Adj\{8\} = \{2, 3, 5, 7, 8\}$$

Node 8 is marked processed. The first unprocessed node in set *comp* is '5'. Therefore, in the next step

$$comp+ = Adj\{5\} = \{2, 3, 4, 5, 7, 8\}$$

When the node 4 is processed, the set *comp* remains unaltered. In the next step, node 7 is processed leading to the following:

$$comp+ = Adj\{7\} = \{2, 3, 4, 5, 6, 7, 8\}$$

Processing of node 6 does not change the set *comp*. At this stage all the nodes in set *comp* are processed. Therefore, we have identified one component of the graph.

Now, a check is made to ascertain if any node is still unprocessed. It turns out that node 1 is still unprocessed. So, a search for component of node 1 is initialized, *i.e.*, $comp = \{1\}$. Since, $Adj\{1\} = \{\phi\}$, adding $Adj\{1\}$, makes no difference to the set *comp* and hence, the other component turns out to be an isolated node 1. At this point, all the nodes of the graph have been processed and therefore, the algorithm terminates with two components *viz.* $\{2, 3, 4, 5, 6, 7, 8\}$ and $\{1\}$. To facilitate graph theoretic computations like finding a component, a class *Graph* has been implemented (refer `graph.h`) in diskette.

2.1.2 Network Level Processing

The network level processing involves aggregation of substations into a network by connecting electrical nodes of different substations through power apparatus like lines, transformers etc. At this stage an admittance network model can be created. The graph of the network is analyzed for components, *i.e.*, islands in power system terminology. The steps involved in finding islands in a network are essentially the same as that of substation level processing. Opening of tie lines with external equivalent can cause islands with load/generation imbalance. Once, the topology of the network is established, the next step is to analyze the meter placement to ascertain whether sufficient data in terms of number and location of measurements is available for estimating the power system state. This processing is known as the observability analysis. During the design stage, observability analysis can be viewed as a meter placement problem. In the online mode, it is a check required to ascertain that estimation problem can be solved.

3. Observability Analysis

A system is said to be observable if the number and placement of measurements is such that the state of the system *viz*, all its complex voltages can be estimated (*e.g.*, by a procedure similar to load flow algorithm). A system in which all the states can not be estimated is said to be unobservable. The measurements made on the system belong to one of the following categories:

- Bus voltage magnitude measurement;
- Active and/or reactive power line flow measurement;
- Transformer tap settings;
- Active and/or reactive power injection measurement.

Measurement of voltage angles is difficult because it requires that all the voltage transducers placed geographically apart in a system to be synchronized to a same reference phasor. To understand the importance of proper meter placement, let us consider an example.

Example 10.4: Figure 10.9 shows a 5-node network along with the placement of meters. Voltage magnitudes at busses 1, 2 and 3 are monitored and so are P+jQ flows on lines 1-3, 2-3 and 2-1. Is the system observable, *i.e.*, is this data sufficient to compute phasor voltages $V_1, V_2, \ldots, V_5$?

Solution: There are a total of 9 measurements and 9 states ($\delta_2, \delta_3, \delta_4, \delta_5$, $|V_1|, |V_2|, |V_3|, |V_4|, |V_5|$) in the system. However, the locations of the measurements is such that state of nodes 4 and 5 cannot be determined uniquely. For example, if line flow 1-3 is (5+j3) MW and flow from 2 to 3 is (4 + j2) MW, then there is no way to decide distribution of 9 MW between lines 3-4 and 3-5. Based upon the transducer placement, it can be seen that the net generation at nodes 1 and 2 can be calculated. The equivalent subnetwork consisting of nodes 1, 2 and 3 can be completely analyzed. However, nodes 4 and 5 are not observable.

One way to tackle the problem of unobservability, say due to failure of a communication link is to use *pseudo*-measurements. Pseudo-measurements are fictitious measurements based upon the historical database, operator's guess etc. Obviously, such measurements are given low weightages during estimation.

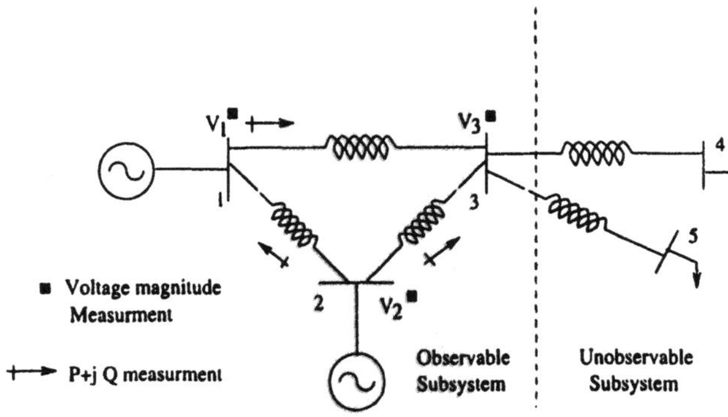

Figure 10.9. 5 Node System for Observability Analysis

If all the data required by load flow analysis is telemetered to energy control center, then even it can also be viewed as a rudimentary state estimator (can you guess it's limitations?). We can say that in load flow analysis based state estimation, the placement of meters *viz* injection measurements at the PQ nodes and P and V transducers in PV-nodes is such that system is just observable. Let in a n-node system, bus-1 be the reference bus and $1, \ldots, g$ be the generator busses. Then, in the equation (10.1), first $n-1$ rows correspond to injected active power measurements at busses $2, \ldots, n$ followed by Q-measurements on PQ busses $g+1, \ldots, n$. The last g rows shown compactly by submatrix I_g (unit matrix of size g) in the Jacobian represent the voltage measurements at the PV busses. The coefficient matrix in (10.1) is Jacobian corresponding to the measurements. The state estimation, (*i.e.*, computing all complex voltages) is possible because the Jacobian is nonsingular, *i.e.*, it has full column rank. Further, the solution in equation (10.1) is uniquely defined because number of measurement and the unknowns are equal. Let us now formally define *redundancy* as a ratio of the number of measurements to number of system states. Then, load flow analysis has a redundancy factor 1. When the system has a redundancy factor 1, noise cannot be eliminated. Also, if fewer measurements are available, then the number of states, *i.e.*, number of columns exceed the number of rows and, the system is underdetermined. Now, it may have full row rank but does not have full column rank. Such a system has either 1) no solution or 2) infinite number of solutions. The lack of a unique solution or no solution implies an unobservable system. In the previous example, we had a situation of infinitely many solutions because there is no way to specify how the power at node 3 from 1-2 and 2-3 should be distributed to node 4 and 5.

$$J_{aug}\Delta x = \Delta z \tag{10.1}$$

where, $J_{aug} =$

$$\begin{bmatrix} \frac{\partial P_2}{\partial \delta_2} & \cdots & \frac{\partial P_2}{\partial \delta_n} & \vdots & \frac{\partial P_2}{\partial V_{g+1}} & \cdots & \frac{\partial P_2}{\partial V_n} & \frac{\partial P_2}{\partial V_1} & \cdots & \frac{\partial P_2}{\partial V_g} \\ \vdots & \ddots & \vdots & \vdots & \vdots & \ddots & \vdots & \vdots & \ddots & \vdots \\ \frac{\partial P_n}{\partial \delta_2} & \cdots & \frac{\partial P_n}{\partial \delta_n} & \vdots & \frac{\partial P_n}{\partial V_{g+1}} & \cdots & \frac{\partial P_n}{\partial V_n} & \frac{\partial P_n}{\partial V_1} & \cdots & \frac{\partial P_n}{\partial V_g} \\ \cdots & \cdots & \cdots & \cdots & \cdots & \cdots & \cdots & \cdots & \cdots & \cdots \\ \frac{\partial Q_{g+1}}{\partial \delta_2} & \cdots & \frac{\partial Q_{g+1}}{\partial \delta_n} & \vdots & \frac{\partial Q_{g+1}}{\partial V_{g+1}} & \cdots & \frac{\partial Q_{g+1}}{\partial V_n} & \frac{\partial Q_{g+1}}{\partial V_1} & \cdots & \frac{\partial Q_{g+1}}{\partial V_g} \\ \vdots & \ddots & \vdots & \vdots & \vdots & \ddots & \vdots & \vdots & \ddots & \vdots \\ \frac{\partial Q_n}{\partial \delta_2} & \cdots & \frac{\partial Q_n}{\partial \delta_n} & \vdots & \frac{\partial Q_n}{\partial V_{g+1}} & \cdots & \frac{\partial Q_n}{\partial V_n} & \frac{\partial Q_n}{\partial V_1} & \cdots & \frac{\partial Q_n}{\partial V_g} \\ & & & & & & & 1 & & \\ & & & & & & & & \ddots & \\ & & & & & & & & & 1 \end{bmatrix}$$

$$\Delta x = \begin{bmatrix} \Delta\delta_2 \\ \vdots \\ \Delta\delta_n \\ \\ \Delta|V_{g+1}| \\ \vdots \\ \Delta|V_n| \\ \cdots \\ \Delta|V_1| \\ \vdots \\ \Delta|V_g| \end{bmatrix} \quad \text{and } \Delta z = \begin{bmatrix} \Delta P_2 \\ \vdots \\ \Delta P_n \\ \cdots \\ \Delta Q_{g+1} \\ \vdots \\ \Delta Q_n \\ \cdots \\ 0 \\ \vdots \\ 0 \end{bmatrix}$$

On the other hand, when redundancy factor is greater than 1, *i.e.*, number of measurements exceed the number of states, and the Jacobian has full column rank, then, it can be shown that the augmented system of equations has a unique solution if noise is zero. If noise is not zero, a more realistic situation, then the system of equations will not be consistent. Then, a unique solution can be specified in a Least Square (LS) minimization framework wherein the sum of the square of residuals $\sum(z_i - f_i(x))^2$ is minimized. When noise is zero, the LS solution corresponds to LSS solution. We conclude that

> **A system is observable if the Jacobian of the measurement vector has full column rank.**

Example 10.5: With bus 1 as reference, the structure of the Jacobian corresponding to measurements in Example 10.4 is shown below:

$$
\begin{array}{c}
P_{13}\\ Q_{13}\\ P_{21}\\ Q_{21}\\ P_{23}\\ Q_{23}\\ |V_1|\\ |V_2|\\ |V_3|
\end{array}
\overset{\begin{array}{ccccccccc}\delta_2 & \delta_3 & \delta_4 & \delta_5 & |V_1| & |V_2| & |V_3| & |V_4| & |V_5|\end{array}}{\left[\begin{array}{ccccccccc}
 & \times & & & \times & & \times & & \\
 & \times & & & \times & & \times & & \\
\times & & & & \times & \times & & & \\
\times & & & & \times & \times & & & \\
\times & \times & & & & \times & \times & & \\
\times & \times & & & & \times & \times & & \\
 & & & & 1 & & & & \\
 & & & & & 1 & & & \\
 & & & & & & 1 & &
\end{array}\right]}
\left[\begin{array}{c}
\Delta\delta_2\\ \Delta\delta_3\\ \Delta\delta_4\\ \Delta\delta_5\\ \Delta|V_1|\\ \Delta|V_2|\\ \Delta|V_3|\\ \Delta|V_4|\\ \Delta|V_5|
\end{array}\right]
=
\left[\begin{array}{c}
\times\\ \times\\ \times\\ \times\\ \times\\ \vdots\\ \\ \vdots\\ \times
\end{array}\right]
\qquad (10.2)
$$

The rank of a sparse matrix depends upon (i) its structure and (ii) the numerical values. In *structural* rank, value of non-zero is not important. Hence, all non-zeros are set to 1. On the other hand, *numerical* rank depends upon both, the matrix structure as well as the numerical values of non zero elements. In equation (10.2), observe that four columns (for $\delta_4, \delta_5, |V_4|, |V_5|$) of the matrix are identically zero and hence, the matrix does not have full column rank. Thus, the system is unobservable.

Now, if flows P_{34} and P_{35} are monitored then, 2-additional rows with the following structure are added and the rank of the coefficient matrix increases by 2, *i.e.*, structural rank becomes 7.

$$
\begin{array}{c} P_{34}\\ P_{35}\end{array}
\overset{\begin{array}{ccccccccc}\delta_2 & \delta_3 & \delta_4 & \delta_5 & |V_1| & |V_2| & |V_3| & |V_4| & |V_5|\end{array}}{\left[\begin{array}{ccccccccc}
 & \times & \times & & & & \times & \times & \\
 & \times & & \times & & & \times & & \times
\end{array}\right]}
\qquad (10.3)
$$

In fact, each new measurement increase the rank by 1. This is the best that can happen, *viz*, a measurement can at most increment rank by 1. Adding corresponding Q-measurements do not increase the *structural* rank, but adding voltage measurements at $|V_4|$ and $|V_5|$ provide full column rank. However, by visual inspection of figure 10.9, one would expect that monitoring $P_{34} + jQ_{34}$ and $P_{35} + jQ_{35}$ should make the system observable. The apparent problem seems to be in our choice of setting all the non-zeros of measurement Jacobian represented by $\times$ to 1. Such a choice makes structure of rows corresponding to measurements P_{ij} and Q_{ij} identical except for those busses that are connected to reference bus. As a consequence, the significance of the additional reactive power information is lost. The problem can be overcome by either computing the numerical rank of the coupled Jacobian or by computing structural rank in a decoupled formulation. The corresponding P-δ and Q-V Jacobian structures with $P_{34} + jQ_{34}$ measurements and $P_{35} + jQ_{35}$ measurements is as shown below:

$$H_1 = \begin{array}{c} \\ P_{13} \\ P_{21} \\ P_{23} \\ P_{34} \\ P_{35} \end{array} \left[\begin{array}{cccc} \delta_2 & \delta_3 & \delta_4 & \delta_5 \\ & \times & & \\ \times & & & \\ \times & \times & & \\ & \times & \times & \\ & \times & & \times \end{array} \right] \quad H_4 = \begin{array}{c} \\ Q_{13} \\ Q_{21} \\ Q_{23} \\ Q_{34} \\ Q_{35} \\ \\ |V_1| \\ |V_2| \\ |V_3| \end{array} \left[\begin{array}{ccccc} |V_1| & |V_2| & |V_3| & |V_4| & |V_5| \\ \times & & \times & & \\ \times & \times & & & \\ & \times & \times & & \\ & & \times & \times & \\ & & \times & & \times \\ \cdots & \cdots & \cdots & \cdots & \cdots \\ \times & & & & \\ & \times & & & \\ & & \times & & \end{array} \right] \tag{10.4}$$

The rank of (5$\times$4) matrix H_1 and (5$\times$5) submatrix of H_4 with all $\times$ set to 1 is 4 and 5 respectively. Therefore, the system is observable. Next, we consider the question of filtering out noise. In what follows, unless stated otherwise, the system is assumed to be observable.

4. Power System State Estimator

All real life measurements have noise. Any measurement made by a physical instrument has a level of accuracy specified by the manufacturer. For example, if a 100 MW meter has an accuracy of 95%, then, it means that a power of 100 MW will be read anywhere between 95 MW to 105 MW, 99% of the time. This brings in an element of *uncertainty* in the measurements.

The commonly used measurements are line currents, voltage magnitudes, active and reactive power flows on lines, transformers, and injection measurements. The equations for these measurements can be differentiated with respect to bus voltage magnitudes and angles. Let $\{z_1, \cdots, z_m\}$ be an ordered set of m measurements and let there be a total of n states in the system. Further, let equations for the measurement be given by a set of differentiable functions $\{f_1(x), \ldots, f_m(x)\}$ where x

is the state vector to be estimated. Thus, the real life model of power system is given by the following equations:

$$\begin{bmatrix} z_1 \\ \vdots \\ z_m \end{bmatrix} = \begin{bmatrix} f_1(\bar{x}) \\ \vdots \\ f_m(\bar{x}) \end{bmatrix} + \begin{bmatrix} v_1 \\ \vdots \\ v_m \end{bmatrix} \tag{10.5}$$

i.e.,

$$[z] = [f(\bar{x})] + [v] \tag{10.6}$$

where, $\bar{x}$ is the true state of the system, and $[v]$ represents the noise vector. Usually, the noise vector has a zero mean. It can be approximated by a Gaussian distribution. It is also referred to as the *white* noise.

Remark 1: We have presumed that there is no correlation between noise in the ith and jth measurment v_i and v_j. Stated more precisely, it means that $\mathbf{E}(v_i v_j) = 0$ and $\mathbf{E}(v_i^2) = \sigma_i^2$ where $\mathbf{E}$ is an expectation operator. The $m \times m$ matrix $\mathbf{E}(vv^T)$ is known as the covariance matrix. It is a positive semidefinite matrix. The diagonal terms define the variance of v_i (of its meter). The off diagonal terms measure the correlation between v_i and v_j. Now, $\mathbf{E}(v_i)$, measures the expected value of noise component v_i. For a zero mean noise, $\mathbf{E}(v) = 0$ and hence from equation (10.5), $\mathbf{E}(z) = f(\bar{x})$.

Since, noise is not a deterministic quantity, in power system state estimation we have to estimate x such that vector $f(x)$ closely fits to measurement vector z. Stated more precisely, power system state estimation is framed as a minimization problem wherein the objective function minimizes a nonnegative measure of residual. The two most commonly used formulations are as follows:

$$\min_x \frac{1}{2}[z - f(x)]^T[z - f(x)] = \sum_{i=1}^{m} \frac{1}{2}(z_i - f_i(x))^2 \tag{10.7}$$

$$\text{or } \min_x \sum_{i=1}^{m} |z_i - f_i(x)| \tag{10.8}$$

The difference $z_i - f_i(x)$ is referred to as the residual for i^{th} measurement. Equation (10.7) minimizes the sum of square of individual measurement residual, the square term ascertaining that the positive residual term is not canceled by a corresponding negative residual term. Since, the sum of the square of error is minimized in (10.7), it is known as a *Least Square estimator (LS)*. On the other hand, in (10.8), the sum of absolute errors is minimized thereby, avoiding positive error terms from canceling with

```
Set k = 1, Choose x^k;
While k=1 or |Δx^k| > ε do
```

$$\min_{\Delta x^k} \sum_{k=1}^{m} |z_i - f_i(x^k) - \nabla f_i^T \Delta x^k|$$

```
Update x^{k+1} = x^k + Δx^k
k=k+1;
```

Figure 10.10. Algorithm for LAV Estimator

negative error terms. It is known as the *Least Absolute Value* (LAV) estimator [Abur and Celik, 1993].

5. LAV Estimator

The LAV estimator minimizes the objective function (10.8). The minimization problem defined by (10.8) is not differentiable. Let x^* be the optimal solution to (10.8). Then, at the optimal:

$$\min_{x} \sum_{i=1}^{m} |z_i - f_i(x)| = \sum_{i=1}^{m} |r_i| = \sum_{i=1}^{m} |z_i - f_i(x^*)|$$

Since, the problem defined by (10.8) is nonlinear, it is solved by an iterative procedure. Convergence is ascertained by evaluating $\max|\Delta x^k| = \max|\Delta x_i^k|$ $i = 1, \ldots, n$. When $\max|\Delta x^k| < \epsilon$, a user specified tolerance, the algorithm terminates. At k^{th} iteration of LAV algorithm, we solve the following linearized problem:

$$\min_{\Delta x^k} \sum_{i=1}^{m} |\Delta z_i^k - \nabla f_i^T \Delta x^k| \tag{10.9}$$

where ∇f_i is the gradient of the function f_i with respect to x. As the objective function (10.9) is not differentiable, the minimization problem is reframed as a linear programming (LP) problem:

$$\min_{\beta^+, \beta^-, \Delta x^k} \sum_{i=1}^{m} \beta_i^+ + \beta_i^- \tag{10.10}$$

$$\text{such that } \beta_i^+ - \beta_i^- = \Delta z_i^k - \nabla f_i^T \Delta x^k \tag{10.11}$$

$$\beta_i^+ \geq 0; \; \beta_i^- \geq 0 \tag{10.12}$$

where $\Delta z_i^k = z_i - f_i(x^k)$. The first task for us is to show that the LP problem defined by (10.10-10.12) is equivalent to the minimization problem (10.9). The solution of a LP problem always lies at a vertex; therefore, β_i^+ and/or β_i^- should be zero. It follows from equation (10.11) and constraint (10.12) that

$$\begin{array}{c} \text{if } r_i > 0,\ \beta_i^+ = r_i,\ \beta_i^- = 0 \\ \text{if } r_i < 0,\ \beta_i^- = -r_i,\ \beta_i^+ = 0 \\ \text{if } r_i = 0,\ \beta_i^+ = \beta_i^- = 0 \end{array}$$

Thus, at any vertex of the LP problem $\sum \beta_i^+ + \beta_i^- = \sum |r_i|$ where, $r_i = \Delta z_i - \nabla f_i^T \Delta x$. Therefore, (10.9) and (10.10-10.12), minimize the same objective function. Thus, if Δx is an optimal of (10.10-10.12), then it also minimizes (10.9) problem.

In the LAV estimator, a LP problem is solved within an iterative solver. Hence, it is computationally expensive. LS estimator to be discussed now, requires lesser computations and is more popular.

6. Weighted LS Estimator

The LS formulation of (10.7) does not consider *relative* accuracy of the meters. For example, if one meter has an accuracy of 99% while another has an accuracy 95%, clearly, the measurement of meter-1 should be given more weightage than the meter-2. Let σ_i represent the standard deviation of meter i. If we introduce a weightage factor of $1/\sigma_i$, with the ith measurement, then accurate meters, (*i.e.*, smaller σ_i) will get a higher weightage. Hence, the power system state estimation problem is framed as a weighted LS problem (WLS) as follows:

$$\min_x \frac{1}{2}\sum_{i=1}^{m} \frac{(z_i - f_i(x))^2}{\sigma_i^2} = \frac{1}{2}[z - f(x)]^T W^{-1}[z - f(x)] \tag{10.13}$$

$$\text{where, } W^{-1} = \begin{bmatrix} 1/\sigma_1^2 & & \\ & \ddots & \\ & & 1/\sigma_m^2 \end{bmatrix} \tag{10.14}$$

Note that the scalar $\frac{1}{2}$ does not affect x^*. It has been introduced for the sake of convenience.

6.1. Optimality Conditions for a LS Estimator

The nonlinear LS problem for power system state estimation is defined as follows:

$$\min_x J(x)$$

$$\text{where } J(x) = \frac{1}{2}[z - f(x)]^T W^{-1}[z - f(x)] = \frac{1}{2}\sum_{i=1}^{m} \frac{(z_i - f_i(x))^2}{\sigma_i^2} \tag{10.15}$$

At an optimal of an unconstrained optimization problem, the gradient must be zero. Therefore, taking a partial derivative of $J(x)$ with respect to x_j we have

$$\frac{\partial J(x)}{\partial x_j} = -\sum_{i=1}^{m} \frac{(z_i - f_i(x))}{\sigma_i^2} \frac{\partial f_i}{\partial x_j} = 0 \; \forall \; j = 1, \cdots, n \tag{10.16}$$

The n-equations given by (10.16) can be elegantly written in matrix notations as follows:

$$\nabla J(x) = - \begin{bmatrix} \frac{\partial f_1}{\partial x_1} & \frac{\partial f_2}{\partial x_1} & \cdots & \frac{\partial f_m}{\partial x_1} \\ \frac{\partial f_1}{\partial x_2} & \frac{\partial f_2}{\partial x_2} & \cdots & \frac{\partial f_m}{\partial x_2} \\ \vdots & \vdots & \ddots & \vdots \\ \frac{\partial f_1}{\partial x_n} & \frac{\partial f_2}{\partial x_n} & \cdots & \frac{\partial f_m}{\partial x_n} \end{bmatrix} \begin{bmatrix} \frac{(z_1 - f_1(x))}{\sigma_1^2} \\ \frac{(z_2 - f_2(x))}{\sigma_2^2} \\ \vdots \\ \frac{(z_m - f_m(x))}{\sigma_m^2} \end{bmatrix} = \begin{bmatrix} 0 \\ 0 \\ \vdots \\ 0 \end{bmatrix} \tag{10.17}$$

The matrix as well as the residual are evaluated at the optimal x^*. It can be easily seen that the matrix in the above equation is the transpose of Jacobian of measurement equations. Further, $z_i - f_i(x^*)$ is the residual (r_i^*) at the optimal solution. The necessary conditions (10.17) for optimality can be written in a compact notation as follows:

$$H^T W^{-1} \Delta z^* = [0] \tag{10.18}$$

i.e., the scaled residual $[W^{-1}\Delta z^*]$ belongs to the null space of a $n \times m$ $(m > n)$ matrix H^T. If $\Delta z^* = 0$, then there is no noise. If $m = n$ and matrix H is nonsingular then $\Delta z^* = 0$ and problem reduces to a equivalent load flow problem. Now, let x^k be a guess to optimal, and $x^* = x^k + \Delta x^k$. One way to devise an iterative procedure for solving minimization problem (10.15) with the help of optimality condition (10.17) is to linearize the equation (10.17) at x^k by the Taylor series expansion. This leads to the following system of equations:

$$[H^k + \Delta H^k]^T W^{-1} [\Delta z^k - H^k \Delta x^k] = [0] \tag{10.19}$$

where H^k is the Jacobian evaluated at x^k and

$$\Delta H^{k^T}(i,j) = \sum_{l=1}^{n} \frac{\partial^2 f_j}{\partial x_l \partial x_i} \Delta x_l$$

Because the entries in ΔH^{k^T} depend upon the correction Δx, they cannot be evaluated *apriori*. A reasonable assumption while devising a computational procedure is based upon the fact that, if x^k is close x^* then the matrix ΔH^k will be close to a zero matrix and hence, ΔH can be neglected. The above qualitative statement can be made precise as follows:

$$||\Delta H^{k^T}(i,j)||_2 = ||\nabla \frac{\partial f_j^T}{\partial x_i} \Delta x||_2 \leq ||\nabla \frac{\partial f_j^T}{\partial x_i}||_2 ||\Delta x||_2$$

In the limit $||\Delta x||_2 \rightarrow 0$, clearly $\Delta H^k(i,j) \rightarrow 0$. Thus, in the near vicinity of the solution, ΔH^k can be neglected. This leads to the following computational procedure.

At an iteration k, solve for Δx^k such that the following equation is satisfied

$$H^{k^T} W^{-1}[\Delta z^k - H^k \Delta x^k] = 0$$

If the system is observable then, H^k has full column rank and so H^{k^T} has a full row rank. It can be shown that $H^{k^T} W^{-1} H^k$ is spd and therefore nonsingular. Hence, the correction Δx^k at each iteration can be obtained by solving a sparse spd LSS as follows:

$$[H^{k^T} W^{-1} H^k] \Delta x^k = H^{k^T} W^{-1} \Delta z^k \qquad (10.20)$$

The matrix $H^T W^{-1} H$ is known as the gain matrix. The formulation of equation (10.20) is known as the *Normal Equations* approach. Equation (10.20) can also expressed as follows:

$$\Delta x^k = [H^{k^T} W^{-1} H^k]^{-1} H^{k^T} W^{-1} \Delta z^k \qquad (10.21)$$

To establish the relationship of equation (10.21) with a linear least squares problem, let us consider the following linearized problem.

$$\min_{\Delta x^k} \frac{1}{2} \left[\Delta z^k - H^k \Delta x^k\right]^T W^{-1} \left[\Delta z^k - H^k \Delta x^k\right]$$

$$= \min_{\Delta x^k} \left[\frac{1}{2} \Delta z^{k^T} W^{-1} \Delta z^k - \Delta x^{k^T} H^{k^T} W^{-1} \Delta z^k + \frac{1}{2} \Delta x^{k^T} H^{k^T} W^{-1} H^k \Delta x^k\right] \qquad (10.22)$$

The first term is a constant and hence vanishes when derivative is taken. Applying the zero derivative condition to above equation leads to the following equation:

$$[H^{k^T} W^{-1} H^k] \Delta x^k = H^{k^T} W^{-1} \Delta z^k \qquad (10.23)$$

Therefore, solving (10.21) is equivalent to solving a LS problem (10.22).

Remark 2*: Let $\hat{H}^k = W^{-1/2} H^k$. The matrix

$$\hat{H}^{k^+} = (H^{k^T} W^{-1} H^k)^{-1} H^{k^T} W^{-1/2}$$

is the generalized Penrose Moore inverse of $W^{-1/2}H^k$. It can be verified that when H^k is full column rank, $\hat{H}^{k^+}$ satisfies the following four-conditions of a pseudo inverse.

$$\hat{H}\hat{H}^+\hat{H} = \hat{H} \tag{10.24}$$

$$\hat{H}^+\hat{H}\hat{H}^+ = \hat{H}^+ \tag{10.25}$$

$$(\hat{H}\hat{H}^+) = (\hat{H}\hat{H}^+)^T \tag{10.26}$$

$$(\hat{H}^+\hat{H}) = (\hat{H}^+\hat{H})^T \tag{10.27}$$

Superscript k has been dropped for convenience. By substituting $\hat{H}^+ = [\hat{H}^T\hat{H}]^{-1}\hat{H}^T$ in (10.27), it can be seen that $\hat{H}^+\hat{H} = I_n$. Hence $\hat{H}^+$ is also called left inverse of $\hat{H}$. However, $\hat{H}\hat{H}^+ = \hat{H}[\hat{H}^T\hat{H}]^{-1}\hat{H}^T$ is not an identity matrix. It is a symmetric semidefinite matrix. This explains the use of the terminology that $\hat{H}^+$ is a pseudo inverse of $\hat{H}$. It can be seen that equation (10.21) can also be written as

$$\Delta x^k = \hat{H}^{k^+}W^{-1/2}\Delta z^k \tag{10.28}$$

Remark 3*: It can be shown that $\Delta x^k = \hat{H}^{k^+}\Delta\hat{z}^k$ is a descent direction for nonlinear LS problem (10.15), where $\Delta\hat{z} = W^{-1/2}\Delta z$. To show that $d^k = \Delta x^k$ is a descent direction, all we need to prove is that

$$d^{k^T}\nabla J(x^k) < 0 \tag{10.29}$$

Substituting the gradient ∇J from (10.17) and d^k from (10.28) in the (10.29) we have

$$\begin{aligned} d_k^T\nabla J(x^k) &= -\Delta\hat{z}^{k^T}\left[\hat{H}^{k^+}\right]^T\hat{H}^{k^T}\Delta\hat{z}^k \\ &= -\Delta\hat{z}^{k^T}\left[\hat{H}^k\left[\hat{H}^{k^T}\hat{H}^k\right]^{-1}\hat{H}^{k^T}\right]\Delta\hat{z}^k \leq 0 \end{aligned} \tag{10.30}$$

If $d^{k^T}\nabla_x J(x^k) = 0$; then, from (10.30) we have $\hat{H}^{k^T}\Delta\hat{z}^k = [0]$. Then, equation (10.17), implies that we have reached the optimal. The descent property of d^k, specified by equation (10.30), when $d^{k^T}\nabla J(x^k) \neq 0$ is used in practice to steer the solution towards the optimal point.

Remark 4*: Let $\bar{x}$ represent true state of the system and x^* the estimate of $\bar{x}$ obtained by a WLS state estimator. If the system is observable then, the following inequality holds true:

$$||\bar{x} - x^*||_2 \leq ||\hat{H}^+||_2||W^{-\frac{1}{2}}||_2||v||_2 \tag{10.31}$$

where the Penrose Moore inverse is given by $\hat{H}^+ = (H^T W^{-1} H)^{-1} H^T W^{-\frac{1}{2}}$ In the presence of noise, the model of the power system is given by the following equation:

$$W^{-\frac{1}{2}} z = W^{-\frac{1}{2}} [f(\bar{x}) + v] \tag{10.32}$$

Let $\bar{x} = x^* + \delta x$. Linearizing the above equation at x^*, we have

$$W^{-\frac{1}{2}} \Delta z^* = W^{-\frac{1}{2}} H \delta x + W^{-\frac{1}{2}} v \tag{10.33}$$

where matrix H is the Jacobian evaluated at x^*. If the system is observable, matrix $\hat{H}$ has full column rank and so the matrix $\hat{H}^T \hat{H}$ is invertible. Therefore, premultiplying (10.33) by $\hat{H}^T$ and simplifying, we have

$$\delta x = \bar{x} - x^* = [H^T W^{-1} H]^{-1} H^T W^{-1} [\Delta z^* - v] \tag{10.34}$$

From equation (10.18) $H^T W^{-1} \Delta z^* = 0$. Therefore,

$$\bar{x} - x^* = -\hat{H}^+ W^{-\frac{1}{2}} v \tag{10.35}$$

By taking the norm of equation (10.35), the statement follows. Hence, we conclude that *in presence of noise there is an uncertainty associated with the computed state.* In the limiting case of noise approaching zero, the estimate approaches the true state. *It should be now clear why we say "power system state estimation problem" and not "power system state computation problem".*

Example 10.6: Consider the three bus system from [Pai, 1979] shown in Figure 10.11. The load flow solution is given as follows.

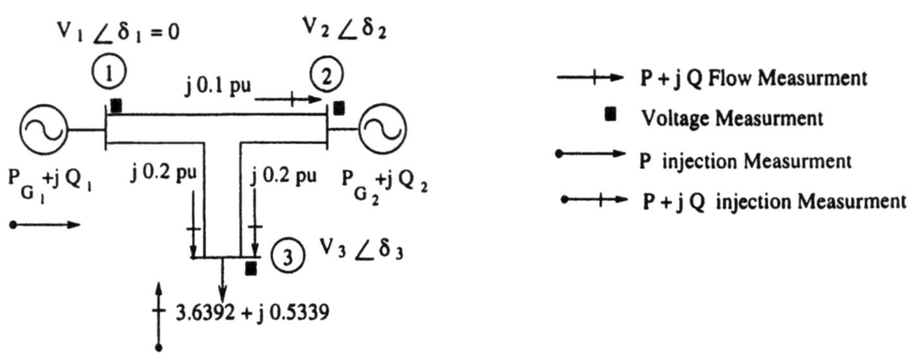

Figure 10.11. 3 Bus Example for State Estimation

$$P_1 = 1.20 \text{ pu}; \quad P_2 = 2.4251 \text{ pu}$$

Table 10.2. Effect of noise on state of the system

% noise	$J(x)$	$\max\lvert r_i\rvert$	$\lvert V_1\rvert$	$\lvert V_2\rvert$	$\lvert V_3\rvert$	δ_2	δ_3
0	0.0016	0.0011	1.0999	1.0986	0.9800	0.0420	-0.3234
2	0.0086	0.0059	1.1007	1.0993	0.9800	0.0419	-0.3244
5	0.0116	0.0072	1.1093	1.1080	0.9905	0.0417	-0.3190

$V_1 = 1.10\angle 0 \quad V_2 = 1.0987\angle 0.0420 \quad V_3 = 0.980 \text{ pu} \angle -0.3234$ (radians)

The flows across the lines are given at the locations shown:

$$P_{12} + jQ_{12} = -0.5084 + j0.0040 \text{ pu}$$

$$P_{23} + jQ_{23} = 1.9251 + j0.2256 \text{ pu}$$

$$P_{13} + jQ_{13} = 1.7140 + j0.3083 \text{ pu}$$

These values represent the true measurements without noise. MATLAB® program (see the diskette) is used to estimate the state of the system with above set of measurements. The program has a facility to add zero mean unit variance normal noise.

- Study the performance of the LS estimator with increasing noise.
- Compute the absolute maximum of residual vector r and the objective function $J(x)$.

Solution: A MATLAB®program (`se.m`) along with this book, implements a state estimator for the 3-bus example. Table 10.2 summarizes the results with the varying noise.

It can be observed that:

- With zero noise, the state estimator replicates the load flow analysis behavior.
- As the % noise increases, the LS residual $J(\hat{x})$ and the absolute maximum residual also increase.
- The estimated state is in close vicinity of the actual state.

6.2. Equality Constrained Power System State Estimation (ECPSSE)

A transmission system has a large number of zero injection busses. Typically, at such busses switching of voltage levels takes place. This additional information can be used to improve quality of estimation by framing a power system state estimation as an equality constrained non-linear LS problem.

$$\min \frac{1}{2}[z - f(x)]^T W^{-1}[z - f(x)] \tag{10.36}$$

subject to constraints

$$c_i(x) = 0;\ i = 1, \ldots, p \tag{10.37}$$

The equality constraint represents either a real zero power injection or a reactive zero power injection. Each iteration of the state estimation solves the following linearized linear equality constrained linear least squares (LSE) problem.

$$\min_{\Delta x^k} \frac{1}{2}[\Delta z^k - H^k \Delta x^k]^T W^{-1}[\Delta z^k - H^k \Delta x^k] \tag{10.38}$$

$$C^k \Delta x^k = \left[\Delta c(x^k)\right] \tag{10.39}$$

where $\Delta z^k = [z^k - f(x^k)]$, $\Delta c^k = [0 - c(x^k)]$, $H = \left[\frac{\partial f}{\partial x}\right]^T$ and $C = \left[\frac{\partial c}{\partial x}\right]^T$. If there are p zero injection busses, then there are $2p$ equations in (10.39). In the next section, we discuss numerical methods for solving LSE problem. The equation (10.38) can also be expressed as a minimization in distance norm, *i.e.*,

$$\min_{\Delta x^k} \frac{1}{2} ||W^{-\frac{1}{2}} \Delta z^k - W^{-\frac{1}{2}} H^k \Delta x^k||_2^2$$

7. Numerical Methods for Equality Constrained Least Squares Estimation

The methods for solving a Equality Constrained Linear Least Squares (LSE) problem can be broadly classified into the following categories [Lawson and Hanson, 1974].

- those using a basis of the null space of the constraint matrix[1];
- methods based on direct elimination of variables using the constraint equations;
- methods based on constraint weighting and
- those using Lagrangian multipliers approach.

The first two methods are not very suitable for large sparse problems. Hence, we limit the discussion to latter two approaches only.

[1] Recall the QP solver discussed in chapter 7.

7.1. Constraint Weighting Approach

By penalizing the equality constraints violations in the objective function the LSE problem can be referred as unconstrained LS minimization problem. Since, the violations (positive or negative) on either side of $Cx - \Delta c = 0$ are equally unacceptable, large quadratic penalties are used. In other words, since telemetry is not required at the zero injection busses; zero power injection is treated as a *virtual* measurement with high accuracy level ($\sigma_i \rightarrow 0$). This results in an unconstrained weighted least squares (WLS) problem. The large sparse WLS problem can be solved by either of the following methods:

- Normal Equations Approach [Schweppe, 1970]

- Method of Peters and Wilkinson [Gu et al., 1983]

- Orthogonalization Approach [Simoes-Costa and Quintana, 1981; Monticelli et al., 1985; Vempati et al., 1991].

7.1.1 Normal Equations (NE) Approach

In this formulation, at each iteration of the power system state estimation algorithm following linearized LS problem is solved :

$$\min_{\Delta x^k} \frac{1}{2}||W^{-1/2}\Delta z^k - W^{-1/2}H^k\Delta x^k||_2^2 + \frac{\rho}{2}||C^k\Delta x^k - \Delta c^k||_2^2 \quad (10.40)$$

The scalar ρ is a large positive real number that heavily penalizes the equality constraints violations. Penalty function ρ should be of the order of $\sqrt{(\text{machep})}$, where machep is the numerical precision of the computer. Solving the problem (10.40) is equivalent to solving the following linear system of equations:

$$[H^{k^T}W^{-1}H^k + \rho C^{k^T}C^k]\Delta x^k = [H^{k^T}W^{-1}\Delta z^k + \rho C^{k^T}\Delta c^k] \quad (10.41)$$

In this approach, the sparse (spd) linear system of equations (10.41) is factorized by Cholesky factorization or square root free LDL^T factorization and solved by Forward Backward substitution. The tools for LSS development in chapter 4 for exploiting sparsity and minimizing fills can be directly used here. The merits of the NE approach are simplicity and minimal implementation complexity. Earlier estimators used NE approach for power system state estimation. However, it was soon realized that the approach lacks reliability because of the numerical illconditioning.

Causes of Numerical Ill-Conditioning in PSSE

Numerical ill-conditioning in the power system state estimation can result from the following:

- Unusual weighting factors: Unusual weighting factors arise because of large penalty factor (ρ) for zero injection measurements and very large weights for inclusion of unreliable pseudo measurements. The disparity in the weights cause numerical illconditioning.
- Numerical squaring introduced by matrix transpose matrix product.
- Large number of injection measurements.
- Connection of long line (large impedance) with a short line (very small impedance).

To overcome these limitations without discarding the simple approach of penalizing equality constraints, numerically stable implementation based on orthogonal factorization approach is developed. Before proceeding further we digress and comment on the requirements of observability analysis in an equality constrained formulation.

Observability in ECPSSE

To proceed with estimation, it is required that matrix $[H^T W^{-1} H + \rho C^T C]$; $\rho > 0$ should be nonsingular in equation (10.41). The superscript k has been dropped for convenience. Since,

$$[H^T W^{-1} H + \rho C^T C] = \left[H^T W^{\frac{-1}{2}} \ \sqrt{\rho} C^T\right] \begin{bmatrix} W^{\frac{-1}{2}} H \\ \sqrt{\rho} C \end{bmatrix} \tag{10.42}$$

it means that for left hand side to be non-singular in (10.42), augmented Jacobian $\begin{bmatrix} W^{\frac{-1}{2}} H \\ \sqrt{\rho} C \end{bmatrix}$ should have full column rank. Note that full column rank of augmented Jacobian does not imply full column rank of H. If in a system, large number of zero injection measurements are present, then the number of transducers required to guarantee observability reduces.

7.1.2 QR Decomposition Approach

At each step of the ECPSSE, unconstrained problem given by (10.43) is solved. Equation (10.40) can also be expressed as follows.

$$\min_{\Delta x} \left\| \begin{bmatrix} \hat{H} \\ \sqrt{\rho} C \end{bmatrix} \Delta x - \begin{bmatrix} \Delta \hat{z} \\ \sqrt{\rho} \Delta c \end{bmatrix} \right\|_2^2 ; \ \rho >> 0 \tag{10.43}$$

where $\hat{H} = W^{-1/2}H$ and $\Delta\hat{z} = W^{-1/2}\Delta z$. If the system is observable then rank of $\begin{bmatrix} \hat{H} \\ \sqrt{\rho}C \end{bmatrix}$ is n. Therefore, the augmented matrix of $\hat{H}$ can be decomposed using QR decomposition as follows:

$$\min_{\Delta x} \left\| Q \begin{bmatrix} R \\ 0 \end{bmatrix} \Delta x - QQ^T \begin{bmatrix} \Delta\hat{z} \\ \Delta\hat{c} \end{bmatrix} \right\|_2^2 \tag{10.44}$$

where $\Delta\hat{c} = \sqrt{\rho}\Delta c$, $\begin{bmatrix} \hat{H} \\ \sqrt{\rho}C \end{bmatrix} = Q \begin{bmatrix} R \\ 0 \end{bmatrix}$.

Q is an $m \times m$ orthogonal matrix, *i.e.*, $QQ^T = Q^TQ = I_m$ and R is $n \times n$ upper triangular matrix such that $R_{ii} > 0 \ \forall \ i = 1, \ldots, n$. Since,

$$||Qy||_2^2 = y^TQ^TQy = y^Ty = ||y||_2^2 \tag{10.45}$$

with $y = \begin{bmatrix} R \\ 0 \end{bmatrix} \Delta x - Q^T \begin{bmatrix} \Delta\hat{z} \\ \Delta\hat{c} \end{bmatrix}$ it follows that (10.44) is equivalent to the following:

$$\min_{\Delta x} \left\| \begin{bmatrix} R \\ 0 \end{bmatrix} \Delta x - \begin{bmatrix} \Delta\tilde{z}_1 \\ \Delta\tilde{z}_2 \end{bmatrix} \right\|_2^2 \tag{10.46}$$

$$\text{where} \begin{bmatrix} \Delta\tilde{z}_1 \\ \Delta\tilde{z}_2 \end{bmatrix} = Q^T \begin{bmatrix} \Delta\hat{z} \\ \Delta\hat{c} \end{bmatrix} \tag{10.47}$$

The subvector $\Delta\tilde{z}_1$ is of dimension n and $\Delta\tilde{z}_2$ is a $(m - n)$ residual vector. It can be seen from (10.46) that whatever be the choice of Δx, residual $\Delta\tilde{z}_2$ can not be nullified. Since, R is $n \times n$ nonsingular upper triangular matrix, it can be used to nullify the $\Delta\tilde{z}_1$ component of $\Delta\tilde{z}$. Hence, the Δx^* for minimization of (10.46) is given by the solution of following equations:

$$R\Delta x = \Delta\tilde{z}_1 \tag{10.48}$$

The minimum value of objective function is given by $||\Delta\tilde{z}_2||_2^2$. The equation (10.48) amounts to a backward substitution. The sparse QR decomposition discussed in chapter 5 can be used for computing the QR decomposition of augmented $\hat{H}$. Computationally, it does not make sense to explicitly create the orthogonal matrix Q and then compute $\Delta\tilde{z} = Q^T\Delta\hat{z}$. Such a Q will not be sparse. Consequently, the benefits of sparsity exploitation cannot be reaped. However, there is a simple remedy to the problem. As discussed in chapter 5, the matrix R is obtained by premultiplying by Givens rotations, *i.e.*,

$$[Q_l^T \ldots Q_2^TQ_1^T] \begin{bmatrix} \hat{H} \\ \sqrt{\rho}C \end{bmatrix} = \begin{bmatrix} R \\ 0 \end{bmatrix} \tag{10.49}$$

Now, from equation (10.44), it can be seen that $\begin{bmatrix} \Delta\hat{z} \\ \Delta\hat{c} \end{bmatrix}$ is also premultiplied by sequence of Givens matrices $[Q_l^T \ldots Q_2 Q_1]$. Thus, it is not necessary to store Q_i at all. Instead when a rotation is applied to matrix $[\hat{H}^T, \sqrt{\rho}C^T]^T$; it should also be applied to right hand side vector $\begin{bmatrix} \Delta\hat{z} \\ \Delta\hat{c} \end{bmatrix}$ and then discarded. This leads to the sparse implementation of QR decomposition based power system state estimation (Figure 10.12). Note that unlike load flow analysis, $\max_i |\Delta z_i|$ mismatch is not checked for convergence because large noise or a bad data results in large $|\Delta z_i|$ at the solution.

In order to reduce the computational effort, fast decoupled state estimators on the lines of FDLF have been developed.

7.2. Fast Decoupled Estimator

In case of a fast decoupled estimator [Soman, 1995], Jacobian H is both decoupled and fixed on the lines of FDLF as follows:

$$H = \begin{bmatrix} H_1 & 0 \\ 0 & H_4 \end{bmatrix} \quad \Delta z = \begin{bmatrix} \Delta z_p \\ \Delta z_q \end{bmatrix} \tag{10.50}$$

Thus, two LS problems one for $P - \delta$ coupling and another for $Q - V$ coupling have to be solved at each iteration. When matrices H_1 and H_4 are fixed, computing $Q_p^T \Delta\hat{z}_p$ and $Q_q^T \Delta\hat{z}_q$ at each iteration requires storage of orthogonal matrix Q, *i.e.*, storage of angle θ of each Givens rotation and the row pair on which operated. An ingenious way to avoid the storage is as follows:

1 Compute matrix R by QR decomposition of $\begin{bmatrix} \hat{H} \\ \sqrt{\rho}C \end{bmatrix}$ and discard Q.

2 Then solve the following NE as follows:

$$R^T R \Delta x = H^T W^{-1} \Delta z + \rho C^T \Delta c$$

This is known as a *hybrid* approach.

Since R has been computed in a numerically stable manner, the approach does not suffer from the numerical problems of NE approach. Using the hybrid approach, a fast decoupled state estimator can be implemented as shown in figure 10.13.

We now discuss the Lagrangian method approach for power system state estimation

1.Read measurement data, tolerance ϵ;
2.$k = 1$, Guess x^k, (flat start);
3.Compute the structure of $\begin{bmatrix} H \\ C \end{bmatrix}$ and then structure of $[H^T H + C^T C]$.
Apply MDA to it and obtain permutation matrix P.
4.Reorder the state variables and columns of $\begin{bmatrix} \hat{H} \\ \sqrt{\rho}C \end{bmatrix}$
5.Perform symbolic factorization of $P^T[H^T H + C^T C]P$ and obtain structure of R. P is a permutation matrix discussed in chapter 4.
6.While $k = 1$ or $\max_i ||\Delta x_i|| < \epsilon$, do
{
a) Compute $\Delta z^k = [z - f(x^k)]; \Delta c^k = [-c(x^k)]$
b) Compute Jacobian $\begin{bmatrix} \hat{H} \\ \sqrt{\rho}C \end{bmatrix}$;
c) Solve $\min \left\| \begin{bmatrix} \hat{H}^k \\ \sqrt{\rho}C^k \end{bmatrix} \Delta x^k - \begin{bmatrix} \Delta \hat{z}^k \\ \sqrt{\rho}\Delta c^k \end{bmatrix} \right\|_2^2$ by QR decomposition.
d) $x^{k+1} = x^k + \Delta x^k$
e) k=k+1
};
7. Proceed for bad data detection.

Figure 10.12. QR Decomposition Based State Estimation Algorithm

7.3. Method of Lagrangian Multipliers

We consider the equality constrained subproblem and solve it using the approach similar to the QP approach discussed in chapter 6 and 7. First we discuss an implementation known as Normal Equations with Constraints approach.

7.3.1 Normal Equations with Constraints (NE/C)

This approach [Aschmoneit et al., 1977] solves a LSE problem defined by (10.38) and (10.39). Solution point must satisfy the KKT optimality conditions. This requires the solution of the following sparse indefinite

```
1. Read data, a) Measurement data b) System topology
   (Y Bus Matrix data) c) Weightage matrix d) Penalty
   ρ >> 0
2. k = 1, Give V^k, δ^k (flat start)
3. Compute structure of [H_1^T H_1 + ρ C_1^T C_1] and
   [H_4^T H_4 + ρ C_4^T C_4], where C_1 and C_4 represent
    the equality constraints for real power
   and reactive power subproblem.
4. Apply MDA to matrices in step-3 and map the new
   order to old order of variables.
5. Perform symbolic factorization to obtain structure
   of R_1 and R_4.
6. Obtain Δz_p^k and Δc_p^k
7. Obtain R_1 and R_4 by sparse QR decomposition of
   [ Ĥ_1 ; √ρ C_1 ] and [ Ĥ_4 ; √ρ C_4 ]
8. While k = 1 or || Δδ^k ; Δ|V|^k || < ε, do
   {
    a) Solve [R_1^T R_1] Δδ^k = [H_1^T W_1^{-1} Δz_p^k + ρ C_1^T Δc_p^k]
    b) δ^{k+1} = δ^k + Δδ^k,
    c) Compute Δz_q^k and Δc_q^k.
    d) Solve [R_4^T R_4] Δ|V|^k = [H_4^T W_4^{-1} Δz_q^k + ρ C_4^T Δc_q^k]
    e) |V|^{k+1} = |V|^k + Δ|V|^k
    f) k = k + 1
    g) Compute || Δδ^k ; Δ|V|^k ||
    }
```

Figure 10.13. Hybrid Approach for Fast Decoupled State Estimator Algorithm

augmented linear system of equations.

$$\begin{bmatrix} H^{k^T} H^k & C^{k^T} \\ C^k & 0 \end{bmatrix} \begin{bmatrix} \Delta x^k \\ \lambda^k \end{bmatrix} = \begin{bmatrix} H^{k^T} \Delta z^k \\ \Delta c^k \end{bmatrix} \tag{10.51}$$

where vector λ^k is the vector of Lagrangian multipliers. Clearly, the unwarranted illconditioning in the problem formulation due to the large penalties to equality constraints is avoided. However,

- the formulation requires solution of a large sparse indefinite system of linear equations with zero diagonal blocks. Therefore, more sophisticated pivoting strategies than in spd solvers to balance numerical stability and exploit sparsity (reduce fills) is a prerequisite for the success of the method;
- implementation complexity increases because of the requirement to handle large sparse indefinite system of matrices;
- undesirable numerical squaring effect due to computing H^TH is encountered.

7.3.2 Hatchel Augmented Matrix Method

Hatchel method is theoretically equivalent to NE/C formulation with the exception that numerical squaring effect of H^TH is avoided. Hachtel augmented matrix method for constrained linear least squares was first used for state estimation by Gjelsvik et. al., [Gjelsvik et al., 1985]. In this method, the residual vector is treated as unknown. Let

$$r^k = \Delta z^k - H^k \Delta x^k \tag{10.52}$$

Then, the solution point satisfying KKT conditions is obtained by solving the following large sparse indefinite system of linear equations.

$$\begin{bmatrix} 0 & 0 & C^k \\ 0 & \alpha I & H^k \\ C^{k^T} & \alpha H^T & 0 \end{bmatrix} \begin{bmatrix} \lambda^k \\ \alpha^{-1} r^k \\ \Delta x^k \end{bmatrix} = \begin{bmatrix} \Delta c^k \\ \Delta z^k \\ 0 \end{bmatrix} \tag{10.53}$$

α is a scalar parameter chosen to improve the numerical conditioning of the resulting matrix. Optimal value of α corresponds to the smallest singular value of H^k. However, this quantity in practice is hard to estimate without significant computations. Therefore, the authors in [Gjelsvik et al., 1985] have used $\alpha = 1$ in computations. [Gjelsvik et al., 1985] use Harwell subroutine MA27AD which selects pivots from the diagonal using minimum degree criteria, with 2×2 blocks, when too small pivots are detected. [Alvarado and Tinney, 1990] and [Nucera and Gilles, 1991] have attempted block partitioned sparse formulations to allow triangular factorization to proceed in positive definite like manner. The significant characteristics of Hatchel method are

- it avoids numerical squaring involved in computing H^TH;

- increases the size of the linear system solver;
- increases implementation complexity;
- requires a proper choice of scalar α for improving numerical stability.

8. Bad Data Processing*

In power systems, telemetered measurements are subjected to noise or error in metering, communication system etc. The zero mean noise is filtered by the basic state estimator. However, removal of bad measurements because of faulty metering require extra processing. Errors in measurements above say 40 standard deviations can be detected by preprocessing with logical or qualitative checks [Wallach, 1986]. However, errors around say five standard deviations may not be detected. Bad data estimators are used to detect and remove the errors falling in this band. If the bad data is not removed, then the quality of estimation deteriorates because of *smearing* effect. Bad data results in large residuals. Consequently, the LS minimization process biases its estimation of state to honor bad measurements, as that would reduce residual and hence the objective function. In contrast, the solution of a LP problem lies at a vertex; the LAV estimators attempts to find those n-linearity independent measurements that give minimum absolute sum of residuals. Therefore, the LAV estimator rejects bad data points. We discuss bad data detection and elimination only in the context of WLS estimator.

The overall process consists of three steps:

1 Detect if bad data problem exists. If not, stop.

2 If yes, identify the bad measurements.

3 Eliminate or compensate bad measurements, and reestimate the power system state and go back to step 1.

The first step in bad data processing is the detection of bad data. Specifically, we answer the following question: *Is there any likelihood of presence of bad data in the set of measurements?* Prominent bad data detection tests include Chi square test and normalized residual r_N test. Various bad data detection tests are summarized below [Handschin et al., 1975; Mili et al., 1985].

8.1. Chi Square (χ^2) Test

This test is based upon the following intuitive idea: *If bad measurements have been used in the estimation, they should reflect in the objective function. Thus, a large value of $J(\hat{x})$ is expected in presence of bad* *data.* The crucial question is how to quantify the large enough criteria. The estimated objective function is given by

$$J(\hat{x}) = [z - h(\hat{x})]^T W^{-1} [z - h(\hat{x})] \tag{10.54}$$

where $\hat{x}$ is the estimated state and $\bar{x}$ is the true state. The minimum value of the objective function is given by

$$J(\bar{x}) = v^T W^{-1} v = [z - h(\bar{x})]^T W^{-1} [z - h(\bar{x})] \tag{10.55}$$

If the noise has normal distribution, then $J(x)$ has χ^2 distribution with $K = m - n$ degrees of freedom. If $(K \geq 30)$, then the normalized random variables $\eta_1(\hat{x}) = \frac{J(\hat{x})-K}{2\sqrt{K}}$ and $\eta_2(\hat{x}) = \sqrt{2J(\hat{x})} - \sqrt{2K}$ both become zero mean, unit variance random numbers from $N(0, 1)$. Consider the following two hypothesis H_0 and H_1:

H_0 : no bad data present;
H_1 : H_0 is not true.

In the χ^2 test, hypothesis H_0 or H_1 are selected based on a threshold parameter γ as follows:

1 if $\eta_1(\hat{x}) < \gamma$ or $\eta_2(\hat{x}) < \gamma$ accept H_0;

2 reject H_0 (accept H_1) otherwise.

Scalar γ is called the detection threshold level. Note that χ^2 test can not identify the location of bad data. Next, we discuss r_W and r_N tests which can be used to both detect and identify bad data locations.

8.2. r_W Test

The intuitive idea behind r_W test is that *if there is bad data present in the measurements, not only should the LS objective function have a large value but it should also reflect as large residuals.* One would presume that a large magnitude of the residual will be seen at a bad measurement location[2]. The ratio of the magnitude of residual to standard deviation of the meter can be used to detect bad data more accurately than the magnitude of the residual alone. For example, if a meter is very accurate, then a moderate residual can also be used to identify a bad data. On the other hand, if it is known that quality of metering is not very good (large standard deviation of the meter), then the corresponding ratio will be small and the measurement will not be suspected easily.

[2]Unfortunately, it is always not true leading to the problem of multiple interacting bad data.

Let $\hat{r}_W = \sqrt{W^{-1}}\hat{r}$, *i.e.*, $\hat{r}_{W,i} = \frac{\hat{r}_i}{\sigma_i}$ $\forall i = 1, \ldots, m$. Then this test uses the following logic for detection.

1 Accept H_0 if $\hat{r}_{W,k} < \gamma$, $k = 1, \ldots, m$

2 Else accept H_1.

Further, the largest normalized residual is the likely bad measurement.

8.3. r_N Test

A more accurate test than the r_W test can be developed if the standard deviation of the residual (squareroot of variance) is used in scaling the residual rather than the variance of the meter.

However, it means that the variance of the residual has to be computed. Let $\hat{r} = z - h(\hat{x})$, $E_r = \mathbf{E}(\hat{r}\hat{r}^T)$, $D = \text{diag}(E_r)$ and $\hat{r}_N = \sqrt{D^{-1}}\hat{r}$. Covariance matrix E_r gives the correlation between r_i and r_j $\forall$ $i = 1, \ldots, m$ and $j = 1, \ldots, m$. The ith diagonal term corresponds to the variance of ith residual. The matrix is symmetric semidefinite matrix. The normalized residual for a measurement corresponds to the ratio of residual to standard deviation of the residual. Then, this test uses the following logic for detection.

1 Accept H_0 if $\hat{r}_{N,k} < \gamma$, $\forall$ $k = 1, \ldots, m$

2 Else accept H_1.

If the same value of γ is used, r_N test is more accurate than r_W test. The variance of the meters (or the noise vector) is known beforehand. However, the r_N test requires the variance of the residual vector which has to be computed.

8.3.1 Bad Data Estimation-Derivations

We now develop necessary theory of bad data estimation for least squares estimators. The validity of the linearized model for bad data estimation is assumed [Handschin et al., 1975]. We now develop a linearized relationship between noise vector and computed residual.

Lemma 1: Let $\bar{x}$ be the true value of the state vector, $\hat{x}$ the estimated value and v the noise in measurements. Then

$$z = h(\bar{x}) + v \tag{10.56}$$

$$\hat{r} = z - h(\hat{x}) \tag{10.57}$$

The linearized sensitivity relationship between $\hat{r}$ and v is given by

$$\hat{r} = \bar{W} v \tag{10.58}$$

where

$$\bar{W} = \left[I_m - H \left[H^T W^{-1} H \right]^{-1} H^T W^{-1} \right] \tag{10.59}$$

and I_m is a $m \times m$ identity matrix. The proof following from remark 4 by substituting (10.35) in $\hat{r} = \Delta z^* = H\delta x + v$. Also, from equation (10.58) it follows that

$$\mathbf{E}(\hat{r}\hat{r}^T) = \bar{W}\mathbf{E}(vv^T)\bar{W}^T = \bar{W}W\bar{W}^T = \left[W - H \left[H^T W^{-1} H \right]^{-1} H^T \right] \tag{10.60}$$

Covariance matrix for the noise $\mathbf{E}(vv^T)$ is a diagonal matrix given by the square of the standard deviation of the meters. The equality constraints are treated as *virtual* measurements with corresponding meters having standard deviation equal to $\frac{1}{\sqrt{\rho}}$. The off diagonal terms in the covariance matrix for noise vector is zero because it is assumed that measurements are independent, *i.e.*, the noise in the two different measurements has no correlation between them. Thus, $\mathbf{E}(vv^T) = W$. It can be seen that the sensitivity matrix $\bar{W}$ (equation 10.59) and the covariance matrix are related by the following relationship.

$$\bar{W} = \mathbf{E}(\hat{r}\hat{r}^T) W^{-1} \tag{10.61}$$

Next, we develop, a relationship to computed the linearized change in residual for a perturbation in the measurement.

Lemma 2: (Sensitivity Relationship)
If the measurement vector is perturbed by Δz^{meas}, then as per linearized analysis, the residual is perturbed by Δr where

$$\Delta r = \mathbf{E}(\hat{r}\hat{r}^T) W^{-1} \Delta z^{meas} \tag{10.62}$$

Proof:

While the measurements are perturbed by Δz^{meas}, let the state be perturbed by Δx.

$$\text{Now } \hat{r} = z - f(x^k)$$

$$\text{The new residual is given } \hat{r} + \Delta r = z + \Delta z^{meas} - f(x^k + \Delta x)$$

$$\Rightarrow \hat{r} + \Delta r = z - f(x^k) + \Delta z^{meas} - H\Delta x$$

$$\Rightarrow \Delta r = \Delta z^{meas} - H\Delta x$$

For a perturbation of Δz^{meas} in right hand side vector from (10.21), x is perturbed by

$$\Delta x = \left[H^T W^{-1} H\right]^{-1} H^T W^{-1} \Delta z^{meas} \tag{10.63}$$

Substituting value of Δx in expression for Δr, we get the following

$$\Delta r = \left[I_m - H\left[H^T W^{-1} H\right]^{-1} H^T W^{-1}\right] \Delta z^{meas} \tag{10.64}$$

$$i.e.,\ \Delta r = \mathbf{E}(rr^T) W^{-1} \Delta z^{meas} \tag{10.65}$$

Equation (10.65), measures the sensitivity of the residual to the perturbation in measurement. This sensitivity can be used to compensate a bad measurement instead of elimination.

8.4. Combined detection-identification test

The approach combines detection and identification phases in bad data processing. After the detection of bad measurement, usual procedure is to delete the measurement with largest normalized residual and reestimate the state of the system. This requires extra computations due to refactorization and checks to ensure that the system neither becomes unobservable nor creates additional critical measurements. Instead, it seems advisable to estimate the error in a measurement and compensate the measurement and reestimate the system state. Such a procedure not only saves computations but avoids creating unobservable states or critical measurements. If the redundancy in the measurements is not high, accuracy of the compensation may not be good.

Let Δz_i be the change in the measurement required to compensate bad data in z_i. Then,

$$z_i^{new} = z_i + \Delta z_i \tag{10.66}$$

Let the new residual for the ith measurement be given by the following equation:

$$r_i^{new} = \hat{r}_i + \Delta r_i \tag{10.67}$$

For simplicity we assume that there is no noise, hence, $\Delta z_j = 0$ for $j \neq i$. Let, $D = \text{diag}\ \mathbf{E}(rr^T)$. From equation (10.65)

$$\frac{\Delta r_i}{\sqrt{D_i}} = \frac{\sqrt{D_i}}{\sigma_i^2} \Delta z_i \tag{10.68}$$

It is desired to determine the change Δz_i that leads to $r_i^{new} = 0$. This implies that the true value of z_i is z_i^{new} if no other bad measurements or noise is assumed. Thus $\Delta r_i = -\hat{r}_i$.Hence,

$$\Delta z_i = -\hat{r}_{N_i} \frac{\sigma_i^2}{\sqrt{D_i}}$$

where $\hat{r}_N = \sqrt{D^{-1}}\hat{r}$

$$z_i^{new} = z_i - \frac{\sigma_i^2}{\sqrt{D_i}} r_{N_i}$$

Now, define b_i as follows

$$b_i = \frac{z_i - z_i^{new}}{\sigma_i} = \frac{\sigma_i}{\sqrt{D_i}} r_{N_i} \tag{10.69}$$

If a measurement is good, r_{Ni} is small and so is b_i. The above equation provides a test for detection of bad measurement. For example, for i^{th} measurement, if the noise is assumed to be normal with zero mean and standard deviation σ_i then, the probability of noise v_i to be within $\pm 3\sigma_i$ is 99%. Therefore, if $|b_i|$ exceeds 3.0, then the corresponding measurement can be flagged for compensation/elimination.

If in addition to gross error $b_i\sigma_i$, zero mean random noise is also considered, then the above expression gives the expected value of b_i. This leads us to the following model algorithm.

8.4.1 Bad Data Processing Algorithm

1 Determine the estimate $\hat{x}$;

2 Calculate the normalized residual $r_{N_j} = \frac{z_j - f(\hat{x})}{\sqrt{D_i}}\ \forall j$;

3 Find i such that $|r_{N_i}|$ is maximum;

4 Compute $b_i = \frac{\sigma_i}{\sqrt{D_i}} r_{N_i}$;

5 Check

(a) $|b_i| > 3.0$, go to step 6;

(b) else, the estimation is completed;

6 Delete/compensate z_i and go to step 1;

The discussion on bad data has been restricted to single bad data or multiple non interacting bad data. There is much more to bad data estimation, *e.g.*, multiple interacting bad data, sparse matrix techniques for computation of variance of residual etc. Section on further reading suggests references for more advanced reading.

Example 10.7: Let only the flow and voltage measurments be used in state estimation in the previous example. Reverse the measurement P_{12} from -0.508 pu to 0.508 pu; implement the bad data detection tests

discussed and check the quality of bad data estimator.
Solution : It is seen that when the P_{12} is reversed, the $J(\hat{x})$ jumps to 0.3262. Thus, bad data is detected by the χ^2 test. The r_N test is implemented for identification of the bad data. With a threshold parameter of 0.01, it is seen that bad data is identified for measurement 2. Hence, the identification is not proper. A closer analysis reveals that, the flow measurements are just sufficient to define the state of the system. Hence, the local redundancy in the measurements is poor. To improve the local redundancy along with the flow P_{12}, we now add an injection measurement at bus-1 (measured valued $P_1 = 1.20$ pu). With this addition, it is seen that identification of bad data corresponds to measurement P_{12}, *i.e.*, correct identification is done.

Similarly, when the flow measurement Q_{23} is reversed (new measured value is -1.9251 pu), it is seen that the bad data detection test is positive but the identification is done for measurement Q_{13}. It is easy to see that problem lies with the low local redundancy. By considering the injection measurements at ($P_3 = -3.6393, Q_3 = -0.5339$) node-3, the correct identification of bad data takes place.

We conclude that (1) bad data does reflect in the objective function $J(x)$, (2) in the absence of local redundancy, wrong identification occurs and (3) improving local redundancy (and not just overall redundancy) improves the quality of bad data identification. The MATLAB®program `se.m` in the diskette implements this example with all the measurements listed so far.

9. Object Oriented Design*

The overall estimation process involves NTP, Observability analysis, State estimation and bad data estimation. We begin the discussion on OOD by first considering the NTP application. To implement NTP, we should model a class substation that should be generic enough to cover all possible substation topologies. It is defined as follows:

```
class SubStn {
    public :
     char name[20], from[20], to[20];
        //name of substation, physical nodes
     AssoArr NameArr;
        // map physical nodes to integers
     spmat<int> *VectCb; // to model a CB
     Set<int> *MergesNodes;
       //set of physical nodes aliased by electrical nodes
     int NodesCreated;
```

```
    SubStn();  // default constructor
    FindConnectivity();// Compute components of
                              // substation graph
    CreateGraph();// create array of closed CB
     :
};
```

Specific topologies like one and half breaker, ring bus can be easily derived (as subclasses) from the class *SubStn*. A CB connects two physical nodes, which can be either on or off. It can be modeled by data type *spmat* < *int* > introduced in chapter 3. The electrical nodes are stored in a set of merged nodes. The substation has a constructor and various other methods.

A class *AllSubstn* is defined to aggregate all the substations primarily to assign unique names to the electrical nodes. It is defined as follows:

```
class AllSubstn {
   protected :
    int TopoCount;  // number of substations in a network
    char **ElectName; // name for electrical nodes
    AssoArr NameToNo, GivenNo;
     // mapping of electrical & substation names
    SubStn * VectSubstn; // array of substations
   public:
    AllSubstn();
    FindConnectivity();
    AssignEleNodes();
    CreateLookup();
     :
};
```

Let us presume that electrical nodes in all of the substations have been identified by FindConnectivity() method of the class *SubStn*. The next task is to assign unique node names (numbers) to all the electrical nodes of each substation, where uniqueness refers to the complete network. As an illustrative example, consider that a physical node say 5 of a substation (say **sub3**) belongs to component-1[3] of that substation graph. It is therefore mapped as an electrical node 1 within the substation. Generally, substation names are defined by geographical locations or voltage levels etc. and are unique. Therefore, a new name to this electrical node is generated by concatenating the substation name with electrical node

[3]Recall that C_{++} index is $0, 1, 2 \ldots$

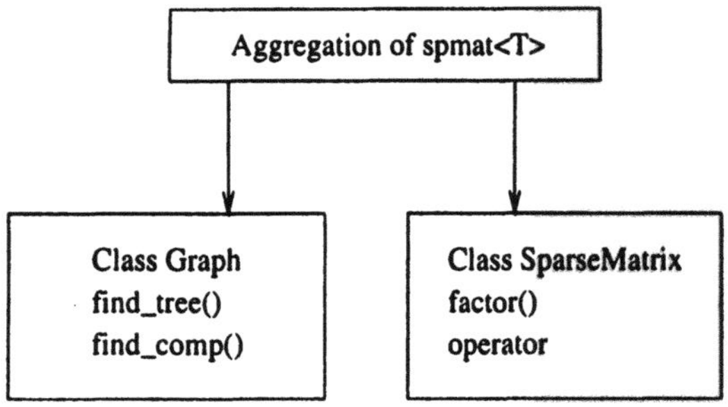

Figure 10.14a. Design-1 For Class Graph

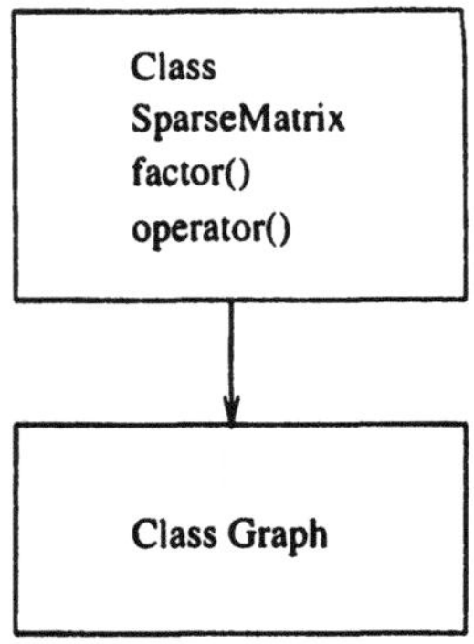

Figure 10.14b. Design-2 For Class Graph

number, *e.g.*, *sub*3_1. Thus, we have established the mapping of physical node number-5 of the substation **sub3** to the electrical node *sub*3_1. The electrical node name *sub*3_1 is unique to the complete network. Now, an associative array is used to assign integer node numbers to the electrical nodes. The circuits connecting the substations are described in the offline database by the substation name and physical nodes number. These can now be mapped to electrical node names (number). Once this mapping is available, the construction of the admittance model YBUS can be carried out. Further checks can be carried out on structure of YBUS to find islands etc. This completes the NTP phase.

Graph theoretic computations in NTP can be implemented by a class *Graph*. A sparse symmetric matrix also has an embedded graph. Conversely, from a weighted undirected graph, a sparse symmetric matrix can be constructed. This close relation between a symmetric matrix and graph should be reflected in the design. Figure 10.14a and 10.14b shows two possible design alternatives. In the figure 10.14a, the graph and matrix share the same data. However, the two objects have a entirely different feel which can be seen in the corresponding methods. For example, *find_tree* has no meaning on a sparse matrix. However, in our implementation, we have used option 10.14b. At this stage, we find it difficult to refine and implement scheme shown in figure 10.14a. The derivation of figure 10.14b has created unnecessarily strong coupling between class *SparseMatrix* and *Graph*. This example also illustrate that OO design should be done with care, even if it means spending more time on the drawing board.

The next step in the overall process is observability analysis. For observability analysis, both graph theoretic and numerical rank computation algorithms can be used. As an OO design problem, we note that there is a correlation between the observability problem and the factorization of Jacobian used in state estimation. This commonality can be exploited in the design. Hence, we prefer the rank determination criteria for observability analysis. Now, two options are possible *viz.* structural rank computation and numerical rank computation. The structural rank computation requires formation of decoupled Jacobian while numerical rank determination can be directly applied to full Jacobian. In this case the authors preference (based on gut feeling) is that a coupled Jacobian approach should be used; the logic being that coupled approach does not involve approximations. Thus, numerical rank determination of the Jacobian becomes the logical choice for observability analysis.

The two questions that have remained unanswered concern the numerical implementation, *viz.* how we determine (numerical) rank accurately and how to handle equality constraints. The standard tool used by numerical analyst is Singular Value Decomposition(SVD) which decomposes a matrix A $(m \times n; m > n)$ as follows

$$A_{m\times n} = P \begin{bmatrix} \sigma_1 & & \\ & \ddots & \\ 0 & & \sigma_n \\ 0 & \cdots & \\ 0 & \cdots & 0 \end{bmatrix} Q^T \tag{10.70}$$

where P is a $m\times m$ orthogonal matrix, Q is a $n\times n$ orthogonal matrix and σ_i s (note that this σ is different than standard deviation used earlier) are singular values arranged in non ascending order. The numerical stability of above decomposition arises from the use of orthogonal rotations (like in QR decomposition). The rank of the matrix is the number of non-zero singular values. The above method of rank determination is very stable, though it cannot be employed in the context of large sparse matrices because the computations simply become prohibitive.

Recall that the QR decomposition also uses orthogonal rotations and hence enjoys good numerical stability characteristics. It can therefore be used to compute the rank of the Jacobian. During decomposition if $R_{ii} < \epsilon$, where ϵ is an numerical approximation to zero, then the corresponding state can be declared as unobservable. Immediately, a pseudo measurements can be added to enforce observability and the algorithm can proceed. QR decomposition is numerically stable even if, large weights for equality constraints are used. It is amenable to sparse matrix computations. Hence, it meets all our requirements. As such,

QR decomposition should be added as a method of class *SparseMatrix*[4]. The class *SparseMatrix*, its methods like MDA, forward backward substitution can be reused. This design is based upon incremental growth through reuse of existing classes.

At a macro level, it seems that the NTP object should return a network object or its variants. A class *PSSE*, which will mainly have methods for estimation, observability analysis and bad data processing, may be derived from the *Network* class. Readers are encouraged to suggest improvements on this design.

10. Summary

Power system state estimation problem has been investigated in detail since, Prof. Schweppe, the leading researcher of Power System Engineering Group at MIT, proposed and developed the idea [Schweppe and Wildes, 1970; Schweppe and Rom, 1970; Schweppe, 1970]. Later on power system state estimation has drawn attention of researchers from universities, research centers and industry.

Power system state estimator is a computational tool that evaluates the state of the system (node voltage magnitude and node angle) based on *redundant* telemetered system measurements. The telemetered measurements contain noise due to limited transducer accuracy and errors introduced in communication to the energy control center. Power System State Estimator filters out noise in measurements by exploiting the redundancy of the measurements. However, certain measurements e.g, if the meters are completely defective, are not included in state estimation. Such flagrantly wrong data, is removed by prefiltering program. Measurements with errors below this value and above say five standard deviations are called *bad data*. They are detected and eliminated by post processing algorithms called bad data estimation algorithms.

Observability analysis algorithm is used to identify the maximal set of nodes in the power system for which successful estimation of the power system states (node voltages and node angles) is possible. A power system is observable if and only if the resulting linearized measurement Jacobian has full column rank. Only, abnormal conditions result in system not being observable.

There are two basic approaches to power system state estimation viz.,

- Least Absolute Value (LAV) estimation and
- Weighted Least Squares (WLS) Estimation.

[4]Distinction should be made from class of symmetric sparse matrix.

The LS estimators are more popular than the LP based estimators. Fast and numerically stable implementation can be developed by using sparse QR decomposition approach. The performance of such an estimator suffers in presence of bad data. Hence, bad data estimators are used in conjunction with the state estimator to detect, estimate and compensate (or eliminate) such measurements. The estimated state of the power system provides the basis for many online network analysis functions like OPF, contingency analysis, operator request load flow etc.

11. Further Reading

[Monticelli, 1999; Monticelli, 2000] covers all the aspects of power system state estimation in quite detail. It is by far the best single source of information. The articles by Schweppe et. al. are recommended for those who want to review the origins of state estimation. [Gu et al., 1983] develops the method of Peters and Wilkinson. [Pandian et al., 1999; Vempati et al., 1991; Simoes-Costa and Quintana, 1981; Monticelli et al., 1985] are recommend for QR decomposition approach. [Gjelsvik et al., 1985] discusses a Lagrangian formulation for power system state estimation.

A comparative study of various bad data identification methods is presented in [Mili et al., 1985]. Observability analysis and bad data processing for state estimation with equality constraints is dealt in [Wu et al., 1988].

12. Review Questions

1 Consider the solution of LSE problem, $\min ||Ax - b||_2^2$, subject to constraints $Cx = d$. The matrix $\begin{bmatrix} A \\ C \end{bmatrix}$ has full column rank and matrix C has full row rank. Show that the null space of A and C intersect trivially.

2 Show that the LSE problem in question 1 has unique solution.

3 For what value of c is the following linear system of equations consistent.

$$\begin{bmatrix} 1 & 0 \\ 0 & 1 \\ 1 & 1 \end{bmatrix} \begin{bmatrix} x_1 \\ x_2 \end{bmatrix} = \begin{bmatrix} 1 \\ 1 \\ c \end{bmatrix}$$

4 Show that a LS problem $\min ||Ax - b||_2^2$ will solve the corresponding linear system of equations if the equations are consistent. If so, why are square LSS usually not solved by LS solvers?

5 Why is the system of figure 10.9 unobservable? Without changing the total number of measurements, suggest a new measurement placement so as to make it observable.

6 Show that in absence of noise, and with same initialization as in load flow analysis, power system state estimation solver tracks an equivalent Newton Raphson load flow problem.

7 Formulate a weighted LAV estimator to incorporate relative accuracy of the transducers. Modify the formula to include the equality constraints.

8 Define a pseudo inverse of A as A^+ which satisfies the following conditions

$$\begin{aligned} AA^+A &= A \\ A^+AA^+ &= A^+ \\ AA^+ &= (AA^+)^T \\ A^+A &= (AA^+)^T \end{aligned}$$

Then, for a $n \times n$ nonsingular matrix show that $A^+ = A^{-1}$

9 Why is it that scaling does not affect solution of LSS while it affects solution of a LS problem?

10 Show that $(HH^+)^2 = HH^+$.

11 Show that the matrix HH^+ is positive semidefinite.

12 Formulate the LAV and WLS state elimination problem considering inequality constraint, *e.g.*, generator reactive power capability curve. Comment on the statement that it will increase the computational complexity and improve the accuracy of the estimator.

Chapter 11

OPTIMAL POWER FLOW

We must make it our goal to find a method of solution of all problems...by means of a single simple method

—D'Alembert, V.M.Tikhomirov, Stories about Maxima and Minima

1. Recap

In the load flow analysis problem, generator voltage magnitudes, active power generation (except for the slack bus), and P and Q loads at all the busses is known beforehand. Consequently, to determine the power system state, we have to solve $2n - g - 1$ equations in as many unknowns, where n is the number of busses in the system and g is the number of generators. Therefore, the resulting system of equations is neither underdetermined nor overdetermined. In contrast, in the state estimation problem, to filter noise, we deliberately work with an overdetermined system (more measurements than unknowns). In both of the above scenario, controller settings, like generator voltages, MW injections, transformer taps, reactive power compensation etc have been decided beforehand. The criteria used for setting these controllers is not relevant to the problem. However, the next logical question that arises is how best to choose the controller settings for a given load condition. Clearly, there exist multiple choices. If the controller settings are also made variable, then a computational problem on the lines of load flow analysis will have to deal with an underdetermined system of equations. For example, if generator voltage magnitudes are not fixed in load flow analysis, as many additional columns (g) will appear in the load flow Jacobian while the number of rows or equations will remain unchanged.

An underdetermined system of equations (few rows than columns) need not have a solution. However, if it has one solution then, it has infinitely many solutions because its null space is non-zero. In absence of noise, usually we really do not have to worry about the possibility of having no solution. Now to choose one solution among the multiple solutions, we require a performance criteria that is optimized. This leads to the Optimal Power Flow (OPF) problem.

2. OPF: An Introduction

Definition: OPF is a steady state constrained optimization tool in which an objective function e.g., active power loss is minimized by setting controllers like generator voltages. The set of constraints in the OPF problem are as follows:

1 At each node, the active and reactive power injected has to be equal to the sum of corresponding load and the flow on the branches connected to it. The list of branches include lines, transformers and shunts. In other words, power balance equation (henceforth referred to as node injection equation) has to be satisfied. These equations represent the strict equality constraints in the optimization problem.

2 The inequality constraints considered are:

   (a) Limits on P and Q capability of generators. This has been discussed in the chapter on load flow analysis.

   (b) Limits on state variables *viz.* voltage magnitude at busses. These limits are typically, known as soft constraints. For example, if we define minimum bus permissible voltage as 0.95 pu, then, a voltage of 0.94 pu is also acceptable from engineering consideration. In short, the limits fall in the category of being *desirable* rather than *mandatory*. These limits are also known as soft limits.

   (c) Limits on other control variables like transformer taps, reactive power injection variables like SVCs, capacitors etc. These are the hard limits in the problem.

   (d) Limits on loading of lines or limits on angular difference between lines.

3 [Kundur, 1994; Dunlop et al., 1979] discusses the specification of line loading limits as a function of voltage class and distance of transmission line. The loading of a transmission line depends upon the following.

   - When the line length is less than 80km, the line is considered as a short line. In such lines, limits are decided by the thermal

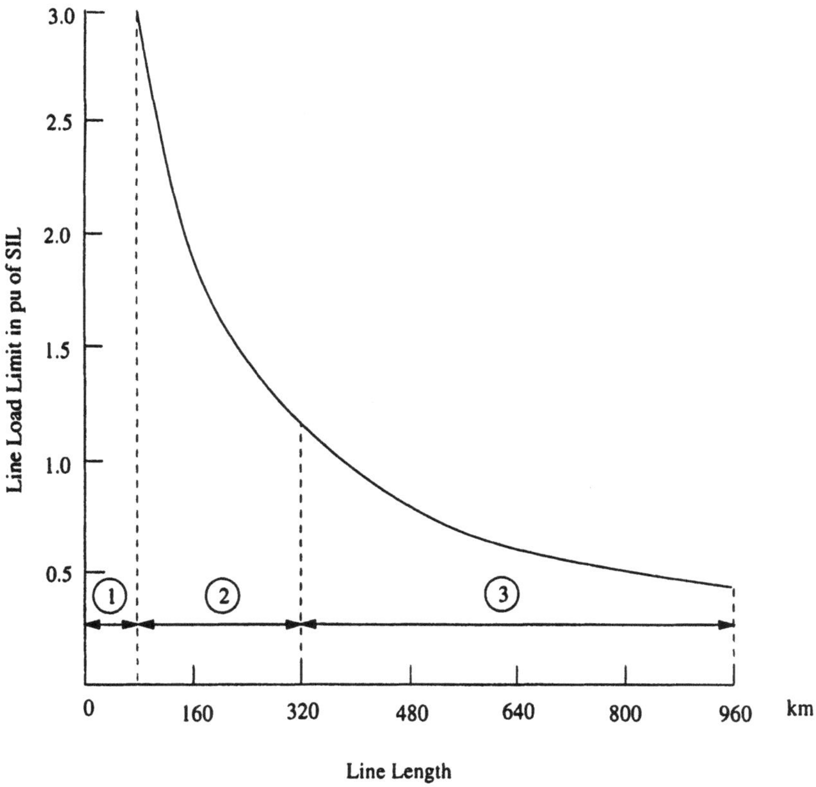

Figure 11.1. Transmission Line Loadability Curve

capacity, *i.e.,* the maximum amperes that the conductors can carry continuously.

- For the lines between 80-320 km, the primary constraint is imposed by the voltage drop limitation.
- For line above 320 km, the limitation is imposed by the stability limits of the line.

Figure 11.1 shows the nature of *St. Clair curves* used for estimating the transmission capability of a line. The capacity is expressed in pu where the base is specified by the Surge Impedance Loading (SIL) of the line. For a 230 kV line, SIL is 140 MW, for a 345 kV line SIL is 420 MW and for a 500kV line SIL is 1000 MW.

Subject to above listed constraints, there are various possible choices for the objective function. Some commonly used objectives are as follows:

1 Total cost of fuel: This corresponds to the Economic Dispatch (ED) problem. The cost function of a generator can be approximated by a

second order quadratic function or by piecewise linear segments.

$$F_i(P_i) = a_i + b_i P_i + \frac{1}{2} c_i P_i^2 \ \$/MW$$

The aim of the ED problem is to run the system economically.

2 Another important OPF formulation is known as minimum emission dispatch problem wherein, goal is to minimize the emissions NO_x, SO_2 etc. The emission dispatch objective can be modeled on similar lines to the fuel cost objective.

3 The other class of OPF problems are related to the system security and control of voltage/reactive power. Various objective functions considered in this category are:

(a) MW or active power loss minimization;

(b) MVAR or reactive power loss minimization;

(c) Maximization of the MVAR reserves;

(d) Minimization of voltage variations;

(e) Function defining improvement of voltage stability etc.

Table 11.1. Control and Constraints for Active Power and Reactive Power OPF Sub Problems [Papalexopoulos et al., 1989]

	CONTROLS	CONSTRAINTS
ACTIVE POWER OPF	MW Generations Economy Interchanges Phase Shifter Positions HVDC Line MW Flows Load Shedding FACTS	Control Variable Limits Voltage Angles Between Specific Busses MW Branch Flows MW Reserve Margins Area MW Interchanges Transmission Corridor MW Transfers Net Area MW Generations
REACTIVE POWER OPF	Generator Voltages or MVARs LTC Tap Positions SVC MVARs Capacitor/Reactor Statuses FACTS	Control Variable Limits Bus Voltage Magnitudes MVAR Generations MVAR Branch Flows MVAR Reserve Margins Area MVAR Interchanges Transmission Corridor MVAR Transfers

4 OPF is also used as a tool for planning purpose. A commonly used objective function in these situations is minimization of cost of compensation to be installed for voltage/reactive power control. A composite objective function involving active power losses and cost of reactive power compensation can also be considered.

5 OPF is also turning out to be an important tool in deregulated scenario [Christie et al., 2000].

Table 11.1 groups the above set of problems into two groups *viz. active* power OPF subproblems and *reactive* power OPF subproblems. This classification is based upon the fact that each controller is designed to affect a specific class of variables or constraints. For example, with phase shifting transformers, we schedule the active power flows across the lines while with the tap changing transformers (also known as on load tap changing transformer (OLTC), under load tap changing transformer (LTC)), we regulate the terminal voltages. Hence, phase shifting transformer is an important control variable in a problem which schedules active power flows e.g., economic dispatch (ED) problem, alleviation of line overloading etc, while transformer tap is an important control variable in a reactive OPF problem e.g., minimize deviation of voltages in a power system from 1.0 pu reference value.

In a well formulated OPF problem, there exists a correlation between the objective function, the controllers used, and the constraints. Conversely, if the problem is not well posed, then the results will not make a good engineering sense. For example, let in ED problem, all generator voltages be fixed to 0.95 pu. Further, they are not included as control variables. Now, if we specify voltage constraints for load busses as 0.95 pu and maximum 1.05 pu then, in a peak load condition scenario, it is unlikely that a feasible solution to OPF will be obtained. Therefore, it can be seen in the table 11.1 that in active OPF subproblems control and constraints pertain to active flow variables. Consequently, for the line loading constraint, only the MW flow constraint is imposed.

In the reactive power OPF subproblem, limits are specified on the MVAR flow across the line. To understand this specification, consider a 230 kV transmission line of 160 Km length. Then, the limit on the MW capacity of the line is about 240 MW. Let us conservatively translate this limit also as the MVA capacity limit of the line. Now, after solving the ED problem, let us say that the line carries 160 MW of active power. Then, in a reactive OPF subproblem, we will specify a MVAR limit of $\sqrt{(240^2 - 160^2)} = 178$ MVAR for the reactive power flow across the line. In essence, in a well formulated OPF problem, there exists correlation between objective function, controls and the constraints.

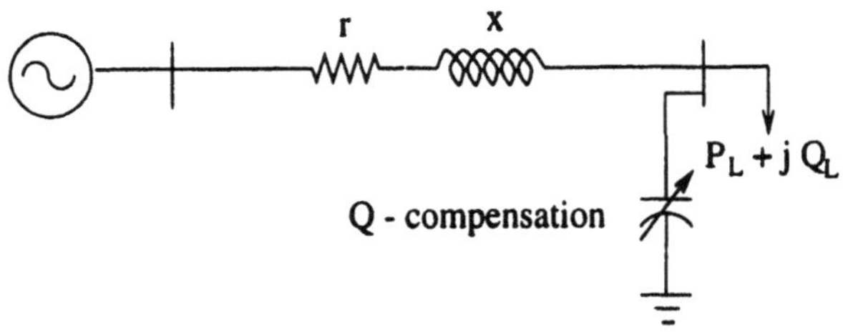

Figure 11.2. A Bus With Reactive Power Compensation

Though, it may be a bit surprising, active power loss minimization problem is considered as a the reactive power OPF subproblem. The reason is that active power loss is sensitive to reactive power control variables. For example, in figure 11.2, if a certain power P_{ij}^0 ($P_{ij} = VI\cos\phi$) has to be delivered from bus i to bus j then, by reducing ϕ, if it is delivered with a smaller magnitude of $|I|$, $|I|^2R$ losses will reduce. Improving power factor mainly involves scheduling of reactive power compensation. An alternative explanation for a meshed system is as follows.

The reactive power flow across a line $i-j$ is proportional to $|V_i|-|V_j|$. Hence, by controlling $|V_i|$ and/or $|V_j|$, the reactive power flow can be controlled. At generator or PV busses, $|V_i|$ can be directly controlled as long as the reactive power limit is not reached. Reduction in the reactive power flow reduces overall current and hence the $|I|^2R$ loss. A smaller reactive power flow also implies a uniform voltage profile. Line charging is another source of MVAR supply. By changing the reference voltage of generator voltages the amount of charging MVAR can be controlled. Thus, under peak load conditions, by increasing the generator reference voltage, the contribution of line charging increases generator MVARs can be reduced. Conversely, under light load conditions by setting a low value for reference voltage, capacitive MVARs and hence MVAR absorption of the generators can be reduced.

There is no hard and fast rule to decouple the active power problem from the reactive power subproblem. For example, losses can also be incorporated in ED problem by assuming some cost coefficient ($/MW) for losses. In the ED problem, the primary constraint is imposed by the transmission capability of the line (limits on P_{ij}, Q_{ij}). For short lines, this corresponds to thermal limit; for long lines, this corresponds to steady state stability limit or alternatively limits on $|\delta_i - \delta_j|$.

OPF solver can also be used for security analysis and control. If a feasible solution exists, *i.e.*, all equality and inequality constraints are satisfied, then the system can be maintained at least in *alert* state.

The various states of power system has been discussed in the previous chapter. By simulating contingencies, the list of constraints can be augmented. Such an OPF solver which accounts for contingencies is known as Security Constrained OPF (SCOPF) solver. If a feasible solution exists to SCOPF, then the system can be maintained in *normal or secure* state. However, additional constraints in SCOPF may increase the minimum objective function value, *e.g.*, losses, total fuel cost etc.

2.1. Mathematical Formulation

Mathematically, the problem can be defined as follows:

$$\text{Problem OPF}: \min_{x,u} \; f(x,u) \tag{11.1}$$

subject to the following constraints:

$$g_i(x,u) = 0; \tag{11.2}$$

$$u_{min_i} \leq u_i \leq u_{max_i}; \tag{11.3}$$

$$x_{min_i} \leq x_i \leq x_{max_i}; \tag{11.4}$$

$$h_{min_i} \leq h_i(x,u) \leq h_{max_i} \tag{11.5}$$

where i is used as a running index. Constraints defined by equations (11.2) are the load flow equations to be satisfied at any operating point. They include both the active and reactive power injection equations. They can be expressed as follows:

$$g_k(x,u) = \left\{ \begin{array}{c} P_i - \Sigma_{j=1}^{n} |V_i||V_j||Y_{ij}| \cos(\delta_i - \delta_j - \theta_{ij}) \\ \text{or} \\ Q_i - \Sigma_{j=1}^{n} |V_i||V_j||Y_{ij}| \sin(\delta_i - \delta_j - \theta_{ij}) \end{array} \right\} \tag{11.6}$$

where i refers to the node number and k is a running index. Constraints defined by (11.3-11.4) are the minimum and maximum permissible control (u) and state variable (x) (bus voltage magnitude and angle) limits. All of these variables are treated as continuous variables. Constraints defined by (11.5) are the security constraints which include the minimum and maximum permissible MVAR loading limits for generators, line loading limits etc.

2.2. Solution Methodologies

An OPF problem is a nonlinear constrained optimization problem[1]. The aim of an OPF solver is to compute a KKT solution to the optimization problem (11.1-11.5) that corresponds to at least a local minimum.

[1]The theory of constrained optimization was discussed in chapter 6.

Computationally, OPF is a much more involved problem than the load flow analysis problem. Basically, the performance of an OPF algorithm depends upon how the power flow equations are handled. Carpentier [Carpentier, 1987] has classified modeling of OPF into the following:

- Compact modeling;
- Sparse modeling.

In compact model formulation, a power flow problem is first solved and then, a reduced optimization problem with only explicit control variables in objective function and constraints is developed. On the other hand, when the bus voltage magnitude and angles are not eliminated while formulating optimization problem, we get a sparse formulation. Since the linearized constraints obtained from equations (11.2-11.5) are highly sparse, optimization problem can be solved using sparse techniques. Resulting algorithms in this family can be subclassified into sparse Sequential Linear Programming (SLP) [Quintana and Mota-Palomino, 1985; Qiu and Shahidehpour, 1987; Iba et al., 1988][2], sparse SQP [Burchett et al., 1984; El-Kady et al., 1986] and Newton method for OPF [Sun et al., 1984]. Since, the power flow equations are not eliminated, explicit linearized power flow equations (11.7) are maintained in the formulation.

$$J_x \Delta x + J_u \Delta u = -g(x^k, u^k) = \begin{bmatrix} \Delta P \\ \Delta Q \end{bmatrix} \tag{11.7}$$

J_x denotes the Jacobian with respect to state variables. It is identical to Jacobian of load flow analysis. J_u denotes the Jacobian with respect to controls. It's computation is similar to the Jacobian in load flow analysis. In this approach, computation of the partial derivatives of functions f, g and h is quite straight forward. However, the optimization program should be able to handle large sparse matrices.

Alternatively, an OPF problem can be classified based upon whether a load flow solution is required before optimization or not [Glavitsch and Bacher, 1991]. An algorithm which requires a solved load flow for optimization belongs to class A while algorithms that compute both load flow and optimal solution simultaneously are said to belong to class B. Broadly, class A algorithms and compact modeling algorithms are synonymous except for the sparse (SLP) methods. Similarly, class B and sparse modeling methods are synonymous except for sparse SLP methods. In the rest of this chapter, we discuss methods belonging to both of these classes.

[2] A sparse SLP technique may require a solved load flow for optimization.

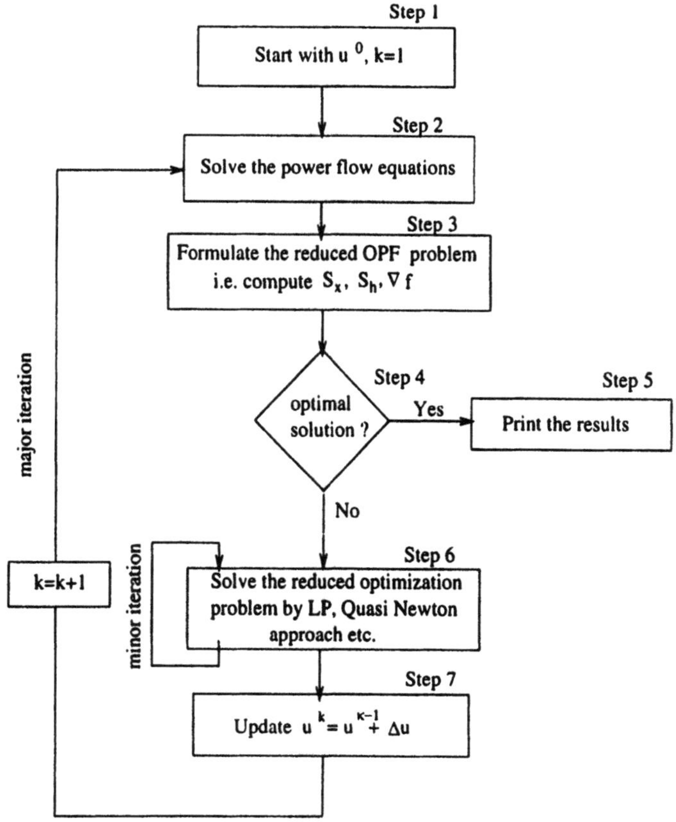

Figure 11.3. Compact Model Based OPF Algorithm

3. Compact Model Formulation

Compact model formulation was developed by [Dommel and Tinney, 1968]. The algorithm solves a reduced optimization problem at each OPF iteration with only control variables. The scheme is illustrated in figure 11.3[3]. In the step 2, load flow solution corresponding to the current setting of control vector is computed. The state vector and dependent constraint magnitudes (load bus voltages and the generator Q-injections) corresponding to exact solution of equality constraints are obtained in this step. The dependent variables are linearized around the current iteration resulting in a reduced model formulation (step 3). Computation of $S_x, S_h, \nabla_u f$ is explained in next section. The resulting

[3]The presentation of Dommel and Tinney algorithm was geared for use of steepest descent method with modifications to account for bound constraints. A more generic formulation which can account for generalized constraints is considered here.

optimization problem is non-sparse mainly due to non-zero elements introduced during the elimination of state variables (PQ node voltages and angles). If the dependent constraints are within the specified bounds and the absolute change in objective function below an user specified tolerance or the reduced gradient zero (norm of reduced gradient less than tolerance), the algorithm terminates (step 4). Otherwise, the linearly constrained optimization problem is solved in step-6.

Let matrices S_x and S_h measure the sensitivity of the state variables x and constraint function h to the controller movement u, *i.e.*, let $\Delta x = S_x \Delta u$ and $\Delta h = S_h \Delta u$. Then, in the compact model formulation based Sequential Linear Programming (SLP) formulation, following optimization problem is solved at step 6:

$$\min_{\Delta u} \nabla_u f^{k^T} \Delta u^k$$

$$\text{such that } x_{min} - x^k \leq \Delta x = S_x \Delta u \leq x_{max} - x^k$$

$$\text{and } h_{min} - h(x^k) \leq \Delta h = S_h \Delta u \leq h_{max} - h(x^k)$$

The point (x^k) is obtained from load flow solution in step-2 with controller setting at u^k. The Matrices S_x, S_h and gradient of objective function is obtained in step 3. In a Quasi-Newton approach, an additional term for modeling the spd approximation of the Hessian may be incorporated. Now, we discuss the various methods for computing the matrices S_x and S_h.

3.1. Computation of $\nabla_u f, S_x$ and S_h matrices

For the purpose of discussion, let us consider the active power loss minimization problem wherein losses are defined as $P_{loss} = \sum_{i=1}^{n} P_i$. Except for slack bus injection, all P injections are fixed. In this case, we presume that they are decided by solving an ED problem separately.

$$\text{Hence, } \Delta P_{loss} = \Delta P_{slack}$$

$$\text{and therefore } \min P_{loss} = \min P_{slack}$$

For such a problem, control variables correspond to (1) generator voltages (2) transformer taps and (3) reactive power injection variables. Third component constitute static VAR compensators, synchronous condensers, shunt capacitors etc. Busses with synchronous condensers can be modeled on similar lines to generators. In the compact model formulation, the generator terminal voltage is treated as a control variable and the limits of Q injection are modeled as dependent constraints[4]

[4] Alternatively, can we specify Q injection of generator in control variable, and the terminal voltage as a dependent state variable? Can the specification of both desirable voltage limit

$h_i(x,u) \leq 0$. The SVC and capacitors can be approximately modeled by equivalent Q compensating variables. In the simplified formulation that we discuss, discrete nature of controllers like shunt capacitor banks, taps etc. is neglected, *i.e.*, all variables are assumed to be continuous. To obtain the reduced problem only in terms of Δu, we need to eliminate state variables x in equation 11.1. Hence, linearizing the power flow equations (11.2) around the power flow solution, we obtain the following[5]:

$$\left[\frac{\partial g}{\partial x}\right]^T \Delta x + \left[\frac{\partial g}{\partial u}\right]^T \Delta u = 0 \tag{11.8}$$

$$\Rightarrow \Delta x = -\left[\frac{\partial g}{\partial x}\right]^{T^{-1}} \times \left[\frac{\partial g}{\partial u}\right]^T \Delta u$$

$$\Delta x = [S_x]\,\Delta u \tag{11.9}$$

and $S_x =$

$$\begin{bmatrix} \frac{d\delta_2}{d|V_1|} & \cdots & \frac{d\delta_2}{d|V_g|} & \vdots & \frac{d\delta_2}{dq_{c_1}} & \cdots & \frac{d\delta_2}{dq_{c_c}} & \vdots & \frac{d\delta_2}{dt_1} & \cdots & \frac{d\delta_2}{dt_l} \\ \vdots & & \vdots & \vdots & \vdots & & \vdots & \vdots & \vdots & & \vdots \\ \vdots & \ddots & \vdots & \vdots & \vdots & \ddots & \vdots & \vdots & \vdots & \ddots & \vdots \\ \frac{d\delta_n}{d|V_1|} & \cdots & \frac{d\delta_n}{d|V_g|} & \vdots & \frac{d\delta_n}{dq_{c_1}} & \cdots & \frac{d\delta_n}{dq_{c_c}} & \vdots & \frac{d\delta_n}{dt_1} & \cdots & \frac{d\delta_n}{dt_l} \\ \cdots & \cdots & \cdots & \cdots & \cdots & \cdots & \cdots & \cdots & \cdots & \cdots & \cdots \\ \frac{d|V_{g+1}|}{d|V_1|} & \cdots & \frac{d|V_{g+1}|}{d|V_g|} & \vdots & \frac{d|V_{g+1}|}{dq_{c_1}} & \cdots & \frac{d|V_{g+1}|}{dq_{c_c}} & \vdots & \frac{d|V_{g+1}|}{dt_1} & \cdots & \frac{d|V_{g+1}|}{dt_l} \\ \vdots & & \vdots & \vdots & \vdots & & \vdots & \vdots & \vdots & & \vdots \\ \vdots & \ddots & \vdots & \vdots & \vdots & \ddots & \vdots & \vdots & \vdots & \ddots & \vdots \\ \frac{d|V_n|}{d|V_1|} & \cdots & \frac{d|V_n|}{d|V_g|} & \vdots & \frac{d|V_n|}{dq_{c_1}} & \cdots & \frac{d|V_n|}{dq_{c_c}} & \vdots & \frac{d|V_n|}{dt_1} & \cdots & \frac{d|V_n|}{dt_l} \end{bmatrix} \tag{11.10}$$

where bus number 1 is the slack bus and $\Delta x = [\Delta\delta_2, \ldots \Delta\delta_n, |V_{g+1}| \ldots |V_n|]^T$ and $\Delta u = [\Delta|V_1|, \ldots \Delta|V_g|, \Delta q_1 \ldots \Delta q_c, \Delta t_1 \ldots \Delta t_l]^T$. The entry $S_x(i,j)$ represents derivative $\frac{dx_i}{du_j}$. The matrix $\left[\frac{\partial g}{\partial x}\right]^T$ corresponds to the load flow Jacobian (J). It is a $2n-g-1 \times 2n-g-1$ sparse matrix. Explicit inversion of J is not the right way to compute S_x. Instead a sparse LU decomposition followed by repeated forward backward substitution should be done. If there are nc control variables, then nc forward back-

and mandatory Q limit on a generator lead to infeasibility in the OPF problem. Argue your case for a lightly loaded system.

[5]Our convention of putting a transpose is in variance with that in [Dommel and Tinney, 1968].

ward solutions are required.

$$
\left[\frac{\partial g}{\partial x}\right]^T = \left[\begin{array}{ccccccc}
\frac{\partial P_2}{\partial \delta_2} & \cdots & \frac{\partial P_2}{\partial \delta_n} & \vdots & \frac{\partial P_2}{\partial |V_{g+1}|} & \cdots & \frac{\partial P_2}{\partial |V_n|} \\
\vdots & & \vdots & \vdots & \vdots & & \vdots \\
\vdots & \ddots & \vdots & \vdots & \vdots & \ddots & \vdots \\
\frac{\partial P_n}{\partial \delta_2} & \cdots & \frac{\partial P_n}{\partial \delta_n} & \vdots & \frac{\partial P_n}{\partial |V_{g+1}|} & \cdots & \frac{\partial P_n}{\partial |V_n|} \\
\cdots & \cdots & \cdots & \cdots & \cdots & \cdots & \cdots \\
\frac{\partial Q_{g+1}}{\partial \delta_2} & \cdots & \frac{\partial Q_{g+1}}{\partial \delta_n} & \vdots & \frac{\partial Q_{g+1}}{\partial |V_{g+1}|} & \cdots & \frac{\partial Q_{g+1}}{\partial |V_n|} \\
\vdots & & \vdots & \vdots & \vdots & & \vdots \\
\vdots & \ddots & \vdots & \vdots & \vdots & \ddots & \vdots \\
\frac{\partial Q_n}{\partial \delta_2} & \cdots & \frac{\partial Q_n}{\partial \delta_n} & \vdots & \frac{\partial Q_n}{\partial |V_{g+1}|} & \cdots & \frac{\partial Q_n}{\partial |V_n|}
\end{array}\right] \qquad (11.11)
$$

$$
\left[\frac{\partial g}{\partial u}\right]^T =
$$

$$
\left[\begin{array}{ccccccccccc}
\frac{\partial P_2}{\partial |V_1|} & \cdots & \frac{\partial P_2}{\partial |V_g|} & \vdots & \frac{\partial P_2}{\partial q_{c_1}} & \cdots & \frac{\partial P_2}{\partial q_{c_c}} & \vdots & \frac{\partial P_2}{\partial t_1} & \cdots & \frac{\partial P_2}{\partial t_l} \\
\vdots & & \vdots & \vdots & \vdots & & \vdots & \vdots & \vdots & & \vdots \\
\vdots & \ddots & \vdots & \vdots & \vdots & \ddots & \vdots & \vdots & \vdots & \ddots & \vdots \\
\frac{\partial P_n}{\partial |V_1|} & \cdots & \frac{\partial P_n}{\partial |V_g|} & \vdots & \frac{\partial P_n}{\partial q_{c_1}} & \cdots & \frac{\partial P_n}{\partial q_{c_c}} & \vdots & \frac{\partial P_n}{\partial t_1} & \cdots & \frac{\partial P_n}{\partial t_l} \\
\cdots & \cdots & \cdots & \cdots & \cdots & \cdots & \cdots & \cdots & \cdots & \cdots & \cdots \\
\frac{\partial Q_{g+1}}{\partial |V_1|} & \cdots & \frac{\partial Q_{g+1}}{\partial |V_g|} & \vdots & \frac{\partial Q_{g+1}}{\partial q_{c_1}} & \cdots & \frac{\partial Q_{g+1}}{\partial q_{c_c}} & \vdots & \frac{\partial Q_{g+1}}{\partial t_1} & \cdots & \frac{\partial Q_{g+1}}{\partial t_l} \\
\vdots & & \vdots & \vdots & \vdots & & \vdots & \vdots & \vdots & & \vdots \\
\vdots & \ddots & \vdots & \vdots & \vdots & \ddots & \vdots & \vdots & \vdots & \ddots & \vdots \\
\frac{\partial Q_n}{\partial |V_1|} & \cdots & \frac{\partial Q_n}{\partial |V_g|} & \vdots & \frac{\partial Q_n}{\partial q_{c_1}} & \cdots & \frac{\partial Q_{g+1}}{\partial q_{c_c}} & \vdots & \frac{\partial Q_n}{\partial t_1} & \cdots & \frac{\partial Q_n}{\partial t_l}
\end{array}\right] \qquad (11.12)
$$

Next, we obtain the gradient of objective function as follows. If P_{sl} is the active power injection at the slack bus, then

$$
\begin{aligned}
\Delta f = \Delta P_{sl} &= \left[\frac{\partial P_{sl}}{\partial x}\right]^T \Delta x + \left[\frac{\partial P_{sl}}{\partial u}\right]^T \Delta u \qquad (11.13) \\
&= \left[\left[\frac{\partial P_{sl}}{\partial x}\right]^T [S_x] + \left[\frac{\partial P_{sl}}{\partial u}\right]^T\right] \Delta u \\
&= \left[\frac{dP_{sl}}{du}\right]^T \Delta u \qquad (11.14)
\end{aligned}
$$

$$\text{where, } \left[\frac{dP_{sl}}{du}\right]^T = \nabla_u f = \left[\left[\frac{\partial P_{sl}}{\partial x}\right]^T [S_x] + \left[\frac{\partial P_{sl}}{\partial u}\right]^T\right] \quad (11.15)$$

$$\text{Similarly, } \Delta h = \left[\frac{\partial h}{\partial x}\right]^T \Delta x + \left[\frac{\partial h}{\partial u}\right]^T \Delta u$$

$$= \left[\left[\frac{\partial h}{\partial x}\right]^T S_x + \left[\frac{\partial h}{\partial u}\right]^T\right] \Delta u \quad (11.16)$$

$$\Delta h = [S_h]\,\Delta u \quad (11.17)$$

where $\Delta h = [\Delta Q_1 \ldots \Delta Q_g]^T$. Though not apparent at first, it can be shown that this approach of computing $\frac{df}{du}$ is identical to computation of $\frac{d\mathcal{L}}{du}$ as discussed in the [Dommel and Tinney, 1968]. From the very nature of the matrix operations, it can be seen that both the matrices S_x and S_h are nonsparse. Instead of an analytical approach followed above, a finite difference approximation can also be used to compute the sensitivities. We now discuss a simple perturbation method to evaluate the sensitivity matrices.

3.1.1 Perturbation Technique

In this method [Soman, 1995], a finite difference approximation of the sensitivity coefficients and the power loss gradient are computed by repeated load flow solutions corresponding to perturbation in the control variables. The philosophy is that we perturb control variables one at a time, then run a load flow to compute sensitivity. After the load flow, the control variable is reset to its initial value. This is repeated for all the control variables. The base case power flow solution at the current iteration provides a close initial guess for load flow with perturbed control variables. Therefore, the convergence of the perturbation load flows is rapid.

The algorithm to evaluate the power loss derivative and the sensitivity matrices (S_x, S_h) is detailed below:

1 Store the base values of state vector x (x^0), reactive power generations Q_i, $\forall\, i = \{1, \ldots g\}$ and the slack bus power injection P_{sl} (P^o_{sl}). Let u^o denote initial control setting.

2 Initialize $ic = 1$;

3 Perturb the ic^{th} control variable, $u_{ic} = u^0_{ic} + \Delta u_{ic}$.

4 Perform a power flow with starting guess for x as x^o. Obtain the corresponding state vector x^{new} and injections P^{new}_{sl} and Q^{new}_j ($j = \{1, \ldots g\}$);

5 Restore control u_{ic}, *i.e.*, $u_{ic} = u_{ic} - \Delta u_{ic}$;

6 Calculate

(a) $\frac{dV_j}{du_{ic}} = \frac{|V_j^{new}|-|V_j^o|}{\Delta u_{ic}} \; \forall \; j = \{g+1 \ldots n\}$

(b) $\frac{dQ_j}{du_{ic}} = \frac{Q_j^{new}-Q_j^o}{\Delta u_{ic}} \; \forall \; j = \{1 \ldots g\}$

(c) $\frac{dP_{sl}}{du_{ic}} = \frac{P_{sl}^{new}-P_{sl}^o}{\Delta u_{ic}}$

7 $ic = ic + 1$

8 If $ic \leq$ number of control variables (nc), go to step 3.

Note that a perturbation of control variable helps in creating one column of the sensitivity matrix S_x and S_h. In this approach, load flow decoupling is used to an advantage without sacrificing the accuracy of computed sensitivities. The perturbations for the generator voltages and the Q-injection variables do not affect the YBUS or B' and B'' matrices. Hence, the load flow solutions for perturbation involve only forward backward substitution. However, perturbation of a transformer tap changes corresponding entries in the YBUS matrix and hence also B''. A direct method to incorporate these changes is to refactorize the relevant matrix (B''). However, such refactorization will be computationally expensive.

When the transformer taps are perturbed, we want to find the solution ΔV to the following system of equations.

$$[B'' + \Delta B''][\Delta V] = [\Delta Q]$$

The LDL^T factorization of B'' is already available. The changes in matrix B'' can be viewed as replacement of two columns of B''. They are basically rank updates of matrix B''. Refactorization can be avoided by using Sherman Morrison formula (refer Chapter 7). Having discussed the reduced optimization problem formulation in sufficient detail, we now discuss a Sequential Linear Programming (SLP) approach which is popularly used with compact or reduced model formulation.

4. SLP based OPF Algorithm

As is shown in figure 11.3, the compact model algorithm involves two iterative loops. In the outer loop (known as major iteration), a constrained optimization problem is formulated until optimality conditions are satisfied. In the SLP approach, this optimization problem corresponds to a LP problem wherein both objective function and the constraints are linear. In the inner loop (known as minor iteration), the LP problem is solved.

For the illustrative loss minimization problem, LP problem at the kth iteration is as follows:

$$\min_{\Delta u^k} \; \Delta P_{sl}^k \;=\; \left[\frac{dP_{sl}^k}{du}\right]^T \Delta u^k \tag{11.18}$$

subject to

$$\max(-ustep, u_{min} - u^k) \leq \Delta u^k \leq \min(ustep, u_{max} - u^k) \tag{11.19}$$
$$x_{min} - x^k \leq S_x^k \Delta u^k \leq x_{max} - x^k \tag{11.20}$$
$$h_{min} - h^k \leq S_h^k \Delta u^k \leq h_{max} - h^k \tag{11.21}$$

Note that the nonlinear objective function and constraints are modeled only by their gradient information. Therefore, it is natural to expect that the region in which the above model holds is small. Hence, a vector *ustep* is used to restrict the maximum permissible movement of the control variables at each LP iteration. The usage of *ustep* vector has a similarity with the formal *trust region methods* in optimization. The choice of step size ($ustep_i$) depends upon the type of control device such as tap, Q-compensation, excitation etc. The final accuracy and the rate of convergence depends upon the number and length of the linear segments considered. Their size can be adaptively changed [Stott et al., 1987].

An inherent merit of the SLP approach is that it can be used to promote discrete scheduling of transformer taps and shunt compensators as the LP solution always lies at the vertex of convex polygon obtained from the linear inequality constraints. Because of the linearization approximation, in the SLP approach, it is recommended that a load flow problem with current control setting be solved before linearization. The advantage of compact modeling philosophy is to reduce programming complexity. Sparsity is handled in load flow program. The optimization problem is non-sparse and has reduced computational complexity.

An infeasible solution for the LP problem implies that problem is terminated in phase-I. Hence, in the context of OPF, Δu^k will correspond to a direction that reduces infeasibility, *e.g.*, violations in load bus voltages etc. After updating the controls, new load flow is repeated and a new LP problem is formed. While it is likely that the initial LP subproblem will not lead to a feasible point, subsequent LP problem may be feasible and hence the objective function is optimized.

An active set method for LP was discussed in chapter-7. It provides a natural and direct framework for handling large LP problems without introducing slack variables for the inequality constraints or formulating a dual LP problem. It is computationally efficient when constraints set size significantly exceeds the control variable set (nc). The method updates an inverse of a matrix of size ($nc \times nc$). For example, for a 319 node power system with 88 controls [Soman et al., 1994], there are about 20 generators. Hence, the vector Δx which equals ΔV_L (voltage changes at load bus) represents 299 constraint equations. In addition, there are 20 equations from constraints on Q-limits of generators. Together $\begin{bmatrix} S_x \\ S_h \end{bmatrix}$ matrix is a 319 × 88 matrix.

SLP is very well discussed in [Bazaraa et al., 1993]. It can have as good convergence characteristic as Newton method near the solution. It is a good choice when number of active constraints nearly equal the dimensionality of the LP problem, *i.e.*, when the solution is at a vertex. The normal operation of a power system is not at a vertex. For example, if $V_{min} = 0.90$ pu and $V_{max} = 1.10$ pu and most of voltages are near 1.0 pu then the solution of problem described by equations (11.1-11.5) is not at a vertex. Hence, SLP technique should not be directly applied to the problem (11.1-11.5). However, in the reduced problem, the solution is likely to be at a vertex defined by control vectors. Hence, SLP is used with compact model formulation.

We now discuss two important methods for large sparse OPF problems. Both these methods use second order information of the objective function and the active constraints. They do not require a converged load flow solution, *i.e.*, they simultaneously compute both the load flow and OPF solution, *say*, from a flat start.

5. OPF by Newton Approach

This method is a generalization of Newton method for equality constrained NLP problem discussed in the chapter 6. It was first applied to OPF by [Sun et al., 1984]. The method computes the optimal solution by applying Newton Raphson algorithm to the nonlinear system of equations obtained from KKT conditions. For the OPF problem (equations 11.1-11.5), Lagrangian function can be defined as follows:

$$\mathcal{L}(x, u, \lambda, \mu) = f(x, u) + \lambda^T g(x, u) + \mu^T h(x, u) \qquad (11.22)$$

where λ is the vector of Lagrangian multipliers associated with power injection equations and μ_is are the Lagrangian multipliers associated with inequality constraints. Recall that we use positive sign for Lagrangian multiplier of an inequality constraint if it is defined as $h_i(x, u) \leq 0$ and minus sign, if it is defined as $h_i(x, u) \geq 0$. This way, the optimality condition ($\mu_i \geq 0$) for all inequality constraints remains unchanged. For strict equality constraints, *e.g.*, the node injection equations, the sign convention used does not matter. Let $z = [x, u, \lambda, \mu]^T$. The KKT conditions demand that at the optimal:

$$\nabla_x \mathcal{L}(z^*) = 0 \tag{11.23}$$

$$\nabla_u \mathcal{L}(z^*) = 0 \tag{11.24}$$

$$\nabla_\lambda \mathcal{L}(z^*) = 0 \Rightarrow g_i(x^*, u^*) = 0 \tag{11.25}$$

$$\nabla_\mu \mathcal{L}(z^*) = 0 \Rightarrow h_i(x^*, u^*) = 0 \;\forall\; i \in \mathcal{A}(z^*) \tag{11.26}$$

$$\mu_i^* \geq 0 \;\forall i \in \mathcal{A}(z^*) \tag{11.27}$$

$$\mu_i^* = 0 \; i \notin \mathcal{A}(z^*) \tag{11.28}$$

where $A(z^*)$ is active set of constraint. In this approach, from an algorithmic viewpoint, there is no distinction between a control variable or a state variable. Hence, unlike the compact model formulation, generator voltage magnitude can also be considered as a state variable. Transformer taps, Q compensation variables, active power injection at generators etc. are treated as control variables. The vector $h(x, u)$ includes limits on line loadings. The linearization of equations (11.23 -11.26) leads to equations of the following form.

$$\begin{bmatrix} \nabla^2_{xx}\mathcal{L}(z^k) & \nabla^2_{xu}\mathcal{L}(z^k) & \frac{\partial g}{\partial x} & \frac{\partial h}{\partial x} \\ \nabla^2_{ux}\mathcal{L}(z^k) & \nabla^2_{uu}\mathcal{L}(z^k) & \frac{\partial g}{\partial u} & \frac{\partial h}{\partial u} \\ \frac{\partial g^T}{\partial x} & \frac{\partial g^T}{\partial u} & 0 & 0 \\ \frac{\partial h^T}{\partial x} & \frac{\partial h^T}{\partial u} & 0 & 0 \end{bmatrix} \begin{bmatrix} \Delta x^k \\ \Delta u^k \\ \Delta \lambda^k \\ \Delta \mu^k \end{bmatrix} = - \begin{bmatrix} \nabla_x \mathcal{L}(z^k) \\ \nabla_u \mathcal{L}(z^k) \\ g(x^k, u^k) \\ h(x^k, u^k) \end{bmatrix} \tag{11.29}$$

where the convention is $\frac{\partial g_i^T}{\partial x} = \left[\frac{\partial g_i}{\partial x_1}, \cdots, \frac{\partial g_i}{\partial x_{2n-1}}\right]$ and

$$\nabla^2_{xx}\mathcal{L}(z^k) = \left[\frac{\partial^2 f}{\partial x^2}\right] + \sum_i \lambda_i \left[\frac{\partial^2 g_i}{\partial x^2}\right] + \sum_i \mu_i \left[\frac{\partial^2 h_i}{\partial x^2}\right]$$

Those constraints that are not active ($\mu_i = 0$) do not have any effect on Hessian. Hence, even though they are a part of Lagrangian function, they are not explicitly shown. Note that Hessian models curvature information of both the objective function and the active constraints The above system of equations is large and sparse and hence the method is amenable to sparsity exploitation. The descent direction is defined by $d^k = [\Delta x^k, \Delta u^k]^T$ and W^k is used to represent coefficient matrix in (11.29).

The key problem in obtaining the above system of equations is that we do not know beforehand which of the inequality constraints are going to be active at the solution. Other than the node injection equations only active inequality constraints should be explicitly considered in equation (11.29). Thus, a procedure is required to refine the set of active constraints such that finally we converge to the active constraints at the optimal. One way to handle inequalities is by the use of penalty functions. Right from the beginning, we know that node injection equality constraints or power flow equations will have to be involved. Hence, they can be easily modeled as explicit equality constraints leading to the following formula.

$$\begin{bmatrix} \nabla^2 \mathcal{L}_{xx} & \nabla^2 \mathcal{L}_{xu} & \frac{\partial g}{\partial x} \\ \nabla^2 \mathcal{L}_{ux} & \nabla^2_{uu} & \frac{\partial g}{\partial u} \\ \frac{\partial g^T}{\partial x} & \frac{\partial g^T}{\partial u} & 0 \end{bmatrix} \begin{bmatrix} \Delta x^k \\ \Delta \lambda_i^k \\ \Delta \lambda^k \end{bmatrix} = - \begin{bmatrix} \nabla_x \mathcal{L} \\ \nabla_u \mathcal{L} \\ g(x^n, u^n) \end{bmatrix} \tag{11.30}$$

Now the inequality constraints as and when they become active can be modeled through quadratic penalty terms augmented to the Lagrangian function. In this way, one can avoid repeated changes in dimension in W. To improve the treatment of infeasibility, the constraints are divided into two types.

1 Soft constraints;

2 Hard constraints.

5.1. Soft Constraints

Soft constraints are those constraints whose limits (min and max) are desirable but not mandatory. Typically, such constraint correspond to voltage limits at PQ busses and possibly line loading limits. An advantage of modeling soft constraint is that infeasibility of the OPF problem can be tackled by exploiting this extra freedom. Soft constraints are handled by augmenting the Lagrangian with quadratic magnitude penalty terms as follows:

$$\psi_i = \rho_i (V_i - V_i^{min})^2; \text{ if } V_i < V_i^{min}$$
$$\psi_i = 0; \text{ if } V_i^{min} \leq V_i \leq V_i^{max}$$
$$\psi_i = \rho_i (V_i - V_i^{max})^2; \text{ if } V_i > V_i^{max}$$

ρ_i is the penalty factor greater than zero. In principle, each constraint can have a separate penalty factor. Similarly, penalty terms can be used for line loading limit violations ($|S_i| \geq S_i^{max}$). The behavior of the penalty term is shown in figure 11.4. If the inequality constraint is not violated, the penalty function has a zero value. On the other hand, as the violation increases, so does the penalty. The magnitude of penalty (ρ_i) can be used to control the softness of the constraint. By a large value of ρ_i, a strict enforcement can be obtained. By adjusting it to a lower value, one can control the importance given to the limit. The voltage penalty term does not adversely affect the sparsity of the formulation

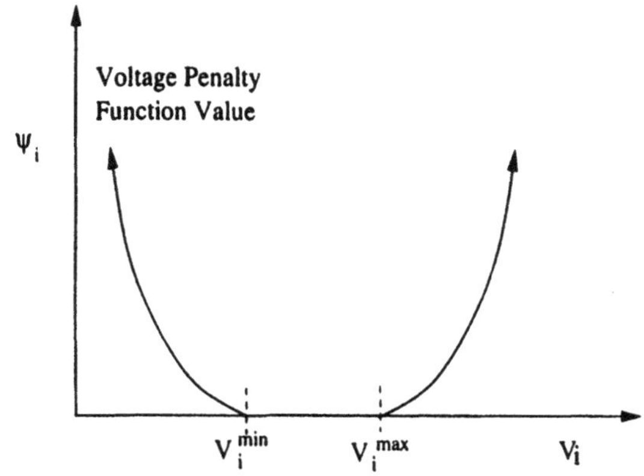

Figure 11.4. Penalty Function for a Bus Voltage

as such; a term changes only the diagonal elements in Hessian. Let $\mathcal{L}^{new} = L^{old} + \rho_i(V_i - V_i^{max})^2$ where at a given iteration k, $V_i > V_i^{max}$. Then

$$\frac{\partial \mathcal{L}^{new}}{\partial V_i} = \frac{\partial \mathcal{L}^{old}}{\partial V_i} + \rho_i(V_i - V_i^{max}) \tag{11.31}$$

$$\frac{\partial^2 \mathcal{L}^{new}}{\partial V_i^2} = \rho_i + \frac{\partial^2 \mathcal{L}^{old}}{\partial V_i^2} \tag{11.32}$$

$$\tag{11.33}$$

However, incorporating the line loading limits through quadratic penalty terms can reduce sparsity of the formulation. An approximate and indirect way to enforce line loading limits for a long line (i-j) is by imposing limit on $|\delta_i - \delta_j|$.

5.2. Hard Constraints

Hard constraints have to be strictly enforced. Typically, hard constraints in OPF formulation are the following:

$$\begin{aligned} \text{Real power Generation limit} : P_G^{min} \leq P_G \leq P_G^{max} \\ \text{Reactive power Generation limit} : Q_G^{min} \leq Q_G \leq Q_G^{max} \\ \text{Reactive Power Compensation Limits} : q^{min} \leq q^{comp} \leq q^{max} \\ \text{Transformer Tap Limits} : t^{min} \leq t \leq t^{max} \end{aligned}$$

Line loading limits can also be treated as hard constraints. Enforcement of hard constraints can be done either through the large penalties or by explicit modeling of equality constraint equations in W. The approach of using explicit equations seems preferable because (1) problems of numerical illconditioning can be avoided and (2) the solution of (11.29) at x^k, u^k will also lead to the Lagrangian multiplier (μ^{k+1}) associated with constraints at (x^{k+1}, u^{k+1}). Let $k+1$ refer to the new iteration count and i to the constraint index, then if $\mu_i^{k+1} < 0$, the constraint can be removed from the active set in the next iteration. However, efficient implementation will require sophisticated factorization update strategies, like Schur's update.

Otherwise, to enforce a hard limit, we can enforce a large quadratic penalty on similar lines to the voltage penalty function discussed above. It can be seen that for the control vector P_G, V_g, q_c, t, penalty terms will affect only the diagonal elements of the Hessian. Thus, structure of the Hessian will not change. However, for a general constraint like Q limit of a generator, we can foresee certain complications. To get an insight, into this, consider that Q_{G_k} of generator k is at it's upper limit $Q_{G_k}^{max}$. Now the augmented Lagrangian will look as follows:

$$\mathcal{L}^{new} = \mathcal{L}^{old} + \frac{1}{2}\rho(Q_{G_k} - Q_{G_k}^{max})^2$$

Now, $Q_{G_k} = \sum V_k V_j Y_{kj} \sin(\delta_k - \delta_j - \theta_{kj})$

$$\frac{\partial \mathcal{L}^{new}}{\partial x_j} = \frac{\partial \mathcal{L}^{old}}{\partial x_j} + \rho(Q_{G_k} - Q_{G_k}^{max})\frac{\partial Q_{G_k}}{\partial x_j}$$

$$\frac{\partial^2 \mathcal{L}^{new}}{\partial x_i x_j} = \frac{\partial^2 \mathcal{L}^{old}}{\partial x_i \partial x_j} + \rho\frac{\partial Q_{G_k}}{\partial x_i}\frac{\partial Q_{G_k}}{\partial x_j} + \rho(Q_{G_k} - Q_{G_k}^{max})\frac{\partial^2 Q_{G_k}}{\partial x_i \partial x_j}$$

Clearly, the sparsity of H deteriorates.

One way to tackle this problem within the gambit of penalty function is to model Q_{G_k} for generators also as explicit variables in vector z. Then the reactive power node injection at the generator bus is given as follows:

$$g(x) = Q_{G_k} - \sum Y_{kj} V_k V_j \sin(\delta_k - \delta_j - \theta_{kj}) = 0$$

It is modeled as explicit equality constraint, with additional variable Q_{G_k}. Thus the matrix has one addition equation and variable each for the Q-injection constraints and hence, W remains a square matrix. Now since Q_{G_k} is modeled as a part of vector z, we have

$$\mathcal{L}^{new} = \mathcal{L}^{old} + \frac{1}{2}(Q_{G_k} - Q_{G_k}^{max})^2 \tag{11.34}$$

$$\frac{\partial \mathcal{L}^{new}}{\partial Q_{G_k}} = \frac{\partial \mathcal{L}^{old}}{\partial Q_{G_k}} + \rho(Q_{G_k} - Q_{G_k}^{max}) \tag{11.35}$$

$$\frac{\partial^2 \mathcal{L}^{new}}{\partial Q_{G_k}^2} = \frac{\partial^2 \mathcal{L}^{old}}{\partial Q_{G_k}^2} + \rho \qquad (11.36)$$

$$\frac{\partial^2 \mathcal{L}^{new}}{\partial x_i \partial x_j} = \frac{\partial^2 \mathcal{L}^{old}}{\partial x_i \partial x_j} \qquad (11.37)$$

Equation (11.25) represents n active and n reactive power injection equations which include the reference bus. Now it can be seen that sparsity structure of Hessian is not disturbed. As per second order KKT necessary conditions, at the optimal $d^T \nabla^2 \mathcal{L} d \geq 0$ for all d belonging to the null space of active constraints and satisfying $h_i(x^k, u^k) + [\frac{\partial h_i^T}{\partial x}, \frac{\partial h_i^T}{\partial u}]^T d \leq 0$ for rest of the constraints. The matrix of the second derivative of $\mathcal{L}$ with respect to x and u (Hessian) need not be positive definite. To ascertain that d^k is always a descent direction, a *spd* approximation $\nabla^2 \mathcal{L} + \gamma I$ can be made where γ is a scalar greater than zero. While initially a large γ will ascertain descent, it may be possible to gradually reduce the penalty to zero.

After a penalty has been imposed to constrain a variable, the sign of its derivative can be used to decide whether the penalty is still needed. The derivative of a penalty on a variable is the negative of its Lagrange multiplier, μ_i, for the enforced limiting value. The derivatives of imposed penalties are also used to determine whether the KKT conditions are satisfied.

The iterative process of formulating and solving equation (11.29) and updating x and u vectors has to be repeated until a certain convergence criteria is satisfied.

5.3. Convergence Criteria

The following condition must be satisfied for the new point (x^{k+1}, u^{k+1}) to be minimum.

1 The mismatches in all node injection equations is within a specified tolerance.

2 The hard inequality constraints are satisfied.

3 The gradient of Lagrangian function is zero. In other words, some norm of the gradient should be below a tolerance.

4 The Lagrangian multiplier μ_i of each binding inequality constraint should be greater than or equal to zero.

In the next section, we now discuss a Sequential Quadratic Programming (SQP) method which is very popular standard tool for solving

NLP problems for moderate sized optimization system, *e.g.*, a power system having a few thousand variables. The implementation complexity is high but gain is a robust implementation.

Example 11.1: Figure 11.5 represents a network with 3 transmission lines, each of 345 kV class and 100 km in length. For simplicity, line resistances are neglected. The load at node 3 is fed by two generators at bus 1 and 2 each having a capacity of 300 MW(3 pu). Following inequality constraints are specified for the problem.

$$\begin{bmatrix} V_{min} \\ V_{min} \\ V_{min} \end{bmatrix} \begin{matrix} \leq \\ \leq \\ \leq \end{matrix} \begin{bmatrix} V_1 \\ V_2 \\ V_3 \end{bmatrix} \begin{matrix} \leq \\ \leq \\ \leq \end{matrix} \begin{bmatrix} V_{max} \\ V_{max} \\ V_{max} \end{bmatrix} \tag{11.38}$$

$$\begin{bmatrix} 0 \\ 0 \end{bmatrix} \leq \begin{bmatrix} P_{G_1} \\ P_{G_2} \end{bmatrix} \leq \begin{bmatrix} 3 \\ 3 \end{bmatrix} \tag{11.39}$$

$$-0.5 \leq q_2 \leq 1.5 \tag{11.40}$$

$$-0.5 \leq q_1 \leq 1.5 \tag{11.41}$$

In this example, we treat all the constraints as hard constraints. A hard constraint is given a weightage of 200. There are no bounds specified for δ_2 and δ_3. The cost coefficients are given by following quadratic equation:

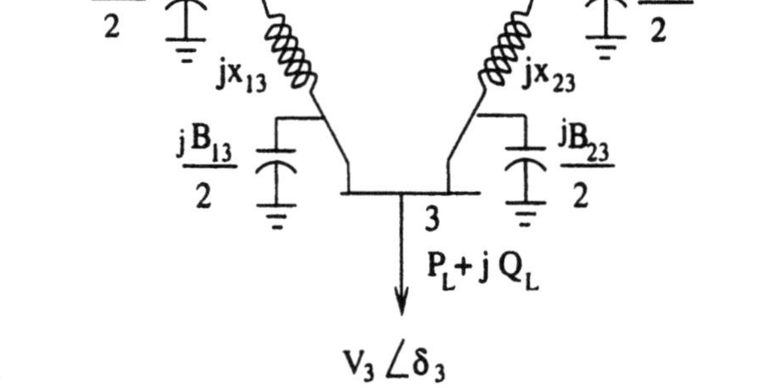

Figure 11.5. Three Bus Test System

$$F_1(P_{G_1}) = \gamma_1 + \beta_1 P_{G_1} + \frac{1}{2}\alpha_1 P_{G_1}^2$$

$$F_2(P_{G_2}) = \gamma_2 + \beta_2 P_{G_2} + \frac{1}{2}\alpha_2 P_{G_2}^2$$

where the generation is specified in pu. Network data is as follows:
$x_{12} = x_{13} = x_{23} = 0.0308$ pu; $B_{12} = B_{13} = B_{23} = 0.54$ pu
$\gamma_1 = 9.6$; $\gamma_2 = 7.6$
$\beta_1 = 4.1$; $\beta_2 = 3.1$; $\alpha_1 = 2.4$; $\alpha_2 = 1.6$
Obtain the optimal economic dispatch solution and load flow solution by Newton method; treat all constraints as hard constraints.
Solution: The Lagrangian function for the problem is as follows:

$$\begin{aligned}\mathcal{L}(P_{G_1}, P_{G_2}, \delta_2, \delta_3, |V_1|, |V_2|, |V_3|, q_1, q_2, \lambda_1, \lambda_2, \lambda_3, \lambda_4, \lambda_5, \lambda_6) = \\ \gamma_1 + \gamma_2 + \beta_1 P_{G_1} + \beta_2 P_{G_2} + \frac{1}{2}\alpha_1 P_{G_1}^2 + \frac{1}{2}\alpha_2 P_{G_2}^2 + \\ \lambda_1 g_1(x) + \lambda_2 g_2(x) + \lambda_3 g_3(x) + \lambda_4 g_4(x) + \lambda_5 g_5(x) + \lambda_6 g_6(x)\end{aligned}$$

The six constraint equations associated with Lagrangian function are node injection equations at the three nodes. The first three equations represent the P-injection equations at nodes 1, 2 and 3 respectively, while the later three equations represent Q-injection equations at nodes 3, 2 and 1 respectively. The ordering of equation is arbitrary and no special significance is attached to it. Explicitly, we can write them as $g_i(x) = 0 \ \forall i = 1, \ldots, 6$ where $g_i(x, u)$ is as follows:

$$\begin{aligned}
g_1(x) &= P_{G_1} + y_{12}V_1V_2\sin\delta_2 + y_{13}V_1V_3\sin\delta_3, \\
g_2(x) &= P_{G_2} - y_{12}V_1V_2\sin\delta_2 - y_{23}V_2V_3\sin(\delta_2 - \delta_3), \\
g_3(x) &= P_L + y_{13}V_1V_3\sin\delta_3 - y_{23}V_2V_3\sin(\delta_2 - \delta_3), \\
g_4(x) &= Q_L - (\frac{B_{13} + B_{23}}{2})V_3^2 + (y_{13} + y_{23})V_3^2 - y_{13}V_1V_3\cos\delta_3 \\
&\quad - y_{23}V_2V_3\cos(\delta_2 - \delta_3), \\
g_5(x) &= q_2 + (\frac{B_{12} + B_{23}}{2})V_2^2 + y_{12}V_1V_2\cos\delta_2 + y_{23}V_2V_3\cos(\delta_3 - \delta_2) \\
&\quad - (y_{12} + y_{23})V_2^2, \\
g_6(x) &= q_1 + (\frac{B_{12} + B_{13}}{2})V_1^2 + y_{12}V_1V_2\cos\delta_2 + y_{13}V_1V_3\cos\delta_3 - \\
&\quad (y_{12} + y_{13})V_1^2,
\end{aligned}$$

Note that unlike load flow analysis, at the generator busses both P_{G_1} and P_{G_2} and also q-injections q_2 and q_1 are not specified. Rather they are variables which will be set by OPF such that cost is minimized and constraints $g_1(x) = 0, \ldots, g_6(x) = 0$ are met. For a 3-bus system, 6 equality constraints result from node injection equations. A load flow analysis problem would have considered only three equations explicitly *viz.* $g_2(x) = 0$; $g_3(x) = 0$ and $g_4(x) = 0$. In load flow analysis, $P_{G_1}, P_{G_2}, q_2, q_1$ will not appear in the vector of unknowns Δz.

Initially, let us set $V_1 = V_2 = V_3 = 1.0$ pu, $\delta_2 = \delta_3 = 0.0$ radians and $\lambda_1 = \lambda_2 \ldots = \lambda_6 = 0$. The system of equations to be solved at each iteration is given by:

$$W^k \Delta z^k = \Delta b^k$$

where $W^k =$

$$\begin{array}{ccccccccccccccc}
P_{G_1} & P_{G_2} & \delta_2 & \delta_3 & V_1 & V_2 & V_3 & q_2 & q_1 & \lambda_1 & \lambda_2 & \lambda_3 & \lambda_4 & \lambda_5 & \lambda_6
\end{array}$$

$$\left[\begin{array}{ccccccccccccccc}
\frac{\partial^2 \mathcal{L}}{\partial P_{G_1}^2} & 0 & 0 & 0 & 0 & 0 & 0 & 0 & 0 & & & & & & \\
 & \frac{\partial^2 \mathcal{L}}{\partial P_{G_2}^2} & 0 & 0 & 0 & 0 & 0 & 0 & 0 & & & & & & \\
 & & \frac{\partial^2 \mathcal{L}}{\partial \delta_2^2} & \frac{\partial^2 \mathcal{L}}{\partial \delta_3 \partial \delta_2} & \frac{\partial^2 \mathcal{L}}{\partial V_1 \partial \delta_2} & \frac{\partial^2 \mathcal{L}}{\partial V_2 \partial \delta_2} & \frac{\partial^2 \mathcal{L}}{\partial V_3 \partial \delta_2} & 0 & 0 & & & & & & \\
 & & & \frac{\partial^2 \mathcal{L}}{\partial \delta_3^2} & \frac{\partial^2 \mathcal{L}}{\partial V_1 \partial \delta_3} & \frac{\partial^2 \mathcal{L}}{\partial V_2 \partial \delta_3} & \frac{\partial^2 \mathcal{L}}{\partial V_3 \partial \delta_3} & 0 & 0 & & & & & & \\
 & & & & \frac{\partial^2 \mathcal{L}}{\partial V_1^2} & \frac{\partial^2 \mathcal{L}}{\partial V_2 \partial V_1} & \frac{\partial^2 \mathcal{L}}{\partial V_3 \partial V_1} & 0 & 0 & & & & & & \\
 & & & & & \frac{\partial^2 \mathcal{L}}{\partial V_2^2} & \frac{\partial^2 \mathcal{L}}{\partial V_3 \partial V_2} & 0 & 0 & & & & & & \\
 & & & & & & \frac{\partial^2 \mathcal{L}}{\partial V_3^2} & 0 & 0 & & & & & & \\
 & & & & & & & 0 & 0 & & & & & 1 & \\
 & & & & & & & 0 & 0 & & & & & & 1 \\
1 & 0 & \frac{\partial g_1}{\partial \delta_2} & \frac{\partial g_1}{\partial \delta_3} & \frac{\partial g_1}{\partial V_1} & \frac{\partial g_1}{\partial V_2} & \frac{\partial g_1}{\partial V_3} & 0 & 0 & 0 & & & & & \\
0 & 1 & \frac{\partial g_2}{\partial \delta_2} & \frac{\partial g_2}{\partial \delta_3} & \frac{\partial g_2}{\partial V_1} & \frac{\partial g_2}{\partial V_2} & \frac{\partial g_2}{\partial V_3} & 0 & 0 & 0 & 0 & & & & \\
0 & 0 & \frac{\partial g_3}{\partial \delta_2} & \frac{\partial g_3}{\partial \delta_3} & \frac{\partial g_3}{\partial V_1} & \frac{\partial g_3}{\partial V_2} & \frac{\partial g_3}{\partial V_3} & 0 & 0 & 0 & 0 & 0 & & & \\
0 & 0 & \frac{\partial g_4}{\partial \delta_2} & \frac{\partial g_4}{\partial \delta_3} & \frac{\partial g_4}{\partial V_1} & \frac{\partial g_4}{\partial V_2} & \frac{\partial g_4}{\partial V_3} & 0 & 0 & 0 & 0 & 0 & 0 & & \\
0 & 0 & \frac{\partial g_5}{\partial \delta_2} & \frac{\partial g_5}{\partial \delta_3} & \frac{\partial g_5}{\partial V_1} & \frac{\partial g_5}{\partial V_2} & \frac{\partial g_5}{\partial V_3} & 1 & 0 & 0 & 0 & 0 & 0 & 0 & \\
0 & 0 & \frac{\partial g_6}{\partial \delta_2} & \frac{\partial g_6}{\partial \delta_3} & \frac{\partial g_6}{\partial V_1} & \frac{\partial g_6}{\partial V_2} & \frac{\partial g_6}{\partial V_3} & 0 & 1 & 0 & 0 & 0 & 0 & 0 & 0
\end{array}\right]$$

$$\Delta b^k = -\begin{bmatrix} \frac{\partial \mathcal{L}}{\partial P_{G_1}} \\ \frac{\partial \mathcal{L}}{\partial P_{G_2}} \\ \frac{\partial \mathcal{L}}{\partial \delta_2} \\ \frac{\partial \mathcal{L}}{\partial \delta_3} \\ \frac{\partial \mathcal{L}}{\partial V_1} \\ \frac{\partial \mathcal{L}}{\partial V_2} \\ \frac{\partial \mathcal{L}}{\partial V_3} \\ \frac{\partial \mathcal{L}}{\partial q_2} \\ \frac{\partial \mathcal{L}}{\partial q_1} \\ g_1(x) \\ g_2(x) \\ g_3(x) \\ g_4(x) \\ g_5(x) \\ g_6(x) \end{bmatrix} \quad \Delta z^k = \begin{bmatrix} \Delta P_{G_1} \\ \Delta P_{G_2} \\ \Delta \delta_2 \\ \Delta \delta_3 \\ \Delta V_1 \\ \Delta V_2 \\ \Delta V_3 \\ \Delta q_2 \\ \Delta q_1 \\ \Delta \lambda_1 \\ \Delta \lambda_2 \\ \Delta \lambda_3 \\ \Delta \lambda_4 \\ \Delta \lambda_5 \\ \Delta \lambda_6 \end{bmatrix} \qquad (11.42)$$

This example problem was solved using MATLAB. In the MATLAB program supported along with this example, the only entries that are marked in W^k are computed. Since, the coefficient matrix is symmetric, the transpose entry need not be recomputed. In the 9×9 principal submatrix of W which represents the Hessian, the last two columns are zero because q_1 and q_2 appear as linear terms in the Lagrangian function. Clearly, Hessian matrix is not full rank. However, in the W-matrix, corresponding two columns are not zero (notice that, $\frac{\partial g_5}{\partial q_2} = 1$ and $\frac{\partial g_6}{\partial q_1} =$

1). Now, the right hand side gradient vector for the Lagrangian function is given by the following equation:

$$\frac{\partial \mathcal{L}}{\partial P_{G_1}} = \beta_1 + \alpha_1 P_{G_1} + \lambda_1;$$

$$\frac{\partial \mathcal{L}}{\partial P_{G_2}} = \beta_2 + \alpha_2 P_{G_2} + \lambda_2;$$

$$\begin{aligned}\frac{\partial \mathcal{L}}{\partial \delta_2} &= (\lambda_1 - \lambda_2) y_{12} V_1 V_2 \cos\delta_2 - (\lambda_2 + \lambda_3) y_{23} V_2 V_3 \cos(\delta_2 - \delta_3)\\ &\quad + (\lambda_4 - \lambda_5) y_{23} V_2 V_3 \sin(\delta_2 - \delta_3) - (\lambda_5 + \lambda_6) y_{12} V_1 V_2 \sin\delta_2;\end{aligned}$$

$$\begin{aligned}\frac{\partial \mathcal{L}}{\partial \delta_3} &= (\lambda_1 + \lambda_3) y_{13} V_1 V_3 \cos\delta_3 + (\lambda_2 + \lambda_3) y_{23} V_2 V_3 \cos(\delta_2 - \delta_3)\\ &\quad + (\lambda_4 - \lambda_6) y_{13} V_1 V_3 \sin\delta_3 + (\lambda_5 - \lambda_4) y_{23} V_2 V_3 \sin(\delta_2 - \delta_3);\end{aligned}$$

$$\begin{aligned}\frac{\partial \mathcal{L}}{\partial V_1} &= (\lambda_1 - \lambda_2) y_{12} V_2 \sin\delta_2 + (\lambda_1 + \lambda_3) y_{13} V_3 \sin\delta_3\\ &\quad + (\lambda_6 - \lambda_4) y_{13} V_3 \cos\delta_3 + (\lambda_5 + \lambda_6) y_{12} V_2 \cos\delta_2\\ &\quad - 2 \times (y_{12} + y_{13}) \lambda_6 V_1 + \lambda_6 (B_{12} + B_{13}) V_1;\end{aligned}$$

$$\begin{aligned}\frac{\partial \mathcal{L}}{\partial V_2} &= (\lambda_1 - \lambda_2) y_{12} V_1 \sin\delta_2 - (\lambda_2 + \lambda_3) y_{23} V_3 \sin(\delta_2 - \delta_3)\\ &\quad + (\lambda_5 - \lambda_4) y_{23} V_3 \cos(\delta_2 - \delta_3) + (\lambda_5 + \lambda_6) y_{12} V_1 \cos\delta_2\\ &\quad - 2 \times (y_{12} + y_{23}) \lambda_5 V_2 + \lambda_5 (B_{12} + B_{23}) V_2;\end{aligned}$$

$$\begin{aligned}\frac{\partial \mathcal{L}}{\delta V_3} &= (\lambda_1 + \lambda_3) y_{13} V_1 \sin\delta_3 - (\lambda_2 + \lambda_3) y_{23} V_2 \sin(\delta_2 - \delta_3)\\ &\quad + (\lambda_5 - \lambda_4) y_{23} V_2 \cos(\delta_2 - \delta_3) + (\lambda_6 - \lambda_4) y_{13} V_1 \cos\delta_3\\ &\quad + 2\lambda_4 \times (y_{13} + y_{23}) V_3 - \lambda_4 (B_{13} + B_{23}) V_3;\end{aligned}$$

$$\frac{\partial \mathcal{L}}{\partial q_2} = \lambda_5;$$

$$\frac{\partial \mathcal{L}}{\partial q_1} = \lambda_6$$

Next, we tabulate the entries of the Hessian matrix. While writing these equations, we make use of the symmetric nature of Hessian. Hence, explicit expression are given for upper triangular matrix only and the order of specification is row wise, *i.e.*, we specify $H(1,1) \ldots H(1,9)$ first, then $H(2,2) \ldots H(2,9)$; $H(3,3) \ldots H(3,9)$ and so on upto $H(9,9)$.

$$\textbf{Row 1}: \frac{\partial^2 \mathcal{L}}{\partial P_{G1}^2} = \alpha_1; \quad \frac{\partial^2 \mathcal{L}}{\partial P_{G2} \partial P_{G1}} = 0$$

$$\frac{\partial^2 \mathcal{L}}{\partial \delta_2 \partial P_{G1}} = \frac{\partial^2 \mathcal{L}}{\partial \delta_3 \partial P_{G1}} = \frac{\partial^2 \mathcal{L}}{\partial V_1 \partial P_{G1}} = \frac{\partial^2 \mathcal{L}}{\partial V_2 \partial P_{G1}} = \frac{\partial^2 \mathcal{L}}{\partial V_3 \partial P_{G1}} =$$

$$\frac{\partial^2 \mathcal{L}}{\partial q_2 \partial P_{G1}} = \frac{\partial^2 \mathcal{L}}{\partial q_1 \partial P_{G1}} = 0$$

$$\textbf{Row 2}: \frac{\partial^2 \mathcal{L}}{\partial P_{G2}^2} = \alpha_2$$

$$\frac{\partial^2 \mathcal{L}}{\partial \delta_2 \partial P_{G2}} = \frac{\partial^2 \mathcal{L}}{\partial \delta_3 \partial P_{G2}} = \frac{\partial^2 \mathcal{L}}{\partial V_1 \partial P_{G2}} = \frac{\partial^2 \mathcal{L}}{\partial V_2 \partial P_{G2}} = \frac{\partial^2 \mathcal{L}}{\partial V_3 \partial P_{G2}} =$$
$$\frac{\partial^2 \mathcal{L}}{\partial q_2 \partial P_{G2}} = \frac{\partial^2 \mathcal{L}}{\partial q_1 \partial P_{G2}} = 0$$

$$\textbf{Row 3}: \frac{\partial^2 \mathcal{L}}{\partial \delta_2^2} = -\lambda_1 y_{12} V_1 V_2 \sin\delta_2 + \lambda_2[y_{12} V_1 V_2 \sin\delta_2 + y_{23} V_2 V_3 \sin(\delta_2 - \delta_3)]$$
$$+\lambda_3[y_{23} V_2 V_3 \sin(\delta_2 - \delta_3)] + \lambda_4[y_{23} V_2 V_3 \cos(\delta_2 - \delta_3)]$$
$$+\lambda_5[-y_{12} V_1 V_2 \cos\delta_2 - y_{23} V_2 V_3 \cos(\delta_2 - \delta_3)] + \lambda_6[-y_{12} V_1 V_2 \cos\delta_2]$$

$$\frac{\partial^2 \mathcal{L}}{\partial \delta_3 \partial \delta_2} = -(\lambda_2 + \lambda_3)[y_{23} V_2 V_3 \sin(\delta_2 - \delta_3)] + (\lambda_5 - \lambda_4) y_{23} V_2 V_3 \cos(\delta_2 - \delta_3)$$

$$\frac{\partial^2 \mathcal{L}}{\partial V_1 \partial \delta_2} = (\lambda_1 - \lambda_2) y_{12} V_2 \cos\delta_2 - (\lambda_5 + \lambda_6) y_{12} V_2 \sin\delta_2$$

$$\frac{\partial^2 \mathcal{L}}{\partial V_2 \partial \delta_2} = [\lambda_1 - \lambda_2] y_{12} V_1 \cos\delta_2 - (\lambda_2 + \lambda_3) y_{23} V_3 \cos(\delta_2 - \delta_3)$$
$$+(\lambda_4 - \lambda_5) y_{23} V_3 \sin(\delta_2 - \delta_3) - (\lambda_5 + \lambda_6) y_{12} V_1 \sin\delta_2$$

$$\frac{\partial^2 \mathcal{L}}{\partial V_3 \partial \delta_2} = -(\lambda_2 + \lambda_3) y_{23} V_2 \cos(\delta_2 - \delta_3) + (\lambda_4 - \lambda_5) y_{23} V_2 \sin(\delta_2 - \delta_3)$$

$$\frac{\partial^2 \mathcal{L}}{\partial q_1 \partial \delta_2} = \frac{\partial^2 \mathcal{L}}{\partial q_2 \partial \delta_2} = 0$$

$$\textbf{Row 4}: \frac{\partial^2 \mathcal{L}}{\partial \delta_3^2} = -(\lambda_1 + \lambda_3) y_{13} V_1 V_3 \sin\delta_3 + (\lambda_2 + \lambda_3) y_{23} V_2 V_3 \sin(\delta_2 - \delta_3)$$
$$+\lambda_4[y_{13} V_1 V_3 \cos\delta_3 + y_{23} V_2 V_3 \cos(\delta_2 - \delta_3)] + \lambda_5[-y_{23} V_2 V_3 \cos(\delta_3 - \delta_2)]$$
$$+\lambda_6[-y_{13} V_1 V_3 \cos\delta_3]$$

$$\frac{\partial^2 \mathcal{L}}{\partial V_1 \partial \delta_3} = (\lambda_1 + \lambda_3) y_{13} V_3 \cos\delta_3 + (\lambda_4 - \lambda_6) y_{13} V_3 \sin\delta_3$$

$$\frac{\partial^2 \mathcal{L}}{\partial V_2 \partial \delta_3} = (\lambda_2 + \lambda_3) y_{23} V_3 \cos(\delta_2 - \delta_3) - (\lambda_4 - \lambda_5) y_{23} V_3 \sin(\delta_2 - \delta_3)$$

$$\frac{\partial^2 \mathcal{L}}{\partial V_3 \partial \delta_3} = (\lambda_1 + \lambda_3) y_{13} V_1 \cos\delta_3 + (\lambda_2 + \lambda_3) y_{23} V_2 \cos(\delta_2 - \delta_3)$$
$$+(\lambda_5 - \lambda_4) y_{23} V_2 \sin(\delta_2 - \delta_3) + (\lambda_4 - \lambda_6) y_{13} V_1 \sin\delta_3$$

$$\frac{\partial^2 \mathcal{L}}{\partial q_2 \partial \delta_3} = \frac{\partial^2 \mathcal{L}}{\partial q_1 \partial \delta_3} = 0$$

$$\textbf{Row 5}: \frac{\partial^2 \mathcal{L}}{\partial V_1^2} = -2\lambda_6(y_{12} + y_{13}) + \lambda_6(B_{12} + B_{13})$$

$$\frac{\partial^2 \mathcal{L}}{\partial V_2 \partial V_1} = (\lambda_5 + \lambda_6) y_{12} \cos\delta_2 + (\lambda_1 - \lambda_2) y_{12} \sin\delta_2$$

$$\frac{\partial^2 \mathcal{L}}{\partial V_3 \partial V_1} = (\lambda_1 + \lambda_3) y_{13} \sin\delta_3 + (\lambda_6 - \lambda_4) y_{13} \cos\delta_3$$

$$\frac{\partial^2 \mathcal{L}}{\partial q_2 \partial V_1} = \frac{\partial^2 \mathcal{L}}{\partial q_1 \partial V_1} = 0$$

$$\textbf{Row 6}: \frac{\partial^2 \mathcal{L}}{\partial V_2^2} = -\lambda_5 2 \times (y_{12} + y_{23}) + \lambda_5 (B_{12} + B_{23})$$

$$\frac{\partial^2 \mathcal{L}}{\partial V_3 \partial V_2} = (\lambda_5 - \lambda_4) y_{23} \cos(\delta_3 - \delta_2) - (\lambda_2 + \lambda_3) y_{23} \sin(\delta_2 - \delta_3)$$

$$\frac{\partial^2 \mathcal{L}}{\partial q_2 \partial V_2} = \frac{\partial^2 \mathcal{L}}{\partial q_1 \partial V_2} = 0$$

$$\textbf{Row 7}: \frac{\partial^2 \mathcal{L}}{\partial V_3^2} = 2 \times (y_{13} + y_{23})\lambda_4 - \lambda_4 (B_{13} + B_{23}); \quad \frac{\partial^2 \mathcal{L}}{\partial q_2 \partial V_3} = \frac{\partial^2 \mathcal{L}}{\partial q_1 \partial V_3} = 0;$$

$$\textbf{Row 8}: \frac{\partial^2 \mathcal{L}}{\partial q_2^2} = \frac{\partial^2 \mathcal{L}}{\partial q_1 q_2} = 0;$$

$$\textbf{Row 9}: \frac{\partial^2 \mathcal{L}}{\partial q_1^2} = 0$$

Finally, we tabulate the elements of Jacobian for node injection equations.

$$\frac{\partial g_1}{\partial P_{G_1}} = 1; \quad \frac{\partial g_1}{\partial P_{G_2}} = 0; \quad \frac{\partial g_1}{\partial \delta_2} = y_{12} V_1 V_2 \cos\delta_2; \quad \frac{\partial g_1}{\partial \delta_3} = y_{13} V_1 V_3 \cos\delta_3$$

$$\frac{\partial g_1}{\partial V_1} = y_{12} V_2 \sin\delta_2 + y_{13} V_3 \sin\delta_3; \quad \frac{\partial g_1}{\partial V_2} = y_{12} V_1 \sin\delta_2; \quad \frac{\partial g_1}{\partial V_3} = y_{13} V_1 \sin\delta_3;$$

$$\frac{\partial g_2}{\partial P_{G_1}} = 0; \quad \frac{\partial g_2}{\partial P_{G_2}} = 1; \quad \frac{\partial g_2}{\partial \delta_2} = -y_{12} V_1 V_2 \cos\delta_2 - y_{23} V_2 V_3 \cos(\delta_2 - \delta_3)$$

$$\frac{\partial g_2}{\partial \delta_3} = y_{23} V_2 V_3 \cos(\delta_2 - \delta_3); \quad \frac{\partial g_2}{\partial V_1} = -y_{12} V_2 \sin\delta_2;$$

$$\frac{\partial g_2}{\partial V_2} = -y_{12} V_1 \sin\delta_2 - y_{12} V_3 \sin(\delta_2 - \delta_3)$$

$$\frac{\partial g_2}{\partial V_3} = -y_{23} V_2 \sin(\delta_2 - \delta_3); \quad \frac{\partial g_3}{\partial P_{G_1}} = \frac{\partial g_3}{\partial P_{G_2}} = 0; \quad \frac{\partial g_3}{\partial \delta_2} = -y_{23} V_2 V_3 \cos(\delta_2 - \delta_3)$$

$$\frac{\partial g_3}{\partial \delta_3} = y_{13} V_1 V_3 \cos\delta_3 + y_{23} V_2 V_3 \cos(\delta_2 - \delta_3); \quad \frac{\partial g_3}{\partial V_1} = y_{13} V_3 \sin\delta_3$$

$$\frac{\partial g_3}{\partial V_2} = -y_{23} V_3 \sin(\delta_2 - \delta_3); \quad \frac{\partial g_3}{\partial V_3} = y_{13} V_1 \sin\delta_3 - y_{23} V_2 \sin(\delta_2 - \delta_3);$$

$$\frac{\partial g_4}{\partial P_{G_1}} = \frac{\partial g_4}{\partial P_{G_2}} = 0$$

$$\frac{\partial g_4}{\partial \delta_2} = y_{23} V_2 V_3 \sin(\delta_2 - \delta_3); \quad \frac{\partial g_4}{\partial \delta_3} = y_{13} V_1 V_3 \sin\delta_3 - y_{23} V_2 V_3 \sin(\delta_2 - \delta_3);$$

$$\frac{\partial g_4}{\partial V_1} = -y_{13} V_3 \cos\delta_3; \quad \frac{\partial g_4}{\partial V_2} = -Y_{23} V_3 \cos(\delta_2 - \delta_3);$$

$$\frac{\partial g_4}{\partial V_3} = 2 \times (y_{13} + y_{23}) V_3 - y_{13} V_1 \cos\delta_3 - y_{23} V_2 \cos(\delta_2 - \delta_3) - (B_{13} + B_{23}) V_3$$

$$\frac{\partial g_5}{\partial P_{G_1}} = \frac{\partial g_5}{\partial P_{G_2}} = 0; \quad \frac{\partial g_5}{\partial \delta_2} = -y_{12} V_1 V_2 \sin\delta_2 + y_{23} V_2 V_3 \sin(\delta_3 - \delta_2)$$

$$\frac{\partial g_5}{\partial \delta_3} = -y_{23} V_2 V_3 \sin(\delta_3 - \delta_2); \quad \frac{\partial g_5}{\partial V_1} = y_{12} V_2 \cos\delta_2$$

$$\frac{\partial g_5}{\partial V_2} = y_{12}V_1\cos\delta_2 + y_{23}V_3\cos(\delta_3-\delta_2) - 2\times(y_{12}+y_{23})V_2 + (B_{12}+B_{23})V_2;$$

$$\frac{\partial g_5}{\partial V_3} = y_{23}V_2\cos(\delta_3-\delta_2);\ \frac{\partial g_5}{\partial q_2} = 1.0;\ \frac{\partial g_5}{\partial q_1} = 0$$

$$\frac{\partial g_6}{\partial \delta_2} = -y_{12}V_1V_2\sin\delta_2;\ \frac{\partial g_6}{\partial \delta_3} = -y_{13}V_1V_3\sin\delta_3$$

$$\frac{\partial g_6}{\partial V_1} = y_{12}V_2\cos\delta_2 + y_{13}V_3\cos\delta_3 - 2\times(y_{12}+y_{13})V_1 + (B_{12}+B_{13})V_1$$

$$\frac{\partial g_6}{\partial V_2} = y_{12}V_1\cos\delta_2;\ \frac{\partial g_6}{\partial V_3} = y_{13}V_1\cos\delta_3$$

$$\frac{\partial g_6}{\partial q_2} = 1.0;\ \frac{\partial g_6}{\partial q_1} = 0$$

Based upon the symmetry argument of the W matrix, it can be seen that with initial conditions defined by $\lambda_5 = \lambda_6 = 0$, we have $\frac{\partial \mathcal{L}}{\partial q_1} = \lambda_6 = 0$, $\frac{\partial \mathcal{L}}{\partial q_2} = \lambda_5 = 0$. Therefore, $\Delta\lambda_5 = \Delta\lambda_6 = 0$, *i.e.*, status $\lambda_5 = \lambda_6 = 0$ remains unaltered in the subsequent iterations. Thus, q_1 and q_2 constraints reactive power injection limits are inactive until one or more of them reaches or exceeds a bound. In that case, the change occurs in right hand side as well as structure of Hessian. Let $q_1^k \geq q_1^{max}$, *i.e.* upper bound has to be enforced. Then the penalty factor is added to Lagrangian function. Therefore,

$$\mathcal{L}^{new} = \mathcal{L}^{old} + \frac{1}{2}\rho(q_1 - q_1^{max})^2$$

$$\frac{\partial \mathcal{L}}{\partial q_1} = \frac{\partial \mathcal{L}^{old}}{\partial q_1} + \rho(q_1 - q_1^{max}) = \lambda_6 + \rho(q_1 - q_1^{max})$$

$$\frac{\partial^2 \mathcal{L}}{\partial q_1^2} = \rho$$

From the structure of W matrix, it can be seen that $\Delta\lambda_6 = \rho(q_1 - q_1^{max})$ and hence $\lambda_6^{k+1} = \lambda_6^k + \Delta\lambda_6 \geq 0$. Initially the Hessian matrix and also W is not full rank. Hence we use an approximation $H^k + \gamma^k I_9$ to H^k where γ is a scalar greater than zero and less than 1. As mentioned earlier, this ensures descent direction initially. As we move towards solution, the scalar is reduced by the power term. Thus, near the optimal the effect of γ is negligible and the actual Hessian is used. In the program we have used $\gamma = 0.30$.

Three cases for system loads are considered;

Case 1: This corresponds to the peak load condition wherein $P_L = 6.0$ pu, $Q_L = 4.0$ pu.

Table 11.2. OPF results with $V_{min} = 0.95$ pu, $V_{max} = 1.05$ pu

P_L	Q_L	P_{G_1}	P_{G_2}	δ_2	δ_3	V_1	V_2	V_3	q_1	q_2	iter
6	4	**2.97**	**3.03**	0.00	-0.08	**1.05**	**1.05**	0.99	**1.51**	**1.51**	6
4	2	1.35	2.65	0.01	-0.05	**1.05**	1.04	1.02	0.25	0.26	4
2	1	0.55	1.45	0.01	-0.02	**0.95**	**0.95**	**0.95**	-0.20	-0.20	4

Table 11.3. OPF results with $V_{min} = 0.90, V_{max} = 1.1$ pu

P_L	Q_L	P_{G_1}	P_{G_2}	δ_2	δ_3	V_1	V_2	V_3	q_1	q_2	iter
6	4	**2.97**	**3.03**	0.00	-0.08	**1.10**	**1.10**	1.04	1.38	1.38	6
4	2	1.35	2.65	0.01	-0.04	**1.10**	1.09	1.07	0.14	0.16	5
2	1	0.55	1.45	0.01	-0.03	0.91	0.91	**0.90**	-0.12	-0.12	4

Case 2: This corresponds to the average load condition wherein $P_L =$ 4.0 pu, $Q_L = 2.0$ pu.
Case 3: This corresponds to the light load condition wherein $P_L = 2.0$ pu $Q_L = 1.0$ pu.

The convergence criteria used is that absolute of maximum mismatch of Δb vector should be below 0.00001. The results as shown in table (11.2-11.3).

Case 1: In table 11.2, the limits on voltages are $V_{min} = 0.95$ and $V_{max} =$ 1.05 pu. It can be seen that under peak load condition, the system is constrained by the maximum limits on both generators P and Q, *i.e.*, generators are working at peak P and Q capacity. The generator voltages are constrained to the maximum value to pump the large reactive power requirement of load. With the limits on voltage changed to $V_{min} =$ 0.90 and $V_{max} = 1.10$ pu, it can be seen in table 11.3 that for peak load case, now the generator voltages is set to its peak 1.10 pu. This results in higher contribution from the line changing MVARs. Hence, the reactive power generations q_1 and q_2 which were earlier constrained at the maximum limits are now well within the bounds. This comparison clearly indicates the significance of choosing proper limits on V_{min} and V_{max} for the nodes, as it can affect both Q-generation and the voltage profile at the optimal solution. Further, it is seen that changing V_{min} and V_{max} has not affected the P solution because the system is lossless. The angles at the busses nearly remain constant.

Case 2: When the system loading is average $(4 + j2)$ pu, it is seen that generators no more operate at the P or Q limits. Since, the system is lossless, the P injections are shared such that incremental cost of generation is same. Comparison of tables 11.2 and 11.3 indicates that with large bounds permitted for V_{min} and V_{max}, higher voltage profile are obtained, infact V_1 and V_2 are fixed to V_{max} in both the cases. Larger V_is implies more contribution from line charging and hence small q-generation.

Case 3: The last case corresponds to the light load condition. Now, the operation of the system is dictated by the minimum limits of the generators. As mentioned, in the load flow chapter, due to end region heating, absorption capacity in practice may be quite limited. It can be seen from table 11.2 that with prescribed voltage band of 0.95 pu to 1.05 pu, the generators are made to absorb reactive power. The system is now dictated primarily by the measurement to hold V_3 at its minimum value 0.95. However, if V_{min} is reduced to 0.90 pu, then it can be seen from table 11.3 that situation improves significantly. The voltage profile is still dictated by V_{min} limit at V_3 node, which is now set to 0.90 pu. Consequently, V_1 and V_2 also reduce and hence the line charging MVAR (capacitive) reduce. The generators now have to absorb less reactive power. Since, the generator reactive power absorption is usually limited due to end region heating effect, this represents a better operating condition from the point of view of maintaining reserve. The P sharing is dictated by the fact that incremental cost of generation should be identical. In practice, to mitigate the problem of reactive power absorption at generators under light load conditions, some generators may be taken out of grid and/or shunt reactors may be used to absorb the surplus line charging MVARs and reduce overvoltages.

In all the cases, the solution is obtained from flat start which substantiates claim that those methods compute load flow and optimal solution simultaneously. As the example system is a lossless system, it can be seen that the optimal solution obtained by just accounting the overall power balance equation P limits on the generators also gives the correct optimal solution for P_{G_1} and P_{G_2}.

Note that the inequality constraints are forced through penalty terms augmented to the Lagrangian. Let us consider a penalty on V_1 for upper limit V_{max} being activated. Then,

$$\mathcal{L}^{new} = \mathcal{L}^{old} + \frac{\rho_i}{2}(V_1 - V_1^{max})^2$$

$$\frac{\partial \mathcal{L}^{new}}{\partial V_1} = \frac{\partial \mathcal{L}^{old}}{\partial V_1} + \rho_i(V_1 - V_1^{max})$$

$$\text{and } \frac{\partial^2 \mathcal{L}^{2new}}{\partial^2 V_1^2} = \frac{\partial \mathcal{L}^{2old}}{\partial V_1^2} + \rho_i$$

Thus, the penalty term only affects the diagonal entry of the Hessian. Readers are encouraged to check out the MATLAB file supported along with this example.

It can be seen that, with current initialization, both matrix H and matrix W are singular at the beginning. The Newton method will have quadratic convergence only near the optimal. Hence, it may not provide descent direction initially. To circumvent this difficulty, a matrix $\gamma_i I_9 (\gamma < 1)$ is added to Hessian (H) to guarantee descent direction initially. This makes the method more robust. The diagonal matrix is successively reduced by a factor of γ. Hence, in the vicinity of optimal, quadratic model of Lagrangian function is used. If γ is very large, the initial iterations will approximate steepest descent direction. The rate of convergence will reduce, *i.e.*, the method will require more iterations. An alternative to Newton approach is the SQP approach.

6. Sequential Quadratic Programming (SQP) Approach

The key to implement a sparse OPF is to solve the augmented equations (11.29). The task is however complicated by the fact that the coefficient matrix is not just large and sparse but also indefinite. Either a good large and sparse indefinite solver is required or sparse QR factorization function may be used. If a sparse QR factorization is used the number of fills may be high. Its suitability to the authors knowledge has not yet been evaluated. [Gould and Toint, 2001c] states that when the number of variables are large (near about and above of few thousands) then one must be suspicious of methods that aim to factorize the coefficient matrix (the KKT matrix of equation 11.29). When such is the case, iterative methods like conjugate gradient methods should be considered.

When the number of active constraints is small, another good alternative is to solve the augmented system of equations (11.29) by range space equations [Gill et al., 1984b].

$$CH^{-1}C^T\lambda = [c(x^k) - H^{-1}\nabla\mathcal{L}]; \tag{11.43}$$

$$Hd = -C^T\lambda - \nabla\mathcal{L} \tag{11.44}$$

where superscript k has been dropped to simplify the notations and

$$H = \begin{bmatrix} \nabla^2_{xx}\mathcal{L} & \nabla^2_{xu}\mathcal{L} \\ \nabla^2_{ux}\mathcal{L} & \nabla^2_{uu}\mathcal{L} \end{bmatrix}; \ \nabla\mathcal{L} = \begin{bmatrix} \nabla_x\mathcal{L} \\ \nabla_u\mathcal{L} \end{bmatrix}$$

and

$$C = \begin{bmatrix} \frac{\partial g}{\partial x} & \frac{\partial h}{\partial x} \\ \frac{\partial g}{\partial u} & \frac{\partial h}{\partial u} \end{bmatrix}^T; \ c(x^k) = \begin{bmatrix} g(x^k, u^k) \\ h(x^k, u^k) \end{bmatrix}$$

The method is appropriate if H is sparse and if C has relatively few rows. For methods which compute the solved power flow and the optimal simultaneously, clearly all the power flow equations have to be modeled in C. Many forward backward substitution are required in (11.44). Hence, the method is not that attractive for large sparse problems like OPF.

The other alternative is that instead of solving (11.29) directly, we solve a QP subproblem to compute descent direction d^k at each iteration. This leads us to the SQP approach. It can be verified that the following QP problem defined by quadratic approximation of Lagrangian function solves the augmented linear system of equations (11.29).

$$\text{Problem QP}: \min_{d^k} d^{k^T} \nabla f(x^k) + \frac{1}{2} d^{k^T} H^k d^k \tag{11.45}$$

subject to $Cd^k = -c(x^k)$ where $d^k = [\Delta x^k, \Delta u^k]^T$.
Notice that gradient of objective function and not Lagrangian function is used in (11.45)[6]. To improve the robustness of the approach, the linearized inequality constraints can also be considered in the QP subproblem. Thus, the QP problem to be solved at each major iteration of SQP is:

$$\min_{d^k} d^{k^T} \nabla f(x^k) + \frac{1}{2} d^{k^T} H^k d^k \tag{11.46}$$

$$\text{such that } h_i(x^k, u^k) + \nabla h_i^T(x^k, u^k) d^k = 0 \ \forall i \in \mathcal{A}(x^k, u^k) \tag{11.47}$$

$$g_i(x^k, u^k) + \nabla g_i^T(x^k, u^k) d^k = 0 \ i = 1, \ldots, 2n \tag{11.48}$$

$$x_{min} - x^k \leq \Delta x^k \leq x_{max} - x^k \tag{11.49}$$

$$u_{min} - u^k \leq \Delta u^k \leq u_{max} - u^k \tag{11.50}$$

$$\text{and } h_i^T(x^k, u^k) + \nabla h_i(x^k, u^k)^T d^k \leq 0 \ \forall i \notin \text{active set } \mathcal{A}(x^k) \tag{11.51}$$

An advantage of the above formulation is that inequality constraints and bounds can also be considered while computing d^k. Each QP subproblem minimizes a quadratic model of a Lagrangian function subject to linearized equality and inequality constraints.

The basic structure of a SQP method involves *major* and *minor* iterations (Figure 11.6). The major iterations generate a sequence of iterates (x^k, u^k, μ^k) that converge to (x^*, u^*, μ^*). At each iteration, a QP subproblem is used to generate a search direction (d^k) towards the next iteration $(x^{k+1}, u^{k+1}, \mu^{k+1})$. A *merit* function is reduced along the each

[6]The justification for the same was discussed in chapter 6 on optimization.

search direction obtained by solving QP subproblem. A merit function is a composite function that stresses both the reduction in objective function and feasibility of nonlinear active and inactive constraint and bounds to ensure convergence from a arbitrary point.

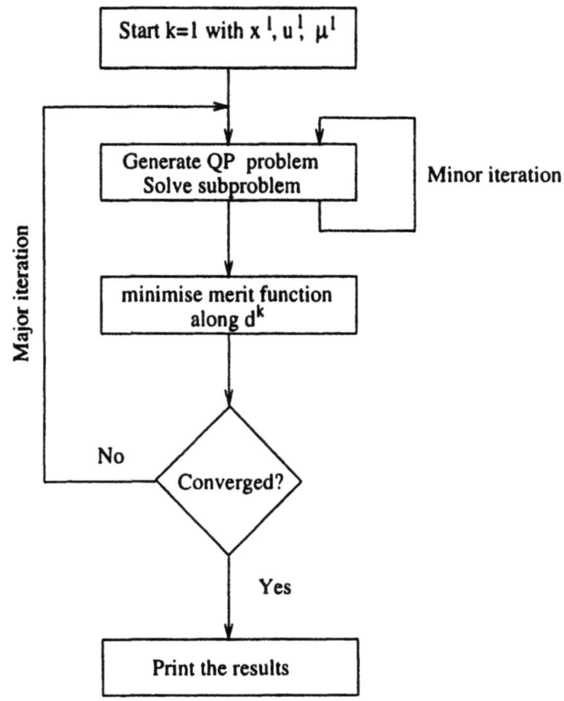

Figure 11.6. SQP Algorithm

Different merit functions can be synthesized. The main idea is that it should not only reduce the objective function but also penalize violation in nonlinear constraints. For example, the power flow equality constraints that we are dealing with are nonlinear and hence even a vector in null space of (11.47, 11.48) can add to infeasibility if a large enough step is taken along it.

$$\begin{bmatrix} x^{k+1} \\ u^{k+1} \end{bmatrix} = \begin{bmatrix} x^k \\ u^k \end{bmatrix} + \alpha^k \begin{bmatrix} \Delta x^k \\ \Delta u^k \end{bmatrix}$$

One commonly used merit function is given by the following expression:

$$F_E(x,u) = f(x,u) + \mu \left[\sum |g_i(x,u)| + \sum \max\{0, h_i(x,u)\} \right]$$

For simplicity all inequality constraints are considered of the form $h_i(x,u) \leq 0$. At the optimal point (x^*, u^*), $g_i(x^*,u^*) = 0$ and $h_i(x^*,u^*) \leq 0$, and hence, $F_E(x^*, u^*) = f(x^*, u^*)$. However, at non-optimal points, merit function penalizes violations in constraints, thereby, emphasizing both descent and feasibility conditions. This helps in developing globally convergent algorithms.

The overall algorithm converges when (1) the norm of the step taken is below a tolerance and (2) power flow mismatches are below a tolerance. (3) no inequality constraint violation exist (4) reduced gradient is zero.

Null space approach to QP solver is discussed in chapter 7. When a new QP problem is framed, the active set from the previous QP problem provides a good guess of the new active set. This is known as the warm start facility. It is because of the warm start facility that effort required in subsequent QP formulations reduce. Even if it is sparse, the reduced Hessian will not be sparse. It may also not be spd. Hence, quasi-Newton approximation of reduced Hessian can be used. Important benefit accrued in the Quasi Newton approach with warm start facility is that the Cholesky factor of the projected Hessian at the solution of one quadratic problem can be used to initiate the next QP problem. This approach has been proved to be very useful for OPF problem[Gill et al., 1984a]. A SQP algorithm for large sparse problems is discussed in [Gill et al., 1997].

7. Summary

There are two philosophies used in OPF solver, *viz.*

1 First compute load flow solution; then frame a reduced optimization problem with only control variables. The results of optimization function are used to update controls. The above sequence can be repeated until absolute change in the objective function reduces below a tolerance. Since, reduced problem formulation requires linearization, the overall method can be classified into Sequential Linear Programming (SLP) or Quasi Newton approach. Another possibility is to linearize the load flow equations but not to eliminate the basic variables (x). Then sparsity can also be exploited within the approach. Typically, the movement of control vector is restricted at each step so as to honor the linearization approximation.

2 More generic OPF solvers directly evaluate the KKT solution to optimization problem

$$viz. \ \min_{x,u} f(x, u)$$

$$\text{such that } g(x, u) = 0$$

$$h(x, u) \leq 0$$

$$\begin{bmatrix} x^{min} \\ u^{min} \end{bmatrix} \leq \begin{bmatrix} x \\ u \end{bmatrix} \leq \begin{bmatrix} x^{max} \\ u^{max} \end{bmatrix}$$

Thus, they compute the load flow and optimal solution simultaneously. Basically, the methods of this class solve a QP problem at each iteration, where the QP problem is obtained by second order expansion of the Lagrangian function and linearization of constraints. This leads to the SQP formulation. Alternatively, the methods can be viewed as application of Newton Raphson solver to the KKT conditions. In this avatar, the method is known as the Newton method. The Newton approach to OPF is easier to understand because of it's intuitive closeness to the load flow solvers. The inequality constraints are usually enforced through quadratic penalty terms. The appropriate choice of the penalty terms is not obvious.

The requirement of sparse matrix computation technology for solving the QP subproblem at each iteration of SQP approach is more involved. For the medium size problem involving a few thousand variables and small degree of freedom, a null space based QP solver is used. A merit of QP solver is that the handling of inequality constraints is directly through computation of Lagrangian multipliers estimates. Further, the computation of descent direction accounts for the linearized inequality constraints. With Quasi Newton approximation for solving the QP subproblem, the effort in computation of explicit projected Hessian is reduced subsequently because of previous iteration provides the initial estimate of projected Hessian (or its inverse). The exploitation of sparsity in linearized constraints is relatively easy.

OPF is used for many applications like (1) economic dispatch (2) loss minimization (3) security constrained OPF and (3) on line applications like EMS. In this chapter, we have not discussed the security constrained OPF formulation (SCOPF). In such problems, the effect of contingencies are also simulated in computing the OPF solution. With this the overall complexity of the algorithm increases. SCOPF is discussed in a nice way in a book by [Barret et al., 1997].

Another OPF application which is not discussed in this chapter is online OPF. For online use it is desired that number and amount of controller actions be restricted. Suboptimal solution is acceptable; also it may not be desirable to compute the controller positions based upon one objective function only. In this case, both security and economy are important. Curtailed number and reduced controller movement optimization algorithms are discussed in [Soman et al., 1994].

8. Further Reading

Economic operation of power system has been topic of interest for a long time. The famous book [Kirchmayer, 1958] was published in 1958. The OPF problem was formulated by [Carpentier, 1962]. The literature on OPF has been periodically reviewed [Happ, 1977]. [Huneault and Galiana, 1991] provides a recent survey of the OPF literature. Reduced model OPF formulation is due to [Dommel and Tinney, 1968]. Year 1984 was a significant year in the history of OPF when two significant computational tools, *viz.* Newton method and SQP method for large sparse problem were developed [Burchett et al., 1984; Sun et al., 1984]. Of late there has been a renewed interested in OPF due to its application in deregulated scenario [Christie et al., 2000; Orfanogianni and Bacher, 2000]. Deficiency of OPF were discussed in [Tinney et al., 1987] and ten years hence challenges in OPF have been discussed in [Momoh et al., 1997]. A good study article on OPF with discussion of security issue is [Stott et al., 1987] and [Glavitsch and Bacher, 1991]. SCOPF is well discussed in [Barret et al., 1997]. The articles by [Saxena et al., 1991; Orfanogianni and Bacher, 2000] discuss requirements of realistic objective function flexibility of OPF code and use of standard optimization packages. Critical cases in OPF have been discussed by [Almeida and Galiana, 1996]. The most recent contribution in this area is a book by [Momoh, 2001].

Chapter 12

POWER SYSTEM DYNAMICS

1. Introduction

A power system is continuously subjected to a variety of disturbances. These may vary from minor events like small and random load changes to major events like faults and line tripping. It is desirable that a power system should be stable (*i.e.*, settle to an acceptable steady state condition) without exceeding equipment ratings. Consequently, the dynamic behavior of a power system has an important bearing on satisfactory system operation. The dynamic behavior of the power system can be described by a set of differential equations. With modeling simplifications they are usually formulated as *differential algebraic* equations (DAEs) as follows:

$$\dot{x} = f(x, y, u, t) \tag{12.1}$$

$$0 = g(x, y, u, t) \tag{12.2}$$

where:
x represents states of the system
y represents variables which are algebraically dependent on x
u represents inputs
t represents time

The equilibrium points (or the steady state conditions) are those values of the states for which the left hand side of equation (12.1) becomes zero. Note that if the equations are formulated in such a way that some inputs are sinusoidally varying then the equilibrium has to be appropriately redefined as the 'sinusoidal steady state'. In general, there could be more than one equilibrium point. A 'disturbance' is modeled by introducing some changes in the above equations. Any disturbance which causes the system to stray away from an equilibrium state, will cause a 'transient'. Note that the pre-disturbance and post-disturbance equilibria may not be the same. In this context, the following questions arise

1 Is the system stable for this disturbance, *i.e.*, do the states settle to an acceptable post-disturbance equilibrium?

2 Are any equipment design limits violated during the transient?

3 Can control and protection design/evaluation be done based on the dynamic analysis?

In this chapter we introduce the reader to some dynamic analysis tools. The dynamic modeling and analysis of a power system is a vast topic in itself and is not covered here. The reader is referred to books by [Kundur, 1994], [Sauer and Pai, 1998], [Padiyar, 1996] for a detailed treatment. We first consider some general-purpose methods for dynamic analysis and then concentrate on some tools specific to power systems.

2. General Analysis Tools for Dynamic Studies

For the purposes of analysis, it is convenient to classify disturbances into small and large disturbances. A system which is small disturbance unstable cannot be operated at all, since, small and random disturbances occur continuously. The major analysis tools for studying the behavior of dynamic systems are:

1 Digital Simulation (Numerical Integration): The output of any digital simulation program is the response of the states at discrete time steps. Digital simulation is the most flexible tool for dynamic analysis. It can be used, for any kind of disturbance, in situations where there are many modeling complexities (*e.g.*, power electronic switching, discrete and continuous (mixed) systems etc.) and where there are several disturbance events.

2 Eigen-Analysis of Linearized differential equations (Small signal analysis): Small disturbance response can be evaluated by a eigen-analysis of the linearized differential equations. Eigen-analysis permits the complete characterization of the system response in terms of different 'modes'. This obviates the need for digital simulation of various disturbances to evaluate small disturbance stability.

A detailed description of digital simulation and linearized analysis are described in the following sub-sections.

2.1. Digital Simulation: Numerical Integration Methods

A system of differential equations (or differential-algebraic equations) with known initial conditions of the states can be simulated by numerical integration of the differential equations. The numerical integration yields the values of the states at discrete time steps (denoted as $x_k, k = 1, 2, 3 \ldots$, and $t_k = hk$, where h is the time step).

A major concern in numerical integration is whether it faithfully captures the actual behavior of the system or not. The errors introduced in any simulation are due to **truncation**, *i.e.*, the approximations involved in discretising the differential equations and **round off** (due to finite precision of number representation in a digital computer). Accumulated round off errors increase when the time step is very small, while accumulated truncation errors decrease. Round off errors can be controlled by selecting higher precision if the step sizes are small.

A major classification of numerical methods for integration is:

A. Explicit methods;

B. Implicit methods.

In explicit methods, the algorithm gives an explicit solution of x_k at the end of kth step. Implicit methods require a solution of algebraic equations for the calculation of x_k. In case of nonlinear systems, iterative solution is required. This requires more computations per step. In general, implicit algorithms are more stable and can tolerate larger step-size. In what follows, the differential equations are assumed to be of the form

$$\dot{x} = f(x, t) \tag{12.3}$$

A few representative integration algorithms are now considered. The reader is referred to [Lapidus and Seufield, 1971] and [Gear, 1971] for an exhaustive treatment.

2.1.1 Single Step Algorithms

Four types of integration methods are discussed below.

1 Forward Euler:

The algorithm is given by

$$x_k = x_{k-1} + hf(x_{k-1}, t_{k-1}) \tag{12.4}$$

2 Backward Euler:

$$x_k = x_{k-1} + hf(x_k, t_k) \tag{12.5}$$

Note that while forward Euler is explicit, the backward Euler's method is implicit.

3 Trapezoidal:

$$x_k = x_{k-1} + \frac{h}{2}[f(x_k, t_k) + f(x_{k-1}, t_{k-1})] \tag{12.6}$$

This is an implicit method and is used extensively in power system studies.

4 Runge-Kutta (fourth order):
This is an explicit algorithm that uses intermediate points in the interval (t_{k-1}, t_k) to calculate state at time t_k.
Defining

$$\begin{aligned} k_1 &= f(x_{k-1}, t_{k-1}) \\ k_2 &= f(x_{k-1} + \frac{h}{2}k_1, t_{k-1} + \frac{h}{2}) \\ k_3 &= f(x_{k-1} + \frac{h}{2}k_2, t_{k-1} + \frac{h}{2}) \\ k_4 &= f(x_{k-1} + hk_3, t_k) \end{aligned}$$

the final state is calculated from

$$x_k = x_{k-1} + \frac{h}{6}[k_1 + 2k_2 + 2k_3 + k_4] \tag{12.7}$$

2.1.2 Multi-Step Algorithms

A general form for an m step algorithm is given by

$$\begin{aligned} x_k &= a_1 x_{k-1} + a_2 x_{k-2} + \ldots a_m x_{k-m} \\ &+ h[b_o f(x_k, t_k) + b_1 f(x_{k-1}, t_{k-1}) + b_2 f(x_{k-2}, t_{k-2}) \\ &+ \ldots b_m f(x_{k-m}, t_{k-m})] \end{aligned} \tag{12.8}$$

where t_{k-i} are evenly spaced with time step h. Different choices of the coefficients a_i and b_i lead to different integration algorithms. If $b_o = 0$, the algorithm is explicit; otherwise, it is implicit. We discuss two commonly used multistep algorithms.

1 Adams-Bashforth: The pth order Adams-Bashforth algorithm is generated by setting $m = p, b_o = 0$ and $a_i = 0$ for $i = 1, 2, \ldots, p$. The

remaining coefficients are chosen such that the algorithm is exact for polynomials of order p. Since $b_o = 0$, this is an explicit algorithm. The first order algorithm is identical to forward Euler's method.

The second order algorithm is given below:

$$x_k = x_{k-1} + \frac{h}{2}[3f(x_{k-1}, t_{k-1}) - f(x_{k-2}, t_{k-2})] \tag{12.9}$$

2 Gear's Algorithm: The pth order Gear's algorithm is obtained by setting $m = p$ and $b_i = 0$ for $i = 1, 2, \ldots, p$. The remaining coefficients are chosen such that the algorithm is exact for polynomials of order p. Since $b_o \neq 0$, it is an implicit method.
First order Gear's algorithm is identical to backward Euler algorithm. The second order algorithm is given below:

$$x_k = \frac{1}{3}[4x_{k-1} - x_{k-2} + 2hf(x_k, t_k)] \tag{12.10}$$

Example 12.1: Compute the response of the following system by (a) Forward Euler's Method and (b) Trapezoidal Method.

$$\dot{x} = v; \quad \dot{v} = -x$$

Initial Conditions: $x(0) = 1$ m, $v(0) = 0$ m/s.
(The above equations represent the dynamics of a spring mass system with mass = 1 kg and spring constant of 1 N/m)
Solution: The exact response (which can be easily verified) of this system is :

$$x(t) = x(0)\cos(t) + v(0)\sin(t) = \cos(t)$$

$$y(t) = -x(0)\sin(t) + v(0)\cos(t) = -\sin(t)$$

i.e., it is an undamped oscillation of frequency 1 rad/s.
(a) Simulation by Forward Euler's method: The differential equations are discretised as follows:

$$x_{k+1} - x_k = hv_k$$

$$v_{k+1} - v_k = -hx_k$$

With $x_0 = 1$, $v_0 = 0$ and $h = 0.2$ s, the response of this numerical integration is as shown in Figure 12.1 .

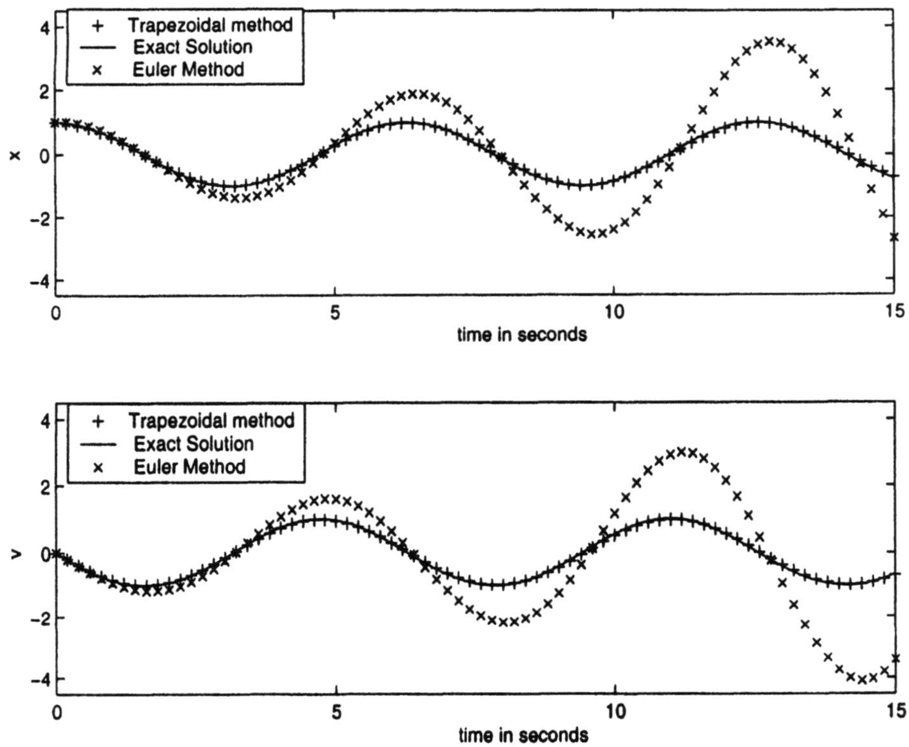

Figure 12.1. Simulation by Euler's method and Trapezoidal method

(b) Simulation by Trapezoidal method:
The differential equations are discretised as follows:

$$x_{k+1} - x_k = \frac{h}{2}(v_k + v_{k+1})$$

$$v_{k+1} - v_k = -\frac{h}{2}(x_k + x_{k+1})$$

Therefore:

$$\begin{bmatrix} x_{k+1} \\ v_{k+1} \end{bmatrix} = \begin{bmatrix} 1 & -h/2 \\ h/2 & 1 \end{bmatrix}^{-1} \begin{bmatrix} 1 & h/2 \\ -h/2 & 1 \end{bmatrix} \begin{bmatrix} x_k \\ v_k \end{bmatrix}$$

With $x_0 = 1$, $v_0 = 0$ and $h = 0.2$ s, the response of this numerical integration is as shown in Figure(12.1).

Remarks:

1 Since the oscillation is of 1 rad/s, $h = 0.2$s appears to be a "reasonably small" time step for numerical integration.

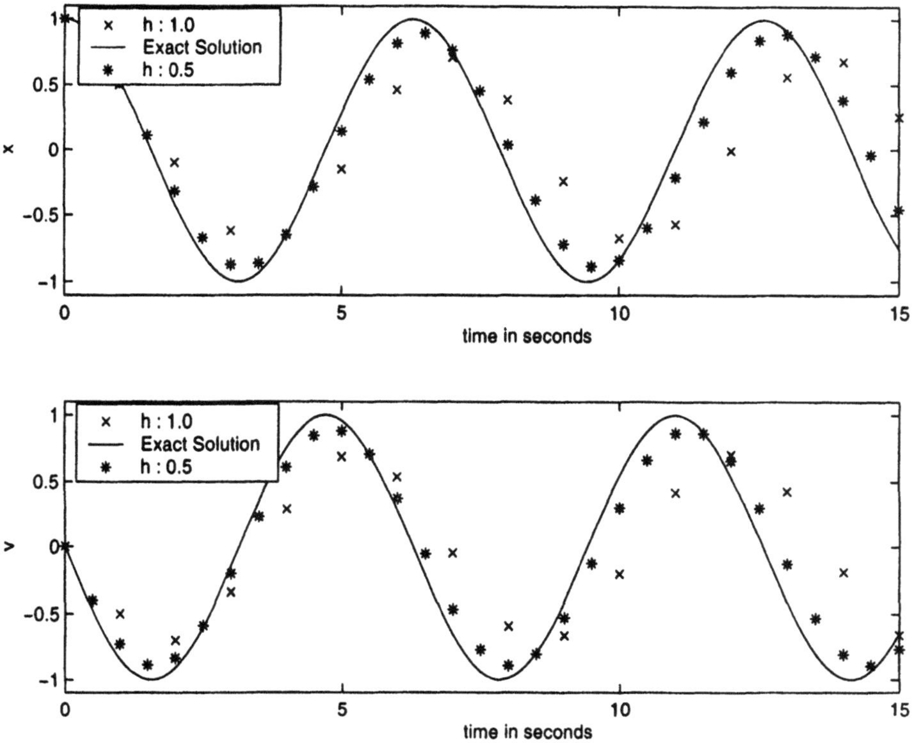

Figure 12.2. Simulation by Trapezoidal method: Effect of time step

2 Forward Euler's method incorrectly determines the system response to be a growing one while Trapezoidal method preserves the stability information of the system (*i.e.*, an undamped oscillation). Forward Euler's method is rarely used in practice due to its poor numerical properties.

3 With an increase in the time step to $h = 0.5$s and 1s, the accuracy of the simulation worsens; however the stability information is still not lost with trapezoidal method (Figure(12.2)). This is an important property of the trapezoidal method which is exploited in the simulation of systems where both slow and fast dynamics are present (*stiff systems*). In stiff systems, if explicit algorithms are used, considerations of numerical stability require very small step sizes even when one is interested in capturing only slow dynamics. Thus, if only slow

transients are of interest, implicit algorithms are computationally efficient as larger step-sizes can be used [1].

2.2. Small Signal Analysis

Consider the system:

$$\dot{x} = f(x,u) \tag{12.11}$$

The stability of an equilibrium point for a small disturbance can be judged by the solution of the linearized system at the equilibrium point x_o.

If

$$x = x_o + \Delta x \tag{12.12}$$

and $\Delta u = 0$, then we obtain

$$\dot{x} = \dot{x}_o + \Delta\dot{x} = f(x_o,u) + \left[\frac{\partial f(x,u)}{\partial x}\right]_{x = x_o} \Delta x \tag{12.13}$$

Therefore,

$$\Delta\dot{x} = [A(x_o,u)]\Delta x \tag{12.14}$$

where A is a $n \times n$ matrix whose elements are functions of x_o and u. The ij^{th} element of [A] is given by

$$A_{ij}(x_o,u) = \frac{\partial f_i}{\partial x_j}(x_o,u) \tag{12.15}$$

For a given x_o and u, the matrix A is constant. The eigenvalues (λ_i) and the corresponding right eigenvectors (p_i) of the matrix A are obtained as solutions of the following equation:

$$Ap_i = \lambda_i p_i \tag{12.16}$$

If the size of A is $n \times n$, then the number of eigenvalues is n. If there are n linearly independent eigenvectors of matrix A (*i.e.*, matrix A is diagonalizable) then the solution of the linearized state equation is given by[2]

$$\Delta x(t) = c_1 e^{\lambda_1 t} p_1 + c_2 e^{\lambda_2 t} p_2 + \cdots + c_n e^{\lambda_n t} p_n \tag{12.17}$$

[1] The reader is encouraged to try other implicit methods like backward Euler method.

[2] if n linearly independent eigenvectors do not exist, then the solution has terms like $te^{\lambda_i t}, t^2 e^{\lambda_i t}$ etc. The reader is referred to [Ogata, 1967] for details.

If a vector q_i corresponding to the eigenvalue λ_i, is related to the right eigenvector p_j as follows [3]:

$$q_i^T p_j = 0 \quad i \neq j \tag{12.18}$$
$$= 1 \quad i = j \tag{12.19}$$

then the extent to which a 'mode' $(e^{\lambda_i t}, p_i)$ is excited is determined by $c_i = q_i^T \Delta x(0)$, where $\Delta x(0)$ is the initial deviation from the equilibrium point.

It can be seen that if $Re[\lambda_i] < 0$ for all λ_i, then for all sufficiently small perturbations from the equilibrium point x_o, the trajectories tend to x_o as $t \to \infty$. Hence, x_o is asymptotically stable. If for even one λ_i, $Re[\lambda_i] \geq 0$ then the system is considered to be unstable.

Note that the eigenvectors yield useful information about the participation of the states in various modes. This information is useful in diagnosing and correcting (by appropriate control) dynamic problems. The computation of eigenvalues and eigenvectors is therefore an important numerical exercise. There are many methods for computing the eigenvalues and eigenvectors. We now consider a general purpose method for computing eigenvalues.

2.2.1 QR Method

The QR method [Golub and Loan, 1989] is a general purpose method for computation of eigenvalues.

The method is based on two important Theorems:

- For an arbitrary real square matrix A, there exists an orthogonal matrix Q and an upper triangular matrix R such that $A = QR$. This decomposition is introduced and discussed in detail in chapter 5.

- If $\{A_0, A_1, \ldots, A_k, \ldots\}$ is a sequence of matrices such that $A_0 = A$ and $A_k = R_{k-1}Q_{k-1}$ and $Q_k R_k = A_k (k = 1, 2, \ldots)$, and the product $Q_0 Q_1 \ldots Q_k$ converges as $k \to \infty$, then A_k converges to an upper triangular matrix with eigenvalues of A as its diagonal elements.

Thus the algorithm for finding eigenvalues proceeds as follows:

1 Perform QR factorization of $A_0 (= A)$. The QR factors are denoted as Q_0 and R_0.

2 Compute $A_1 = R_0 Q_0$.

[3] q_i also satisfies the equation $q_i^T A = \lambda_i q_i^T$ and is a left eigenvector corresponding to λ_i

3 Perform QR factorization of A_1. The QR factors are denoted as Q_1 and R_1.

4 Repeat the above steps till convergence.

Example 12.2: Find the eigenvalues of the matrix A given below by QR factorization
Solution:

$$A = \begin{bmatrix} 1 & 5 \\ 7 & 9 \end{bmatrix}$$

Step1: $[Q_1\ R_1] = QR(A)$

$$Q_1 = \begin{bmatrix} -0.1414 & -0.9899 \\ -0.9899 & 0.1414 \end{bmatrix};\ R_1 = \begin{bmatrix} -7.0711 & -9.6167 \\ 0 & -3.6770 \end{bmatrix}$$

$[A_1] = [R_1] \times [Q_1]$

$$A_1 = \begin{bmatrix} 10.5200 & 5.6400 \\ 3.6400 & -0.5200 \end{bmatrix}$$

Step 2: $[Q_2\ R_2] = QR(A_1)$

$$Q_2 = \begin{bmatrix} -0.9450 & -0.3270 \\ -0.3270 & 0.9450 \end{bmatrix};\ R_2 = \begin{bmatrix} -11.1319 & -5.1599 \\ 0 & -2.3356 \end{bmatrix}$$

$[A_2] = [R_2] \times [Q_2]$

$$A_2 = \begin{bmatrix} 12.2072 & -1.2363 \\ 0.7637 & -2.2072 \end{bmatrix}$$

Step 3: $[Q_3\ R_3] = QR(A_2)$

$$Q_3 = \begin{bmatrix} -0.9980 & -0.0624 \\ -0.0624 & 0.9980 \end{bmatrix};\ R_3 = \begin{bmatrix} -12.2311 & 1.3717 \\ 0 & -2.1257 \end{bmatrix}$$

$[A_3] = [R_3] \times [Q3]$

$$A_3 = \begin{bmatrix} 12.1216 & 2.1327 \\ 0.1327 & -2.1216 \end{bmatrix}$$

Step 4: $[Q_4\ R_4] = QR(A_3)$

$$Q_4 = \begin{bmatrix} -0.9999 & -0.0109 \\ -0.0109 & 0.9999 \end{bmatrix};\ R_4 = \begin{bmatrix} -12.1223 & -2.1094 \\ 0 & -2.1448 \end{bmatrix}$$

$[A_4] = [R_4] \times [Q_4]$

$$A_4 = \begin{bmatrix} 12.1447 & -1.9765 \\ 0.0235 & -2.1447 \end{bmatrix}$$

Step 5: $[Q_5\ R_5] = QR(A_4)$

$$Q_5 = \begin{bmatrix} -1.0000 & -0.0109 \\ -0.0019 & 1.0000 \end{bmatrix}; \quad R_5 = \begin{bmatrix} -12.1447 & -1.9807 \\ 0 & -2.1409 \end{bmatrix}$$

$[A_5] = [R_5] \times [Q_5]$

$$A_5 = \begin{bmatrix} 12.1408 & 2.0041 \\ 0.0041 & -2.1408 \end{bmatrix}$$

If accuracy upto the 2nd decimal place is acceptable, the algorithm can be terminated here. Therefore, eigenvalues of A are 12.1408 and -2.1408.

3. Modeling In Power System Dynamic Studies

The power system is made up of a variety of dynamic sub-systems like transmission network, transformers, HVDC and FACTS devices, generators, exciters, governors etc. The transients associated with these sub-systems can be classified into transients which are fast and others which are relatively slow (see Figure(12.3)). A reduction in computational expense can be obtained without significant loss of accuracy if the following approximations are made:

1 If fast transients are of interest and the response over only a short time period is desired, then "freeze" the slowly varying states at their pre-disturbance value.
For example, in the study of network transients, control /firing systems associated with HVDC or FACTS etc. it is common to neglect the change in the slow electro-mechanical states such as generator speed or angle.

2 If slow transients are of interest and the response over a large time period is desired, then fast transients are neglected.
For example, in the study of transient stability, the natural transients of the network is neglected.

Example 12.3: The equations for a generator modeled with one field winding and one damper winding in the d-axis, and 2 dampers on the q-axis are given in the Park reference frame by [Padiyar, 1999]:

1 d-axis Flux Equations:

$$\frac{d\psi_h}{dt} = \frac{1}{T_d''}\left[-\psi_h + \psi_d\right] \tag{12.20}$$

$$\frac{d\psi_f}{dt} = \frac{1}{T_d'}\left[-\psi_f + \psi_d + \frac{x_d' E_{fd}}{(x_d - x_d')}\right] \tag{12.21}$$

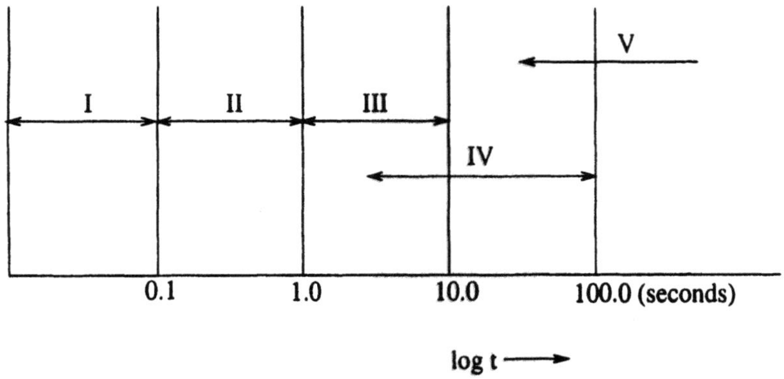

Region	Time (in seconds)	Phenomena
I	$0<t<0.1$	Electromagnetic transients and switching surges
II	$0.1<t1$	Synchronous machine dynamics and its controls
III	$1<t<10$	Governors
IV	$4<t<100$	Load frequency control
V	$t>60$	Boiler Dynamics

Figure 12.3. Transient Spectrum [Pai, 1979]

2 q-axis Flux Equations:

$$\frac{d\psi_g}{dt} = \frac{1}{T_q'}[-\psi_g + \psi_q] \tag{12.22}$$

$$\frac{d\psi_k}{dt} = \frac{1}{T_q''}[-\psi_k + \psi_q] \tag{12.23}$$

3 Stator Flux Equations:

$$\frac{1}{\omega_B}\frac{d\psi_d}{dt} = -(1+S_m)\psi_q - R_a i_d - v_d \tag{12.24}$$

$$\frac{1}{\omega_B}\frac{d\psi_q}{dt} = (1+S_m)\psi_d - R_a i_q - v_q \tag{12.25}$$

4 Currents are algebraically related to the states as follows:

$$\psi_q = x_q'' i_q + \frac{(x_q' - x_q'')}{x_q'}\psi_k + \frac{(x_q - x_q')}{x_q}\frac{x_q''}{x_q'}\psi_g \tag{12.26}$$

$$\psi_d = x_d'' i_d + \frac{(x_d' - x_d'')}{x_d'}\psi_h + \frac{(x_d - x_d')}{x_d}\frac{x_d''}{x_d'}\psi_f \tag{12.27}$$

5 Rotor Mechanical Equations in the synchronously rotating frame:

$$2H\frac{dS_m}{dt} = T_m - T_e = T_m - (\psi_d i_q - \psi_q i_d) \quad (12.28)$$

$$\frac{d\delta}{dt} = \omega_B\ S_m \quad (12.29)$$

Where T_m = mechanical input torque, S_m = generator slip and E_{fd} is the steady state open circuit voltage at the terminals of the synchronous machine. The various constants and their typical values are given in Table 12.1 below:

Table 12.1. Typical values of various constants

$T_d^{'} = 0.4$s	$T_d^{''} = 0.026$s	$T_q^{'} = 0.107$s	$T_q^{''} = 0.046$s	$\omega_B = 2\pi \times 50$ rad/s
$R_a = 0.002$	$x_d = 1.79$	$x_d^{'} = 0.169$	$x_d^{''}$=0.135	
x_q=1.71	$x_q^{'} = 0.228$	$x_q^{''}$=0.2	H = 5 MJ/MVA	

Remark: The flux equations are in the Parks reference frame (*i.e.*, rotor d-q reference frame). When interfacing the generator voltages and currents with the network, it is necessary that all the currents and voltages in the system are expressed in a common reference frame. It is convenient to use the synchronously rotating (D-Q) frame, which is used to define the rotor mechanical equations as the common reference frame. The conversion between d-q and D-Q variables is given by:

$$\begin{bmatrix} f_D \\ f_Q \end{bmatrix} = \begin{bmatrix} \cos\delta & \sin\delta \\ -\sin\delta & \cos\delta \end{bmatrix} \begin{bmatrix} f_d \\ f_q \end{bmatrix} = T_1 \begin{bmatrix} f_d \\ f_q \end{bmatrix} \quad (12.30)$$

or

$$f_Q + jf_D = (f_q + jf_d)e^{j\delta} \quad (12.31)$$

Note that the matrix T_1 is a Givens rotation matrix to rotate a vector by an angle δ.

- **Approximations for study of fast transients for a time-scale of 0-20 cycles.**

  The field flux and electromechanical variables can be assumed to be constant for this study, *i.e.*,

$$\psi_f = \psi_f(0); \quad S_m = S_m(0); \quad \delta = \delta(0) \quad (12.32)$$

- **Approximations for conventional short circuit calculations**

1 For a time scale of 0-3 cycles the following assumptions are made:

(a) The states $\psi_g, \psi_h, \psi_f, \psi_k, \delta$ and S_m are assumed to be frozen at the pre-disturbance values.

(b) $x_d^{''} = x_q^{''} = x^{''}$.

(c) Stator transients are neglected, *i.e.*, $(\frac{1}{\omega_B}\dot{\psi}_d = 0, \frac{1}{\omega_B}\dot{\psi}_q = 0)$. Also it is assumed that $S_m = 0$ and $R_a = 0$.
Therefore, from equations (12.24) and (12.25), we obtain:

$$\psi_d = v_q; \quad \psi_q = -v_d \tag{12.33}$$

Consequently, if we define

$$E_q^{''} = \frac{(x_d^{'} - x_d^{''})}{x_d^{'}}\psi_h + \frac{(x_d - x_d^{'})}{x_d}\frac{x_d^{''}}{x_d^{'}}\psi_f \tag{12.34}$$

$$E_d^{''} = -\left\{\frac{(x_q^{'} - x_q^{''})}{x_q^{'}}\psi_k + \frac{(x_q - x_q^{'})}{x_q}\frac{x_q^{''}}{x_q^{'}}\psi_g\right\} \tag{12.35}$$

Then the equation which defines the generator is:
(from equations(12.33) -(12.35), (12.26) and (12.27))

$$(E_q^{''} + jE_d^{''}) - jx^{''}(i_q + ji_d) = (v_q + jv_d) \tag{12.36}$$

In the synchronously rotating frame the equations become:

$$(E_Q^{''} + jE_D^{''}) - jx^{''}(i_Q + ji_D) = (v_Q + jv_D) \tag{12.37}$$

The above equation defines the sub-transient model of a synchronous machine for short circuit studies (*i.e.*, a constant voltage behind a sub-transient reactance). Note that $E_Q^{''}$ and $E_D^{''}$ are constant as a result of assumption (1a).

2 For a time scale of 3-50 cycles the following assumptions are made:

(a) The states ψ_f, ψ_g, δ, and S_m are assumed to be frozen at the pre-disturbance values.

(b) $x_d^{'} = x_q^{'} = x^{'}$.

(c) Stator transients are neglected (see equation (12.33)).

(d) Transients in the damper windings ψ_h and ψ_k are neglected, *i.e.*, $T_d^{''}\dot{\psi}_h = 0$ and $T_q^{''}\dot{\psi}_k = 0$. The differential equations (12.20) and (12.23) are replaced by the algebraic equations given below:

$$\psi_h = \psi_d; \quad \psi_k = \psi_q \tag{12.38}$$

Consequently, if we define

$$E_q' = \frac{(x_d - x_d')}{x_d}\psi_f; \quad E_d' = -\frac{(x_q - x_q')}{x_q}\psi_g \tag{12.39}$$

Then the equation which defines the generator is :

$$(E_q' + jE_d') - jx'(i_q + ji_d) = (v_q + jv_d) \tag{12.40}$$

In the synchronously rotating frame the equations become:

$$(E_Q' + jE_D') - jx'(i_Q + ji_D) = (v_Q + jv_D) \tag{12.41}$$

The above equations defines the transient model of a synchronous machine for short circuit studies (*i.e.*, a constant voltage behind a transient reactance). Note that E_Q' and E_D' are constant as a result of assumption (2a).

- **Approximations for study of slow transients for a time-scale of 0-10 seconds (transient stability studies)**

  The stator transients are neglected as in conventional short circuit calculations (see equation(12.33)). Consequently we obtain:

$$\begin{bmatrix} E_q'' \\ E_d'' \end{bmatrix} - \begin{bmatrix} 0 & -x_d'' \\ x_q'' & 0 \end{bmatrix} \begin{bmatrix} i_q \\ i_d \end{bmatrix} = \begin{bmatrix} v_q \\ v_d \end{bmatrix} \tag{12.42}$$

  The algebraic equations expressed in the synchronously rotating frame of reference are [4]:

$$\begin{bmatrix} E_Q'' \\ E_D'' \end{bmatrix} - [T_1] \begin{bmatrix} 0 & -x_d'' \\ x_q'' & 0 \end{bmatrix} [T_1]^{-1} \begin{bmatrix} i_Q \\ i_D \end{bmatrix} = \begin{bmatrix} v_Q \\ v_D \end{bmatrix} \tag{12.43}$$

  The differential equations are used as in equations (12.20)-(12.23), (12.28) and (12.29).

- **Study of Sub Synchronous Resonance** For the study of sub-synchronous resonance stator transients are not neglected. As such, equations (12.20-12.27) are used. Moreover, the mechanical systems are modeled in detail to include the torsional dynamics of the turbine-generator system.

In the next section, we introduce some important power system computational tools which exploit some modeling simplifications which were discussed in this section.

[4]if $x_d'' \neq x_q''$, the equations cannot be expressed as one complex circuit equation.

4. Power System Dynamic Computation Tools

While the general tools of digital simulation and eigen-analysis are widely used in power system dynamic computations, some modifications and enhancements have been proposed to improve the computational efficiency. This section describes the basic features of three special types of programs:

1 Electromagnetic Transients Simulation Program. This is used for the simulation of fast network transients (e.g., Circuit recovery voltage, switching and lightning transients, power electronic switchings and fast control systems associated with HVDC, FACTS etc.).

2 Transient Stability Program. This is used for the simulation of electromechanical transients (e.g., stability of synchronous generator following faults or line tripping).

3 Small Signal Analysis Program. This is used for Eigen-analysis of electromechanical transients (e.g, study of damping of power oscillations , design of Power System Stabilizer and other controllers).

4.1. Electromagnetic Transient Simulation

Digital simulation of any dynamic system requires the formulation of the system differential equations. In an electrical network made of R-L-C elements, the differential equations representing the individual dynamic elements are coupled and this coupling is determined by the network topology. In a large network, the process of writing down all the differential equations can be tedious since network topology has to be considered in the choice of state variables.

The method of [Dommel, 1969] for component modeling is a systematic approach to this problem. In this method, lumped inductors or capacitors are modeled as resistances in parallel with current sources dependent on the previous history.

For example, the governing equation for the voltage across a capacitor is:

$$v(t) = \frac{1}{C}\int_{t-h}^{t} i(\tau)d\tau + v(t-h) \qquad (12.44)$$

By applying Trapezoidal rule of integration, the above equation reduces to

$$v(t) = \frac{h}{2C}[i(t) + i(t-h)] + v(t-h) \qquad (12.45)$$

This reduces to

$$v(t) = \frac{h}{2C}i(t) + [\frac{h}{2C}i(t-h) + v(t-h)]$$

$$i(t) = \frac{2C}{h}v(t) + I_C(t-h) \tag{12.46}$$

where,

$$I_C(t-h) = \left[-i(t-h) - \frac{2C}{h}v(t-h)\right]$$

Equation (12.46) can be represented by an equivalent circuit shown in Figure(12.4). Similarly, an inductor can be represented by a circuit

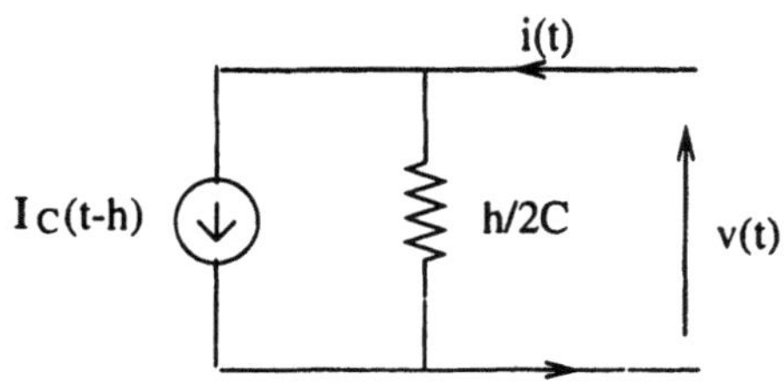

Figure 12.4. Equivalent circuit of a capacitor

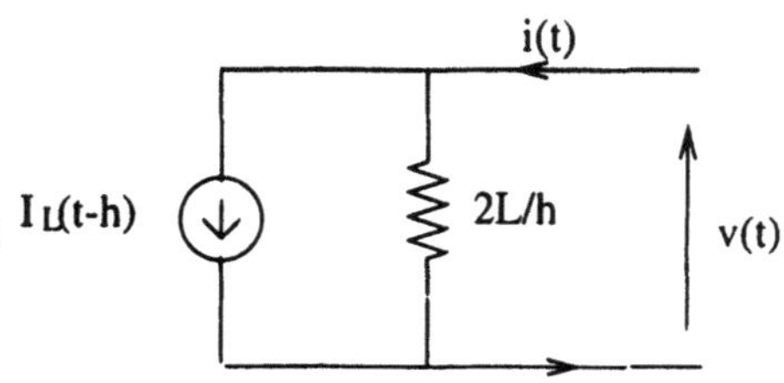

Figure 12.5. Equivalent circuit of an inductor

as shown in Figure (12.5) where $I_L(t-h) = i(t-h) + \frac{h}{2L}v(t-h)$. In a network L and C elements can be replaced by their equivalents. The network then becomes resistive network, where equations can be written as:

$$[G]v(t) = I_s(t) - I(t-h) \tag{12.47}$$

where, $I_s(t)$= vector of current source, $I(t-h)$= vector of the history terms and $[G]$ is the conductance matrix, the formation of which is straightforward. The simulation of the network boils down to the solution of a set of linear algebraic equations.

Remarks:

- Usually, all three phases are considered in the formation of the conductance matrix G. Therefore the structure of G is similar to a three phase bus admittance matrix. Sparsity exploitation techniques discussed in chapter 3 and 4 can be used for the linear system solution.

- Network changes (like a fault or change in parameters of an element) result in low rank changes in G and can be efficiently handled by techniques like Sherman- Morrison formula, or Schur update discussed in chapter 7.

- The above method is prone to problem of numerical oscillations when the derivative of the current changes suddenly (*e.g.*, when power electronic switches are used in the circuit). A solution to this problem is to use snubber circuits across the switches. [Gole et al., 1997] have also suggested a modification of the basic algorithm which overcomes this problem.

4.2. Transient Stability Simulation

A transient stability program refers to a digital simulation program for simulation of electro-mechanical transients following a large disturbance. For transient stability studies a power system is represented as dynamic subsystems (like generators) connected to a static network (network and stator transients are neglected). The behavior can be described by a set of differential-algebraic equations. The algebraic equations arise from the neglect of transients in the fast-subsystems, like the stator windings of generators and the network elements.

4.2.1 Representation of Dynamic Subsystems

The ith dynamic - subsystem can be mathematically represented as follows:

$$\dot{x}_i = f_i(x_i, V_i, u_i) \tag{12.48}$$

$$I_i = g_i(x_i, V_i) \tag{12.49}$$

where
I_i are the components (i_{Qi}, i_{Di}) of the current injected by the device in the synchronously rotating frame of reference.
V_i are the components (v_{Qi}, v_{Di}) of the subsystem bus voltage in the synchronously rotating frame of reference.
x_i represents the states of the dynamic subsystem.
u_i represents inputs to the dynamic subsystem.

Examples 12.4: Modeling of few subsystems like generators, SVC and constant power type loads are discussed below.

1 Generator : Refer to example (12.3) in the previous section (see approximations for the study of slow transients).

2 Model of Static VAR Compensator (SVC) :
The block diagram of the control system of an SVC is given in Figure (12.6). The SVC 'plant' is represented by a first order transfer function [IEEE et al., 1990].
The equations are :

$$\dot{B}_i = k_i(V_{ref} - |V|) \tag{12.50}$$

where, $|V| = \sqrt{(v_Q^2 + v_D^2)}$
Compute $B'_{ref} = k_p(V_{ref} - |V|) + B_i$

If $B_{min} < B'_{ref} < B_{max}$ then $B_{ref} = B'_{ref}$
If $B'_{ref} > B_{max}$ then $B_{ref} = B_{max}$
If $B'_{ref} < B_{min}$ then $B_{ref} = B_{min}$

$$\dot{B}_{SVC} = \frac{1}{T_P}[-B_{SVC} + B_{ref}] \tag{12.51}$$

The algebraic equations can be written as follows:

$$(i_Q + j i_D) = -jB_{SVC}(v_Q + jv_D) \tag{12.52}$$

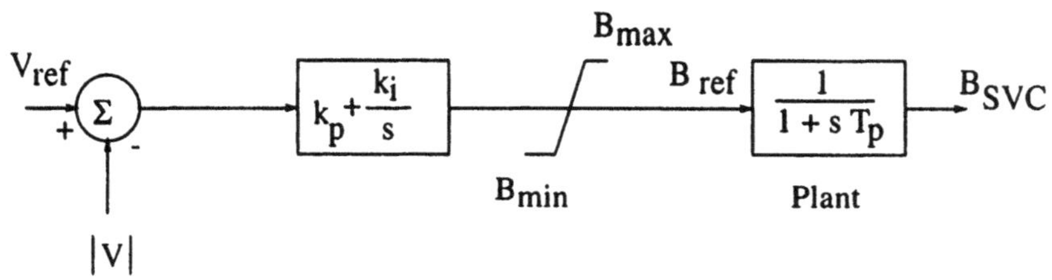

Figure 12.6. Block diagram of SVC with voltage regulator

3 <u>Model of a Static Constant Power load</u>:
Current injected by a static constant power load is given by an equation:

$$(i_Q + j i_D) = -\frac{(P_L - jQ_L)}{(v_Q - jv_D)} \tag{12.53}$$

Note that there is a *nonlinear* dependence of injected current on the bus voltage.

4.2.2 Representation of the Network

The network is represented by algebraic equations as follows:

$$I = [Y]V \tag{12.54}$$

Note that I is a vector of bus current injections and V is a vector of bus voltages. Alternatively, the real and reactive power balance equations may be used instead of the above current balance equations. However, since the network admittance matrix is constant (except when there is a network change), the current-balance formulation is preferred for simulation.

Note:

- The formulation given above assumes that the dynamic subsystems are connected in shunt to the network. For series connected devices such as TCSC (Thyristor Controlled Series Compensator) the equations need to be reformulated to allow for series voltage injections in addition to shunt current injections.

- All the network variables (node voltages and injected currents) are expressed in a common synchronously rotating frame of reference.
- The matrix Y is the positive sequence bus admittance matrix. Unbalance is usually not considered in transient stability programs except when unbalanced faults take place.

There are two approaches to the numerical solution of the differential-algebraic equations :

1 Partitioned Method;

2 Simultaneous Implicit Method.

4.2.3 Partitioned Solution

All the algebraic equations of the system can be clubbed together as follows:

$$I = [Y]V \tag{12.55}$$

$$I_i = g_i(x_i, V_i) \tag{12.56}$$

where, $i = 1, 2, \ldots,$ (number of subsystems)
The differential equations are as follows:

$$\dot{x}_i = f_i(x_i, V_i, u_i) \tag{12.57}$$

where, $i = 1, 2, \ldots,$ (number of subsystems)

In the partitioned method of solution, the algebraic equations and differential equations are solved alternately. The differential equations are integrated using an implicit or an explicit method. The algebraic variables (bus voltages) used in the numerical integration are obtained from the solution of the algebraic equations where, the *previously* computed value of the states are used in the calculation of injected currents. Therefore, the method suffers from interface errors.

The solution of algebraic equations is a major component of the computational procedure. Therefore extrapolation of the bus voltages can be used when it is required to solve the algebraic equations more than once in an integration step.

The network algebraic equations in the current-balance form are linear. Computations are substantially reduced if *all* algebraic equations are linear and have parameters not varying over time (except when there are changes in the network) since it would require only one LU factorization for each network condition. This is generally not the case because of the following reasons:

1 Non-linearity due to subsystems like constant power loads. This will call for an iterative solution of the algebraic equations. Gauss - Jacobi method can be conveniently utilized for this purpose. This involves using an initial guess of the bus voltage to compute the current injected by the load. This injected current is used to solve for the bus voltages using the rest of the *linear* equations. These bus voltages are in-turn used to recompute the injected currents and the procedure is repeated till convergence is attained.

2 Sub-transient saliency of generators, $x_d^{''} \neq x_q^{''}$. This results in equations with time varying parameters i.e., (from equation 12.43)

$$\begin{bmatrix} i_Q \\ i_D \end{bmatrix} = -[X''(\delta)]^{-1} \begin{bmatrix} v_Q \\ v_D \end{bmatrix} + [X''(\delta)]^{-1} \begin{bmatrix} E_Q^{''} \\ E_D^{''} \end{bmatrix} \quad (12.58)$$

Note that $[X'']$ is a function of δ which changes with time. Solution of the linear algebraic equations will require LU decomposition at every time step. This can be avoided by the use of an iterative procedure or by representing the effect of saliency using a dummy coil [Padiyar, 1996].

The advantage of partitioned method is its simplicity and inherent modularity.

4.2.4 Simultaneous Implicit Solution

In the simultaneous implicit method, the algebraic equations and differential equations are solved together. This is done by discretising the differential equations as follows using an implicit method like trapezoidal rule:

$$x_{ik+1} = x_{ik} + \frac{h}{2}\left(f_i(x_{ik}, V_{i_k}, u_{i_k}) + f_i(x_{i_{k+1}}, V_{i_{k+1}}, u_{i_{k+1}})\right) \quad (12.59)$$

where, $i = 1, 2, \ldots,$ (number of subsystems)
The algebraic equations are given by:

$$I_{i_{k+1}} = g_i(x_{i_{k+1}}, V_{i_{k+1}}) \quad (12.60)$$

$$I_{k+1} = [Y]V_{k+1} \quad (12.61)$$

where, $i = 1, 2, \ldots,$ (number of subsystems)
The discretised differential equations (which are also algebraic equations), and equations (12.60) and (12.61) are solved together to obtain x_{k+1}, V_{k+1} and I_{k+1}. The algebraic equations can be solved by Newton method.

The following points are important from a computational perspective:

1 The solution of algebraic equations requires an iterative solution at every time step. The iterative solution by Newton method requires a linear system solution using the system Jacobian at every iteration. To avoid this, the system Jacobian may be calculated only once during the time step and the same Jacobian is used for the iterations. The Jacobian may also be kept constant over a few time steps. This can substantially reduce the computational burden.

2 The implicit method allows the use of larger time steps.

While both electromagnetic transients program and transient stability program do time domain simulation, small signal analysis requires large scale eigenvalue computation. The issues involved are outlined in the following section.

4.3. Small Signal Analysis program: Large Scale Eigenvalue Computation

An eigenvalue computation is preceded by the following:

1 Computation of equilibrium point
The computation of the equilibrium point is obtained by setting the left hand side of the DAEs in equation (12.1) to zero and solving the resulting algebraic equations. Usually, a standard load-flow study is first done. The states and the inputs are then *back-calculated* from the bus voltages and powers.

2 Linearization of nonlinear equations
This involves the evaluation of partial derivatives of the nonlinear functions at the equilibrium point (see equation (12.15)). The partial derivatives may be calculated analytically or numerically.

For the study of small signal stability of the electromechanical modes of the system, the differential-algebraic equations (12.55-12.57) are linearized around an equilibrium point. The resulting equations (assuming $\Delta u = 0$ and eliminating ΔI) can be written as

$$\Delta \dot{x} = A\Delta x + B\Delta V \quad (12.62)$$
$$0 = C\Delta x + D\Delta V \quad (12.63)$$

Eliminating the algebraic variable ΔV, we obtain

$$\Delta \dot{x} = \left(A - BD^{-1}C\right)\Delta x \quad = A_{red}\Delta x \tag{12.64}$$

The stability of the system can be judged from the eigenvalues of A_{red}. The following points are to be noted:

1 The matrix A is block diagonal.

2 The matrix D is derived using the bus admittance matrix and is usually sparse.

3 A_{red} is not sparse. Therefore, for large systems ΔV is not explicitly eliminated from the equations.

4 For large and sparse matrices, only a small set of eigenvalues is desired. It is not feasible to use the QR method for this purpose. Instead, special iterative schemes for the computation of selected eigenvalues have been proposed in the power system stability literature [Byerly et al., 1982; Wang et al., 1988; Perez et al., 1982; Uchida and T.Nago, 1988].

Consider the following scheme for iterative computation of an eigenvalue, λ_i:

1 Suppose that the equations (12.62) and (12.63) are partitioned and re-written as follows:

$$\begin{bmatrix} \Delta \dot{x}_r \\ \Delta \dot{x}_{nr} \\ 0 \end{bmatrix} = \left[\begin{array}{c|c|c} a_r & a_{rnr} & b_r \\ \hline a_{nrr} & A_{nr} & B_{nr} \\ \hline c_r & C_{nr} & D \end{array}\right] \begin{bmatrix} \Delta x_r \\ \Delta x_{nr} \\ \Delta V \end{bmatrix} \tag{12.65}$$

where Δx_r is a scalar and $\Delta x^T = [\Delta x_r \ \Delta x_{nr}^T]$.

2 Assume that the initial conditions are such that only one eigenvalue, λ_i is excited. Therefore, $\Delta x(t) = c_i e^{\lambda_i t} p_i$, $\Delta \dot{x}(t) = \lambda_i \Delta x(t)$.

$$\begin{bmatrix} \lambda_i \Delta x_r \\ \lambda_i \Delta x_{nr} \\ 0 \end{bmatrix} = \left[\begin{array}{c|c|c} a_r & a_{rnr} & b_r \\ \hline a_{nrr} & A_{nr} & B_{nr} \\ \hline c_r & C_{nr} & D \end{array}\right] \begin{bmatrix} \Delta x_r \\ \Delta x_{nr} \\ \Delta V \end{bmatrix} \tag{12.66}$$

3 Thus, we get an algebraic equation

$$\lambda_i \Delta x_r = \left\{ a_r - [a_{rnr} \ \ b_r] \begin{bmatrix} A_{nr} - \lambda_i I & B_{nr} \\ C_{nr} & D \end{bmatrix}^{-1} \begin{bmatrix} a_{nrr} \\ c_r \end{bmatrix} \right\} \Delta x_r \tag{12.67}$$

Therefore,

$$\lambda_i = \left\{ a_r - [a_{rnr}\ b_r] \begin{bmatrix} A_{nr} - \lambda_i I & B_{nr} \\ C_{nr} & D \end{bmatrix}^{-1} \begin{bmatrix} a_{nrr} \\ c_r \end{bmatrix} \right\} \tag{12.68}$$

The algebraic equation can be solved by Newton method or by a fixed point iteration scheme. The convergence of the solution is dependent on appropriate choice of the state variable Δx_r (the state should have high participation ($\frac{\partial \lambda_i}{\partial a_r}$) in λ_i [Perez et al., 1982]). The algorithm can be modified and enhanced to include a larger set of 'relevant' states and eigenvalues instead of a scalar Δx_r. This is the basis of the Selective Modal Analysis technique describe in [Perez et al., 1982].

5. Summary

In this chapter we have briefly reviewed some dynamic analysis tools for digital simulation and eigenvalue analysis. Some special techniques applicable to power system dynamic studies are also reviewed.
The basic features common to all power system dynamic analysis techniques are:

- Modeling simplifications based on transients of interest.
- Systematic procedure for formulation of equations keeping in view the modularity.
- Solution of equations by exploiting sparsity.

These are brought out by a discussion of the basic techniques in network simulation, transient stability simulation and large scale eigenvalue computations.

6. Further Reading

The reader is referred to the books [Sauer and Pai, 1998], [Padiyar, 1996] and [Kundur, 1994] for an exhaustive coverage of modeling. Computational techniques are also covered in these books. For modeling HVDC and FACTS devices the reader is referred to the books [Arrillaga and Bruce, 1998] and [Song and Johns, 1999] respectively. The techniques in the study of Sub-synchronous Resonance are covered in [Padiyar, 1999]. For a detailed coverage on voltage stability and its analysis techniques, reader is recommended to refer [Kundur, 1994; Padiyar, 1996; Gao, 1992; Taylor, 1994; Cutsem and Vounas, 1998].

Appendix A
Calculation of the Fault Currents

In short circuit analysis, the following bus faults are usually considered:

1 Single Line to Ground Fault (SLG);

2 Line to Line Fault (LL);

3 Double Line to Ground fault (LLG);

4 Three Phase fault.

The various fault current equations are now summarized.

0.1. Three phase Fault

A 3ϕ fault is a balanced fault. Thus, only positive sequence network is considered. The data requirements are nearly identical to load flow analysis. The fault currents are given by the following equations.

$$I_1 = \frac{V}{Z_1} \text{ (Solid Fault)} \tag{A.1}$$

$$I_1 = \frac{V}{Z_1 + Z_F} \text{ (Fault Through impedance } Z_F\text{)} \tag{A.2}$$

It is common practice in fault studies on subtransmission and transmission systems to assume negligible fault impedance. However, on distribution systems many faults like tree contact can present a large Z_F.

0.2. SLG Fault

A phase-a-to-ground fault is represented by connecting the three sequence networks together in series. The positive sequence, negative and zero sequence fault current I_1 I_2 I_0 are given by the following equations.

$$I_1 = I_2 = I_0 = \frac{V}{Z_1 + Z_2 + Z_0} \tag{A.3}$$

$$I_1 = I_2 = I_0 = \frac{V}{Z_1 + Z_2 + Z_0 + 3Z_F} \tag{A.4}$$

$$I_{aF} = I_1 + I_2 + I_0 = 3I_1 = 3I_2 = 3I_0 \tag{A.5}$$

The first equation (A.3) corresponds to the solid fault while the second equation (A.4) corresponds to the fault with impedance. V is the prefault voltage of the bus and Z denotes the short circuit impedance. The subscripts 1,2 and 0 denote positive, negative and zero sequence.

0.3. LL Fault

Zero sequence network is not required for LL faults. Thus, like 3ϕ faults the data requirements are almost identical to load flow analysis. Usually, $Z_1 = Z_2$. For the LL fault the relevant equations are

$$I_1 = -I_2 = \frac{V}{Z_1 + Z_2} \quad or$$

$$I_1 = -I_2 = \frac{V}{Z_1 + Z_2 + Z_F} \tag{A.6}$$

0.4. LLG fault

For the LLG fault the relevant equations are:

$$I_1 = \frac{V}{Z_1 + \frac{Z_2 Z_0}{Z_2 + Z_0}} \tag{A.7}$$

$$I_1 = \frac{V}{Z_1 + \frac{Z_F}{2} + \frac{(Z_2 + \frac{Z_F}{2})(Z_2 + \frac{Z_F}{2} + 3Z_{FG})}{Z_2 + Z_0 + Z_F + 3Z_{FG}}} \tag{A.8}$$

$$I_2 = -I_1 \frac{Z_0}{Z_2 + Z_0} \quad or$$

$$I_2 = -I_1 \frac{Z_0 + \frac{Z_F}{2} + 3Z_{FG}}{Z_2 + Z_0 + Z_F + 3Z_{FG}} \tag{A.9}$$

$$I_0 = -I_1 \frac{Z_2}{Z_2 + Z_0} \quad or$$

$$I_0 = -I_1 \frac{Z_2 + \frac{Z_F}{2}}{Z_2 + Z_0 + Z_F + 3Z_{FG}} \tag{A.10}$$

Z_F is the fault impedance between the lines while Z_{FG} is the fault impedance to ground. With the knowledge of the sequence currents, the line currents can be computed as follows

$$I_a = I_0 + I_1 + I_2 \tag{A.11}$$

$$I_b = I_0 + a^2 I_1 + a I_2 \tag{A.12}$$

$$I_c = I_0 + a I_1 + a^2 I_2 \tag{A.13}$$

The constant a equals $e^{j\frac{2\pi}{3}}$.

References

Abur, A. and Celik, M. K. (1993). Least Absolute Value State Estimation With Equality and Inequality Constraints. *IEEE Transactions on Power Systems*, 8(2):680–686.

Aho, A. V., Hopcroft, J. E., and Ullman, J. D. (1974). *The Design and Analysis of Computer Algorithms.* Addison-Wesley Longman, Inc.

Ajjarapu, V. and Christy, C. (1992). The Continuation Power Flow: A Tool For Steady State Voltage Stability Analysis . *IEEE Transactions on Power Systems*, 7(1):416–423.

Almeida, K. C. and Galiana, F. D. (1996). Critical Cases in the Optimal Power Flow. *IEEE Transactions on Power Systems*, 11(3):1509–1516.

Alvarado, F. L. and Tinney, W. F. (1990). State Estimation Using Augmented Blocked Matrices. *IEEE Transactions on Power Systems*, 5(3):911– 921.

Alvarado, F. L., Tinney, W. F., and Enns, M. K. (1991). Sparsity in Large Scale Computations. In *Control and Dynamic Systems - Advances in Theory and Applications Vol 41: Analysis and Control System Techniques for Electric Power Systems Part 1 of 4, Edited By C.T. Leondes*, pages 207–272. Academic Press, Inc.

Anda, A. A. and Park, H. (1994). Fast Plane Rotations With Dynamic Scaling. *SIAM Journal of Matrix Analysis and Applications*, 15(1):162–174.

Anderson, P. M. (1973). *Analysis of Faulted Power Systems.* Iowa State University Press, Ames Iowa.

Arrillaga, J. and Bruce, S. (1998). *AC-DC Power System Analysis.* IEE, London.

Aschmoneit, F. C., Peterson, N. M., and Adrian, E. C. (1977). State Estimation With Equality Constraints. In *PICA Conference*, pages 427–430.

Ashcraft, C. and Liu, J. W. (1999). SPOOLES : An Object Oriented Sparse Matrix Library. Technical report, www.netlib.bell-labs.com/netlib/linalg/spooles.

Ashcraft, C. and Liu, J. W. H. (1996). A Software Package for Ordering Sparse Matrices. Technical report, www.solon.cma.univie.ac.al/~neum/software.html.

Barret, J. P., Bornard, P., and Meyer, B. (1997). *Power System Simulation.* Chapman & Hall, London.

Bazaraa, M. S., Sherali, H. D., and Shetty, C. M. (1993). *Nonlinear Programming.* John Wiley and Sons, Inc., New York.

Bertsekas, P. D. (1982). *Constrained optimization and Lagrange Multiplier Methods.* Academic Press, Boston.

Blackburn, J. L. (1987). *Protective Relaying.* Marcel Dekker, New York.

Blackburn, J. L. (1993). *Symmetrical Components for Power System Engineering.* Marcel Dekker, New York.

Booch, G. (1994). *Object-Oriented Analysis and Design with Applications -Second Edition.* Addison-Wesley Publishing Company.

Brown, H. (1975). *Solution of Large Networks by Matrix Methods.* Wiley Interscience, New York.

Budd, T. A. (1994). *Classic Data Structures in C++* . Addison-Wesley Publishing Company, U.S.A.

Bunch, J. R. and Kaufman, L. (1977). Some Stable Methods for Calculating Inertia and Solving Symmetric Linear Systems. *Mathematics of Computation*, 31:163–179.

Bunch, J. R. and Parlett, B. N. (1971). Direct Methods for Solving Symmetric Indefinite Systems of Linear Equations. *SIAM Journal of Numerical Analysis*, 8:639–655.

Burchett, R. C., Happ, H. H., and Vierath, D. R. (1984). Quadratically convergent optimal power flow. *IEEE Transactions on Power Apparatus And Systems*, PAS-103(11):3267–3275.

Burchett, R. C., Happ, H. H., and Wirgau, K. A. (1982). Large Scale Optimal Power Flow. *IEEE Transactions on Power Apparatus And Systems*, PAS-101:3722–3732.

Byerly, R. T., Bennon, R. J., and Sherman, D. E. (1982). Eigenvalue Analysis of Synchronizing Power Flow Oscillations in Large Electric Power Systems. *IEEE Transactions on Power Apparatus And Systems*, PAS-101(1):235–243.

Carpentier, J. (1962). Contribution a. 'l' etude du Dispatching Economique. *Bulletin de la Societe Francaise des Electriciens*, 3:431–447.

Carpentier, J. (1987). Towards a Secure and Optimal Automatic Operation of Power Systems. In *PICA Conference*, pages 2–37.

Chattopadhyay, B., Sachdev, M. S., and Sidhu, T. S. (1996). An On-line Relay Coordination Algorithm for Adaptive Protection using Linear Programming Technique. *IEEE Transactions on Power Delivery*, 11(1):165–171.

Christie, R. D., Wollenberg, B. I., and Wangensteen, I. (2000). Transmission management in the deregulated enviornment. *Proceedings of the IEEE*, 88(2):170–195.

Chvatal, V. and Freeman, W. H. (1983). *Linear Programming.* W.H.Freeman and Company, New York.

Cutsem, T. V. and Vounas, C. (1998). *Voltage Stability of Electric Power System* . Kluwer Academic Publisher, Boston.

Damborg, M. J. and Venkata, S. S. (1984). Specifications of Computer Aided Design of Transmission Protection Systems. Technical report, EPRI EL-3337 (Final Report),, Department of Electrical Engineering, University of Washington.

Dobrian, F., Kumfert, G., and Pothen, A. (1999). The Design of Sparse Direct Solvers Using Object-Oriented Techniques : ICASE Final Report, NASA/CR-1999-209558 Report No. 99-38, September 1999. Technical report, Institute for Computer Applications in Science and Engineering, NASA Langley Research Center, Hampton, VA 23681-2199, www.icase.edu.

Dommel, H. W. (1969). Digital Computer Solution of Electromagnetic Transient In Single And Multiphase Networks. *IEEE Transactions on Power Apparatus And Systems*, PAS-88(4):388–399.

Dommel, H. W. and Sato, N. (1972). Fast Transient Stability Solutions. *IEEE Transactions on Power Apparatus And Systems*, PAS-91:1643–1650.

Dommel, H. W. and Tinney, W. F. (1968). Optimal Power Flow Solutions. *IEEE Transactions on Power Apparatus And Systems*, PAS-87(10):1866–1875.

Dongarra, J., Lumsdaine, A., Niu, X., Pozo, R., and Remington, K. (1994). A Sparse Matrix Library in C_{++} for High Performance Architectures. Technical report, www.citeseer.nj.nec.com/dongarra94sparse.html.

Duff, I. S. (1974). Pivot Selection and Row Ordering in Givens Reduction on Sparse Matrices. *Computing*, 13:239–248.

Duff, I. S., Erisman, A. M., and Reid, J. K. (1986). *Direct Methods for Sparse Matrices*. Clarendon Press, Oxford.

Dunlop, R. D., Gutman, R., and Marchenko, P. P. (1979). Analytical Development of Loadability Characteristics for EHV and UHV Transmissions Lines. *IEEE Transactions on Power Apparatus And Systems*, 98:606–617.

El-Kady, M. A., Bell, B. D., Carvalho, V. F., Burchett, R. C., Happ, H. H., and Vierath, D. R. (1986). Assessment of Real-Time Optimal Voltage Control. *IEEE Transactions on Power Systems*, 1(2):98–107.

Elgerd, O. I. (1982). *Electric Energy Systems Theory: An Introduction - 2nd Edition*. Tata McGraw-Hill, New Delhi.

Facchine, F., Judice, J., and Soares, J. (1998). An Active Set Newton Algorithm For Large-Scale Nonlinear Programs With Box Constraints. *SIAM Journal of Optimization*, 8:158–186.

Fletcher, R. (1987). *Practical Methods of Optimization, Vol. 2, Constrained Optimization*. John Wiley & Sons Ltd., Chichester.

Fortescue, C. L. (1918). Method of Symmetrical Co-ordinates Applied to the Solution of Polyphase Networks. *AIEE Transactions*, 37:1027–1140.

Fowler, M. and Scott, K. (2000). *UML Distilled*. Addison-Wesley Publishing Company, U.S.A.

Fuerte-Esquivel, C. and Acha, E. (1997). A Newton Type Algorithm For The Control Of Power Flow In Electrical Power Networks. *IEEE Transactions on Power Systems*, 12(4):1474–1480.

Gao, B. (1992). *Voltage Stability Analysis of Large Power Systems*. University of Toronto, Toronto.

Gear, C. W. (1971). *Numerical Initial Value Problems in Ordinary Differential Equations*. Prentice Hall, Englwood Cliff.

Gentleman, W. M. (1973). Least Squares Computations by Givens Transformations Without Square Roots. *J. Inst. Maths Applics*, 12:329–336.

Gentleman, W. M. (1975). *Row Elimination for Solving Sparse Linear Systems and Least Squares Problems*. Lecture Notes in Mathematics, 506, Springer-Verlag, New York.

George, A. and Heath, M. T. (1980). Solution of Sparse Linear Least Squares Problems Using Givens Rotations. *Linear Algebra And Its Applications*, 34:69–83.

George, A. and Liu, J. W. (1980). A Fast Implementation of the Minimum Degree Algorithm using Quotient Graphs. *ACM Transactions On Mathematical Software*, 6(3):337–358.

George, A. and Liu, J. W. H. (1981). *Computer Solution of Large, Sparse, Positive Definite Systems*. Prentice Hall.

George, A. and Liu, J. W. H. (1989). The Evolution of Minimum Degree Ordering Algorithm. *SIAM Review*, 31(1):1–19.

George, A. and Liu, J. W. H. (1997). An Object Oriented Approach to the Design of a User Interface for a Sparse Matrix Package. Technical report, Department of Computer Science, York University.

Gill, P. E., Murray, W., and Saunders, M. A. (1997). SNOPT: An SQP Algorithm For Large-Scale Constrained Optimization. (NA–97–2).

Gill, P. E., Murray, W., Saunders, M. A., and Wright, M. H. (1984a). Software And Its Relationship To Methods. *SIAM journal of Numerical Optimization.*

Gill, P. E., Murray, W., Saunders, M. A., and Wright, M. H. (1984b). Sparse Matrix Methods In Optimization. *SIAM Journal of Scientific and Statistical Computing*, 5(3):562–589.

Gill, P. E., Murray, W., and Wright, M. H. (1981). *Practical Optimization.* Associated Press.

Gill, P. E., Murray, W., and Wright, M. H. (1991). *Numerical Linear Algebra and Optimization, Volume 1.* Addison Wesley Publishing Company.

Givens, W. (1958). Computation of The Plane Unitary Rotations Transforming A General Matrix To Triangular Form. *J.Soc. Ind. Appl. Math.*, 6:26–50.

Gjelsvik, A., Aam, S., and Holten, L. (1985). Hatchels Augmented Matrix Method - A Rapid Method Improving Numerical Stability In Power System State Estimation. *IEEE Transactions on Power Apparatus And Systems*, PAS-104(11):2987–2993.

Glavitsch, H. and Bacher, R. (1991). Optimal Power Flow Algorithms. In *Control and Dynamic Systems - Advances in Theory and Applications Vol 41: Analysis and Control System Techniques for Electric Power Systems Part 1 of 4, Edited by C.T. Leondes*, pages 135–206. Academic Press, Inc.

Gole, A. M., Fernando, I. T., Irwin, G. D., and Nayak, O. B. (June 1997). Modeling of Power Electronic Apparatus: Additional Interpolation Issues . *Proceedings of the Intern. Conf. on Power System Transients (IPST)*, pages 23–28.

Golub, G. H. and Loan, C. F. V. (1989). *Matrix Computations.* John Hopkins Series in the Mathematical Sciences, The John Hopkins University Press, Baltimore, MD.

Gotham, D. J. and Heydt, G. (1998). Steady State Power Flow Control and Power Flow Studies for Systems with FACTS Devices. *IEEE Transactions on Power Systems*, 13(1).

Gould, N. I. M. and Toint, P. L. (2001a). A Quadratic Programming Bibliography. *RAL Numerical Analysis Group Internal Report.*

Gould, N. I. M. and Toint, P. L. (2001b). Large Scale Geodetic Least Squares Adjustment by Dissection and Orthogonal Decomposition.

Gould, N. I. M. and Toint, P. L. (2001c). Numerical Methods For Large-Scale Non-Convex Quadratic Programming. *First International Conference on Industrial and Applied Mathematics in (the) Indian Subcontinent, Amritsar, India.*

Grainger, J. J. and William D. Stevenson, J. (1994). *Power System Analysis.* McGraw Hill, New York.

Gu, J. W., Clements, A., Krumpholz, G. R., and Davis, P. W. (1983). The Solution of Ill-Conditioned Power System State Estimation Problems Via The Method of Peters and Wilkinson. *IEEE Transactions on Power Apparatus And Systems*, PAS-102(10):3473–3480.

Hakavik, B. and Holen, A. T. (1994). Power System Modelling And Sparse Matrix Operations Using Object Oriented Programming. *IEEE Transactions on Power Systems*, 9(2):1045–1052.

Hammarling, S. (1974). A Note On The Modifications To The Givens Plane Rotation. *J. Inst. MAth. Appl.*, 13:215–218.

Handschin, E., Schweppe, F. C., Kohlas, J., and Flechter, A. (1975). Bad Data Analysis for Power System State Estimation. *IEEE Transactions on Power Apparatus And Systems*, 94(2):329–336.

Happ, H. H. (1977). Optimal power dispatch - a comprehensive survey. *IEEE Transactions on Power Apparatus And Systems*, PAS-96:841–854.

Herstein, I. N. (1975). *Topics in Algebra.* Wiley Eastern Limited, New Delhi.

Heydt (1986). *Computer Analysis Methods for Power Systems*. Macmillan Publishing Company, New York.

Huneault, M. and Galiana, F. D. (1991). A Survey of the Optimal Power Flow Literature. *IEEE Transactions on Power Systems*, 6(2):762–770.

Iba, K., Suzuki, H., Suzuki, K., and Suzuki, K. (1988). Practical Reactive Power Allocation/Operation Planning Using Successive Linear Programming. *IEEE Transactions on Power Systems*, 3(2):558–566.

IEEE Special Stability Controls Working Group (Feb.1990). Static VAR Compensator Modeling for Powerflow and Dynamic Performance Simulation. *IEEE Trans. Power System*, 9(1):229–239.

Jacobson, I., Booch, G., and Rumbaugh, J. (1999). *The Unified Software Development Process*. The Addison-Wesley Longman Inc.

Kirchmayer, L. K. (1958). *Economic Operation of Power Systems*. Wiley Eastern Limited.

Kundur, P. (1994). *Power System Stability and Control*. McGraw Hill Inc, New York.

Kusic, G. L. (1986). *Computer Aided Power Systems Analysis*. Prentice Hall of India, New Delhi.

Lapidus, L. and Seufield, J. H. (1971). *Numerical Solution of Ordinary Differential Equations*. Academic Press, New York.

Lawson, C. L. and Hanson, R. J. (1974). *Solving Least Squares Problems*. Prentice Hall, Englewood Cliffs, NJ.

Liacco, T. E. D. (1978). System Security: The Computer's Role. *IEEE Spectrum*, 15:43–50.

Liu, J. W. (1985). Modification of the Minimum Degree Algorithm by Multiple Elimination. *ACM Transactions On Mathematical Software*, 11(2):141–153.

Liu, J. W. (1990). The Role of Elimination Trees in Sparse Factorization. *SIAM Journal of Matrix Analysis and Applications*, 11(1):134–172.

Markowitz, H. M. (1957). The Elimination Form of The Inverse And Its Applications to Linear Programming. *Management Sci.*, 3:255–269.

Meyers, S. (1996). *More Effective C++* . Addison-Wesley Professional Computing Series, Reading, Massachusetts 01867.

Mili, L., Cutsem, T. V., and Pavella, M. R. (1985). Bad Data Identification Methods in Power System State Estimation - a Comparative Study. *IEEE Transactions on Power Apparatus And Systems*, PAS-104(11):3037–3049.

Momoh, J. A. (2001). *Electric Power System Applications of Optimization*. Marcel Dekker AG, Switzerland.

Momoh, J. A., Koessler, R. J., Bond, M. S., Scott, B., Sun, D., Papalexopoulos, A., and Ristanovic, P. (1997). Challenges to Optimal Power Flow. *IEEE Transactions on Power Systems*, 12(1):444–454.

Monticelli, A. (1999). *State Estimation in Electric Power Systems: A Generalized Approach*. Kluwer Academic Publishers, Boston.

Monticelli, A. (2000). Electric Power System State Estimation. In *Proceedings of the IEEE*, volume 88, pages 262–271.

Monticelli, A., Murari, C. A. F., and Wu, F. F. (1985). A Hybrid State Estimator: Solving Normal Equations By Orthogonal Transformations. *IEEE Transactions on Power Apparatus And Systems*, PAS-105:3460–3466.

Morales, J. L., Nocedal, J., Waltz, R. A., Liu, G., and Goux, J. P. (1980). Assessing the Potential of Interior Methods for Nonlinear Optimization.

Neyer, A. F., Wu, F. F., and Imhof, K. (1990). Object Oriented Programming for Flexible Software : Example of a Load Flow. *IEEE Transactions on Power Systems*, 5(3):689–695.

Noroozian, M. and Andersson, G. (1993). Power Flow Control by Use of Controllable Series Components. *IEEE Transactions on Power Delivery*, 8(3):1420–1429.

Nucera, R. R. and Gilles, M. L. (1991). A Blocked Sparse Matrix Formulation For The Solution of Equality Constrained State Estimation. *IEEE Transactions on Power Systems*, 6(1):214–221.

Ogata, K. (1967). *State Space Analysis of Control Systems*. Prentice Hall, Englewood Cliff.

Orfanogianni, T. and Bacher, R. (2000). Increased OPF Development Efficiency by Integration of General Purpose Optimization and Derivative Computation Tools. *IEEE Transactions on Power Systems*, 15(3):987–993.

Osterlund, L.-O. G. (2000). Component Technology. *IEEE Computer Applications in Power*, (1):17–25.

Padiyar, K. R. (1996). *Power System Dynamics, Stability and Control*. Interline Publishers, Bangalore.

Padiyar, K. R. (1999). *Analysis of Subsynchronous Resonance in Power Systems*. Kluwer Academic Publishers, Boston.

Pai, M. A. (1979). *Computer Techniques in Power System Analysis*. Tata McGraw Hill Inc, Delhi.

Pandian, A., Parthasarathy, K., and Soman, S. A. (1999). Towards Faster Givens Rotations Based Power System State Estimator. *IEEE Transactions on Power Systems*, 14(3):837–843.

Pandit, S. (2001). *Object Oriented Power System Analysis*. PhD thesis, Dept. of Electrical Engineering, Indian Institute of Technology, Bombay, Mumbai India.

Pandit, S., Soman, S., and Khaparde, S. (2001a). Object-Oriented Network Topology Processor. *IEEE Computer Applications in Power*, pages 42–46.

Pandit, S., Soman, S. A., and Khaparde, S. A. (2000a). Object-Oriented Design for Power System Applications. *IEEE Computer Applications in Power*, 13(4):43–47.

Pandit, S., Soman, S. A., and Khaparde, S. A. (2001b). Design Of Generic Direct Sparse Linear System Solver in *C++* For Power System Analysis . *Accepted for publication in IEEE Transactions on Power Systems.*

Pandit, S., Soman, S. A., Khaparde, S. A., and Pandian, A. (2000b). An OOP based Real Time Reactive Power Control Algorithm for Large Power Systems. *IEEE Winter Meeting*, 06-08-04.pdf, Singapore.

Papalexopoulos, A. D., Imparato, C. F., and Wu, F. F. (1989). Large-Scale Optimal Power Flow:Effects of Initialization, Decoupling and Discretization. *IEEE Transactions on Power Systems*, 4(2):748–756.

Parlett, B. N. (1998). *Symmetric Eigenvalue Problem*. Society for Industrial and Applied Mathematics (SIAM), Philadelphia, PA.

Perez, I. J., Verghjese, G. C., and Schweppe, F. C. (1982). Selective Modal Analysis With Application to Electric power systems, PART-I Heuristic Introduction, PART-II The Dynamic Stability Problem. *IEEE Transactions on Power Apparatus And Systems*, PAS-101(9):3117–3134.

Prata, S. (1997). *The Waite Group's C++ Primer Plus*. Galgotia Publishing Pvt Company, New Delhi.

Press, W. H., Teukolsky, S. A., Vellerling, W. T., and Flannery, B. P. (1992). *Numerical Recipes in C*. Cambridge University Press, Daryaganj, New Delhi.

Qiu, J. and Shahidehpour, S. M. (1987). A New Approach to Minimizing Power Losses and Improving Voltage Profile. *IEEE Transactions on Power Systems*, 2(2):287–295.

Quintana, V. H. and Mota-Palomino, R. (1985). Sparse Linear Reactive Power Dispatch. In *Proc. IFAC Symposium on Planning and Operation of Electric Energy Systems*, pages 405–410.

Robey, T. H. and Sulsky, D. L. (1994). Row Ordering For Sparse QR Decomposition. *SIAM Journal of Matrix Analysis and Applications*, 15(4):1208–1255.

Rose, D. J. (1972). A Graph Theoretic Study of the Numerical Solution of Sparse Positive Definite Systems of Linear Equations. In *Graph Theory and Computing*, pages 183–217.

Rumbaugh, J., Blaha, M., Premerlani, W., Eddy, F., and Lorensen, W. (1998). *Object Oriented Modelling and Design*. Prentice Hall of India, New Delhi.

Sachdev, M. S. and Medicherla, T. K. P. (1977). A Second Order Load Flow Technique. *IEEE Transactions on Power Apparatus And Systems*, 97:1586.

Sauer, P. and Pai, M. A. (1998). *Power System Dyamics and Stability*. Prentice Hall, Upper Saddle River, NJ.

Sauer, P. M. and Pai, M. A. (1995). Power System Steady State Stability and the Load Flow Jacobian. *IEEE Transactions on Power Systems*, 10(1):125–131.

Saxena, P., Mukerji, R., and Neugebauer, W. (1991). Information-Oriented Architecture Adds New Flavor to Optimal Power Flow. *IEEE Computer Applications in Power*, pages 25–30.

Schweppe, F. C. (1970). Power System State Estimation, Part III: Implementation. *IEEE Transactions on Power Apparatus And Systems*, PAS-89:130–135.

Schweppe, F. C. and Rom, D. B. (1970). Power System State Estimation, Part II: Approximate Model. *IEEE Transactions on Power Apparatus And Systems*, PAS-89:126–130.

Schweppe, F. C. and Wildes, J. (1970). Power System State Estimation, Part I: Exact Model. *IEEE Transactions on Power Apparatus And Systems*, PAS-89:120–125.

Serbin, S. M. (1980). On Factoring a Class of Complex Symmetric Matrices without Pivoting. *Mathematics of Computation*, 35(152):1231–1234.

Sergio, P. (1984). *Sparse Matrix Technology*. Academic Press, London.

Sherman, J. and Morrison, W. J. (1949). Adjustment of An Inverse Matrix Corresponding to Changes in the Elements of a Given Column Or A Given Row of the Original Matrix. *Ann. Math. Stat.*, 20:621.

Simoes-Costa, A. and Quintana, V. H. (1981). An Orthogonal Row Processing Algorithm For Power System Sequential State Estimation. *IEEE Transactions on Power Apparatus And Systems*, PAS-100:3791–3800.

Singh, L. P. (1986). *Advanced Power System Analysis and Dynamics*. Wiley Eastern Limited, New Delhi, second edition.

Smith, D. H. (1990). *IEEE Recommended Practice for Industrial and Commercial Power Systems Analysis, IEEE Std 399-1990*. The IEEE Inc., 345 East 47th Street, New York, NY 10017, USA.

Soman, S. A. (1995). On Large Sparse Equality Constrained Linear Least Squares With Applications in Energy Management Systems. Technical report, Electrical Engineering Department, Indian Institute of Science, Bangalore, India.

Soman, S. A., Parthasarathy, K., and Thukaram, D. (1994). Curtailed Number and Reduced Controller Movement Optimization Algorithms for Real Time Voltage/Reactive Power Control. *IEEE Transactions on Power Systems*, 0(0):2035–2042.

Song, Y. H. and Johns, A. T. (1999). *Flexible AC Tansmission Systems (FACTS)*. IEE, Herts,UK.

Stagg, G. W. and El-Abiad., H. (1968). *Computer Methods in Power Systems Analysis*. McGraw Hill Inc, New York.

Stewart, G. W. (1973). *Introduction to Matrix Computations*. Acdemic Press, San Diego.

Stott, B. and Alsac, O. (1974). Fast Decoupled Load Flow. *IEEE Transactions on Power Apparatus And Systems*, 93:859–869.

Stott, B., Alsac, O., and Monticelli, A. (1987). Security Analysis and Optimization. In *Proceedings of IEEE, Special issue on computers in power systems operations*, pages 1623–1644.

Stroustrup, B. (1991). *The C++ Programming Language-Second Edition*. Addison-Wesley Publishing Company, AT & T Bell.

Stroustrup, B. (1997). *The C++ Programming Language-Third Edition*. Addison-Wesley Publishing Company, AT & T Bell.

Sun, D. I., Ashley, B., Brewer, B., Hughes, A., and Tinney, W. F. (1984). Optimal Power Flow By Newtons Approach. *IEEE Transactions on Power Apparatus And Systems*, PAS-103(10):2864–2880.

Taylor, C. (1994). *Power System Voltage Stability*. McGraw Hill, New York.

Tinney, W. F., Bright, J. M., Demaree, K. D., and Hughes, B. (1987). Some Deficiencies in Optimal Power Flow. In *IEEE-PICA Conf. Proc, Montreal Canada*, pages 164–169.

Tinney, W. F. and Walker, J. W. (1967). Direct Solutions of Sparse Network Equations by Optimally Ordered Triangular Factorization. *Proceedings of the IEEE*, 55(11):1801–1809.

Uchida, N. and T.Nago (1988). A New Eigen-Analysis Method of Steady State Stability for Large power systems : S Matrix Method. *IEEE Transactions on Power Systems*, 3(3):706–714.

Vempati, N., Slutsker, I. W., and Tinney, W. F. (1991). Enhancements to Givens Rotations For Power System State Estimation. *IEEE Transactions on Power Systems*, 6(2):842–849.

Wallach, Y. (1986). *Calculation and Programs For Power System Networks*. Prentice-Hall, New Jersey.

Wang, D. Y., Roger, G., Poretta, B., and Kundur, P. (1988). Eigenvalue Analysis of Very Large Power Systems. *IEEE Transactions on Power Systems*, 3(3):472–480.

Ward, J. and Hale, H. W. (1956). Digital Computer Solution of Power-Flow Problems. *AIEE Transactions, Part III Power Apparatus and Systems*, 75:398–404.

Weiss, M. A. (1999). *Data Structures and Algorithm Analysis in C++* . Addison-Wesley Publishing Company, U.S.A.

Wood, A. J. and Wollenberg, B. F. (1996). *Power Generation Operation and Control*. Wiley Interscience, second edition.

Woodbury, M. (1950). *Inverting Modified Matrices*. Memorandum 42, Statistics Research Group, Princeton, New Jersey.

Wu, F. F., Wen-Hsiung, Liu, E., and Lun, S.-M. (1988). Observability Analysis and Bad Data Processing for State Estimation With Equality Constraints. *IEEE Transactions on Power Systems*, 3(2):541–547.

Yannakakis, M. (1981). Computing the Minimum Fill-in is NP-complete. *SIAM J. Alg. Disc. Meth.*, 2:77–79.

Zhu, J. and Jossman, P. (1999). Application of Design Patterns for Object Oriented Modeling of Power Systems. *IEEE Transactions on Power Systems*, 14(2):532–537.

Zhu, J. and Lubkeman, D. (1997). Object Oriented Developements of Software Systems for Power System Simulations. *IEEE Transactions on Power Systems*, 12(2):1002–1008.

Index

CD-ROM Disclaimer

MIX
Papier aus verantwortungsvollen Quellen
Paper from responsible sources
FSC® C105338

If you have any concerns about our products,
you can contact us on
ProductSafety@springernature.com

In case Publisher is established outside the EU,
the EU authorized representative is:
Springer Nature Customer Service Center GmbH
Europaplatz 3, 69115 Heidelberg, Germany

Printed by Libri Plureos GmbH
in Hamburg, Germany